Topics in Applied Physics   Volume 70

Topics in Applied Physics   Founded by Helmut K. V. Lotsch

Dye Lasers: 25 Years

Edited by Michael Stuke

With 151 Figures

Springer-Verlag Berlin Heidelberg GmbH

Dr. Michael Stuke
Max-Planck-Institut für Biophysikalische Chemie
W-3400 Göttingen, Fed. Rep. of Germany

ISBN 978-3-662-30876-9   ISBN 978-3-540-46605-5 (eBook)
DOI 10.1007/978-3-540-46605-5

© Springer-Verlag Berlin Heidelberg 1992

Originally published by Springer-Verlag Berlin Heidelberg New York in 1992.
Softcover reprint of the hardcover 1st edition 1992

Typesetting: Macmillan India Ltd., India

57/3140-543210 – Printed on acid-free paper

Preface

The remarkable role of the dye laser in the advancement of science and technology since its invention 25 years ago is ranging now from tests of fundamental quantum physics and diagnostics of regioselective photochemistry in complexes to contactless test techniques for the characterization of ultrafast optoelectronic devices for future high-bit-rate optical communication systems.

The 18 chapters of this book are written by the pioneers of the dye laser field.

The fascinating developments for fundamental science and new practical devices origin from the unique properties of the dye laser: its wavelength can be tuned with precise control directly from the ultraviolet to beyond one micrometer, and ultranarrow bandwidth operation with a precision below 10^{-14} can be obtained. Ultrashort light pulses down to six femtosecond duration can be generated − a few optical cycles. The electric field strength corresponding to the available focussed intensity is comparable to the Coulomb field in atoms.

New concepts in laser resonator design and a precise understanding of ultrashort pulse formation enable the design of stable oscillators. Pulse front distortion in optical elements and the propagation through amplifiers are now understood in depth and can be controlled and used in hybrid dye/excimer laser amplifier systems generating powers in the terawatt range.

The tunable narrowband dye laser is the light source that spectroscopists, used to hollow cathode discharge lamps, have probably dreamed about for decades. High-resolution molecular spectroscopy can now give rotationally resolved electronic spectra of large molecular systems leading to a microscopic understanding of the processes of solvation and complex formation. The atomic motion can be precisely manipulated by means of dye laser radiation: atomic optics. The spectroscopy of isolated atoms and ions immersed in superfluid helium allows the investigation of point defects in that environment.

Precise femtosecond synchronization of energetic pulses − optical and/or electrical, petawatt powers and focussed optical intensities exceeding $10^{20} \, \mathrm{Wcm}^{-2}$ open the door to new phenomena in high power nonlinear optics and X-ray and plasma physics. It is a pleasure to thank the inventors of this unique light source for the creation in 1966, and for raising to her beauty − 25 years young!

Göttingen, February 1992 *Michael Stuke*

Contents

Contributors

Axmann, A.
Max-Planck-Institut für Festkörperforschung, Heisenbergstrasse 1,
W-7000 Stuttgart 80, Fed. Rep. of Germany

Beau, M.
Physikalisches Institut der Universität Heidelberg, Philosophenweg 12,
W-6900 Heidelberg 1, Fed. Rep. of Germany

Beutel, V.
Fachbereich Physik, Universität Kaiserslautern,
W-6750 Kaiserslautern, Fed. Rep. of Germany

Bhale, G.
Fachbereich Physik, Universität Kaiserslautern,
W-6750 Kaiserslautern, Fed. Rep. of Germany

Böhmer, E.
Department of Chemistry, University of Southern California,
Los Angeles, CA 90089-0482, USA

Bor, Z.
Department of Optics and Quantum Electronics, József Attila University,
Dóm tér 9. H-6720 Szeged, Hungary

Chen, Y.
Department of Chemistry, University of Southern California,
Los Angeles, CA 90089-0482, USA

Cybo-Ottone, A.
Centro Elettronica Quantistica e Strumentazione Elettronica del CNR,
Istituto di Fisica del Politecnico, Piazza L. da Vinci 32, I-20133 Milano, Italy

De Silvestri, S.
Centro Elettronica Quantistica e Strumentazione Elettronica del CNR,
Istituto di Fisica del Politecnico, Piazza L. da Vinci 32, I-20133 Milano, Italy

Demtröder, W.
Fachbereich Physik, Universität Kaiserslautern,
W-6750 Kaiserslautern, Fed. Rep. of Germany

Eckel, H.-A.
Fachbereich Physik, Universität Kaiserslautern,
W-6750 Kaiserslautern, Fed. Rep. of Germany

Elsaesser, T.
Physics Department E 11, Technische Universität München,
W-8000 München 2, Fed. Rep. of Germany

Glownia, J.
    IBM Research, Yorktown Heights, NY 10598, USA
Gress, J.
    Fachbereich Physik, Universität Kaiserslautern,
    W-6750 Kaiserslautern, Fed. Rep. of Germany
Hofmann, T.
    Department of Electrical and Computer Engineering,
    Rice University, Houston, TX 77251-1892, USA
Horváth, Z.L.
    Department of Optics and Quantum Electronics, József Attila University,
    Dóm tér 9. H-6720 Szeged, Hungary
Kaiser, W.
    Physics Department E 11, Technische Universität München,
    W-8000 München 2, Fed. Rep. of Germany
Kaschke, M.
    IBM Research, Yorktown Heights, NY 10598, USA
Klingenstein, M.
    Max-Planck-Institut für Festkörperforschung, Heisenbergstrasse 1,
    W-7000 Stuttgart 80, Fed. Rep. of Germany
Kuhl, J.
    Max-Planck-Institut für Festkörperforschung, Heisenbergstrasse 1,
    W-7000 Stuttgart 80, Fed. Rep. of Germany
Kuhn, M.
    Fachbereich Physik, Universität Kaiserslautern,
    W-6750 Kaiserslautern, Fed. Rep. of Germany
Lambsdorff, M.
    Max-Planck-Institut für Festkörperforschung, Heisenbergstrasse 1,
    W-7000 Stuttgart 80, Fed. Rep. of Germany
Letokhov, V.S.
    Institute of Spectroscopy, USSR Academy of Sciences,
    142092 Troitsk, Moscow Region, USSR
von der Linde, D.
    Institut für Laser- und Plasmaphysik, Universität-GH-Essen,
    W-4300 Essen 1, Fed. Rep. of Germany
Magni, V.
    Centro Elettronica Quantistica e Strumentazione Elettronica del CNR,
    Istituto di Fisica del Politecnico, Piazza L. da Vinci 32, 20133 Milano, Italy
Misewich, J.
    IBM Research, Yorktown Heights, NY 10598, USA
Moglestue, C.
    Max-Planck-Institut für Festkörperforschung, Heisenbergstrasse 1,
    W-7000 Stuttgart 80, Fed. Rep. of Germany
Neusser, H.J.
    Institut für Physikalische und Theoretische Chemie,
    Technische Universität Müchen, Lichtenbergstrasse 4, W-8046 Garching,
    Fed. Rep. of Germany

Nisoli, M.
Centro Elettronica Quantistica e Strumentazione Elettronica del CNR,
Istituto di Fisica del Politecnico, Piazza L. da Vinci 32, I-20133 Milano, Italy
zu Putlitz, G.
Physikalisches Institut der Universität Heidelberg, Philosophenweg 12,
W-6900 Heidelberg 1, Fed. Rep. of Germany
Richter, W.
TU Berlin, Fachbereich 4, Physik, Hardenbergstrasse 36,
W-1000 Berlin 12, Fed. Rep. of Germany
Riedle, E.
Institut für Physikalische und Theoretische Chemie,
Technische Universität München, Lichtenbergstrasse 4, W-8046 Garching,
Fed. Rep. of Germany
Rosenzweig, J.
Max-Planck-Institut für Festkörperforschung, Heisenbergstrasse 1,
W-7000 Stuttgart 80, Fed. Rep. of Germany
Schäfer, F. P.
Max-Planck-Institut für biophysikalische Chemie, Postfach 2841,
W-3400 Göttingen-Nikolausberg, Fed. Rep. of Germany
Schlag, E. W.
Institut für Physikalische und Theoretische Chemie,
Technische Universität München, Lichtenbergstrasse 4, W-8046 Garching,
Fed. Rep. of Germany
Sharp, T.E.
Department of Electrical and Computer Engineering, Rice University,
Houston, TX 77251-1892, USA
Shin, S. K.
Department of Chemistry, University of Southern California,
Los Angeles, CA 90089-0482, USA
Sorokin, P.
IBM Research, Yorktown Heights, NY 10598, USA
Svanberg, S.
Department of Physics, Lund Institute of Technology, P.O. Box 118,
S-221 00 Lund, Sweden
Svelto, O.
Centro Elettronica Quantistica e Strumentazione Elettronica del CNR,
Istituto di Fisica del Politecnico, Piazza L. da Vinci 32, 20133 Milano, Italy
Szabó, G.
Department of Optics and Quantum Electronics, JATE University,
Szeged, Hungary
Szatmári, S.
Max-Planck-Institut für biophysikalische Chemie, Abt. Laserphysik,
Postfach 2841, W-3400 Göttingen, Fed. Rep. of Germany
Tittel, F.K.
Department of Electrical and Computer Engineering,
Rice University, Houston, TX 77251-1892, USA

Walkup, R.
  IBM Research, Yorktown Heights, NY 10598, USA
Walther, H.
  Sektion Physik der Universität München and Max-Planck-Institut
  für Quantenoptik, W-8045 Garching, Fed. Rep. of Germany
Weber, T.
  Institut für Physikalische und Theoretische Chemie, Technische Universität
  München, Lichtenbergstrasse 4, W-8046 Garching, Fed. Rep. of Germany
Wilhelmi, B.
  Zentralinstitut für Optik und Specktroskopie Berlin, Rudower Chaussee 6,
  O-1199 Berlin, Fed. Rep. of Germany
Wilson, W.L.
  Department of Electrical and Computer Engineering, Rice University,
  Houston, TX 77251-1892, USA
Wisoff, P.J.
  Department of Electrical and Computer Engineering, Rice .University,
  Houston, TX 77251-1892, USA
Wittig, C.
  Department of Chemistry, University of Southern California,
  Los Angeles, CA 90089-0482, USA

1. Femtosecond Broadband Absorption Spectroscopy of Photodissociating Molecules

J. Glownia, J. Misewich, R. Walkup, M. Kaschke and P. Sorokin

With 12 Figures

In this chapter we describe the results of some recent work that we have done in the area of ultrafast measurements. Specifically, we summarize what we have learned in trying to understand the spectra we have recorded of gas-phase molecules in the process of dissociating into smaller fragments. It was, of course, the discovery of the CPM (colliding-pulse, mode-locked) dye laser by *Fork* et al. [1.1] that first gave scientists ready access to laser pulse widths on the order of 100 fs. Such pulses make it possible to probe, in real time, events that occur on a subpicosecond time scale, such as various intramolecular reactions.

Conceptually, the simplest intramolecular reaction to study is gas-phase photodissociation. One can obtain enough molecules dissociating in synchronism so that they can be detected by a probe pulse, by relying upon a prompt photodissociative process that can be initiated with a femtosecond pump-laser pulse. A femtosecond probe pulse is then immediately launched into the gas cell, to propagate behind the pump pulse with a delay that can be precisely adjusted. The simplest case to consider is obviously that of a diatomic molecule dissociating into two atoms. As shown schematically in Fig. 1.1, a femtosecond pump-laser pulse can be used to excite an initially bound diatomic molecule to a repulsive state. As the wavepacket, which represents the probability distribution of particles excited to state $|1\rangle$ by the pump pulse, propagates down state $|1\rangle$, spreading in time, it can be monitored in real time by the precisely timed probe pulse, which acts on an allowed transition from state $|1\rangle$ to a higher electronic state, in a manner that will be described in detail.

The potential curves shown in Fig. 1.1 approximately represent electronic states that are probed following the 308-nm photolysis of TlI, a system we have studied to some considerable degree [1.2, 3, 5]. (The potential curve for state $|2\rangle$ has been shifted down to make the variations in the potentials more apparent.) In drawing these curves, we have tried to incorporate pre-existing relevant information about this particular system. For example, the shape of the molecular continuum absorption band [1.6] through which the photodissociative process is accessed enables us to estimate the slope of state $|1\rangle$ in the region where vertical transitions from the molecular ground state intersect curve $|1\rangle$, the so-called Franck–Condon region. (In the figure, the initial distribution of TlI molecules in the ground electronic state is shown for $T = 0°$ K. The experiments, however, were generally performed at 650–800° K, i.e. at temperatures for which sufficient TlI vapor pressure exists. At these temperatures, the initial distributions are, of course, much wider.)

Topics in Applied Physics, Vol. 70
Dye Lasers: 25 Years Ed.: Dr. Michael Stuke
© Springer-Verlag Berlin Heidelberg 1992

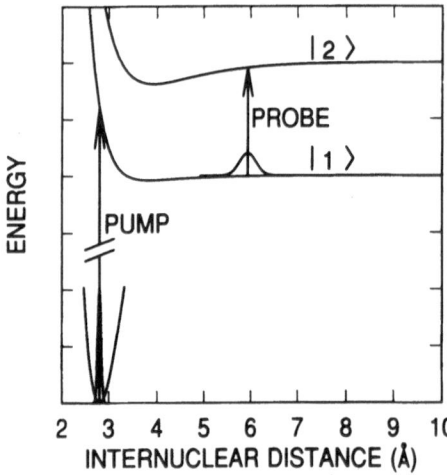

Fig. 1.1. Schematic diagram of photodissociation by a pump pulse, followed by time-resolved interrogation of the wave packet propagating on state |1⟩ by a probe pulse

From an internuclear distance of, say, 6 Å on out, one would normally expect the shapes of the potential curves |1⟩ and |2⟩ to be dominated by the so-called van der Waals interaction, varying as the inverse sixth power of the internuclear separation in each case, but with different individual coefficients (C_6) for the two states. For the TlI system, estimates for these van der Waals coefficients may be obtained from collisional broadening studies [1.7] of Tl atoms in high-pressure Xe buffer gas, since Xe atoms are about as big and as polarizable as I atoms, and the van der Waals interaction between a Tl atom and another atom varies with the polarizability of the latter. At small internuclear distances one assumes that both potential curves are dominated by exponential repulsive walls. From knowledge of the dissociation energy of the TlI molecule, one is able to state how much potential energy is lost by the system during the "half collision" initiated by absorption of a 308-nm photon. The result must, of course, be consistent with measurements [1.8] of the kinetic energy distributions of photofragments produced by 308-nm photolysis.

With the assumption of a definite form for the potential curve for state |1⟩, one can calculate how wave packets propagate on state |1⟩ [1.5]. In Fig. 1.1, the probability density on state |1⟩ is shown for a 0.4-ps delay after excitation by a delta-function pump pulse, an approximation to the shortest pump pulse possible. In Fig. 1.2, the probability densities for TlI molecules excited to the same potential curve |1⟩ are shown for delay times of 0.4 and 0.8 ps. Results for a delta-function pump pulse and molecules initially at $T = °0 \, \text{K}$ are shown as dotted lines. Results for a 160-fs FWHM sech^2 pump pulse and molecules initially at 800°K are shown as solid lines. From this figure, one sees that it takes somewhat less than a picosecond for the wavepacket to travel from its initial position on state |1⟩ out to an internuclear distance of about 10 Å, where the atoms have almost no longer any interaction with one another. Since the duration of this entire dissociative "half collision" is somewhat less than a picosecond, it is evident why one needs a time resolution at least on the order of

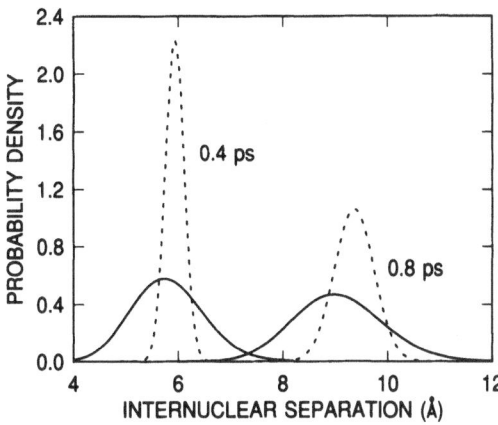

Fig. 1.2. Propagating wave packets on state $|1\rangle$ (in Fig. 1.1) seen at two delay times after the pump pulse. Dotted lines: delta-function pump pulse, $0\,^{\circ}$K initial temperature. Solid lines: 160-fs pump pulse, 800 K initial temperature

100 fs in order to perform transition-state spectroscopy of photodissociating molecules.

1.1 Probing the Dissociation Process: Tll Transient Spectra

The dissociation process has thus far been probed in two ways. In the best known approach, that taken by the group of *Zewail* at Caltech [1.9], the probe pulse is an ultrashort tunable dye laser pulse, and it is used to excite molecules from state $|1\rangle$ to state $|2\rangle$, the fluorescence from state $|2\rangle$ then being monitored from the side of the cell. The probe pulses have temporal durations of roughly 100 fs, and spectral bandwidths of roughly 150 cm^{-1} (greater, if the pulses are not bandwidth-limited). In these experiments, the probe laser frequency is tuned such that resonance with the molecular transition is achieved at some value, R^*, of the internuclear separation. Fluorescence from state $|2\rangle$ will be observed only if the wavepacket on state $|1\rangle$ happens to arrive at R^* when the probe laser pulse is turned on. By varying the pump–probe delay in conjunction with tuning of the probe pulse, one can effectively map out how the frequency of the molecular transition being probed, $\omega_M(R)$, varies as a function of R. This is one of the major goals of transition-state spectroscopy applied to systems such as these. The other goal is the determination of the transition dipole moment, $d(R)$, as a function of R.

In the approach taken by our IBM group, an ultrashort UV or visible pulse is used to generate an ultrashort white-light continuum pulse [1.3]. This broadband probe pulse is propagated through the sample, behind the pump pulse, and the spectral power density of the transmitted probe beam is measured with a grating spectrograph. Thus, the entire transient absorption spectrum can be recorded as a function of pump–probe delay. Before we had actually attempted this experiment, we speculated on what the appearance of the

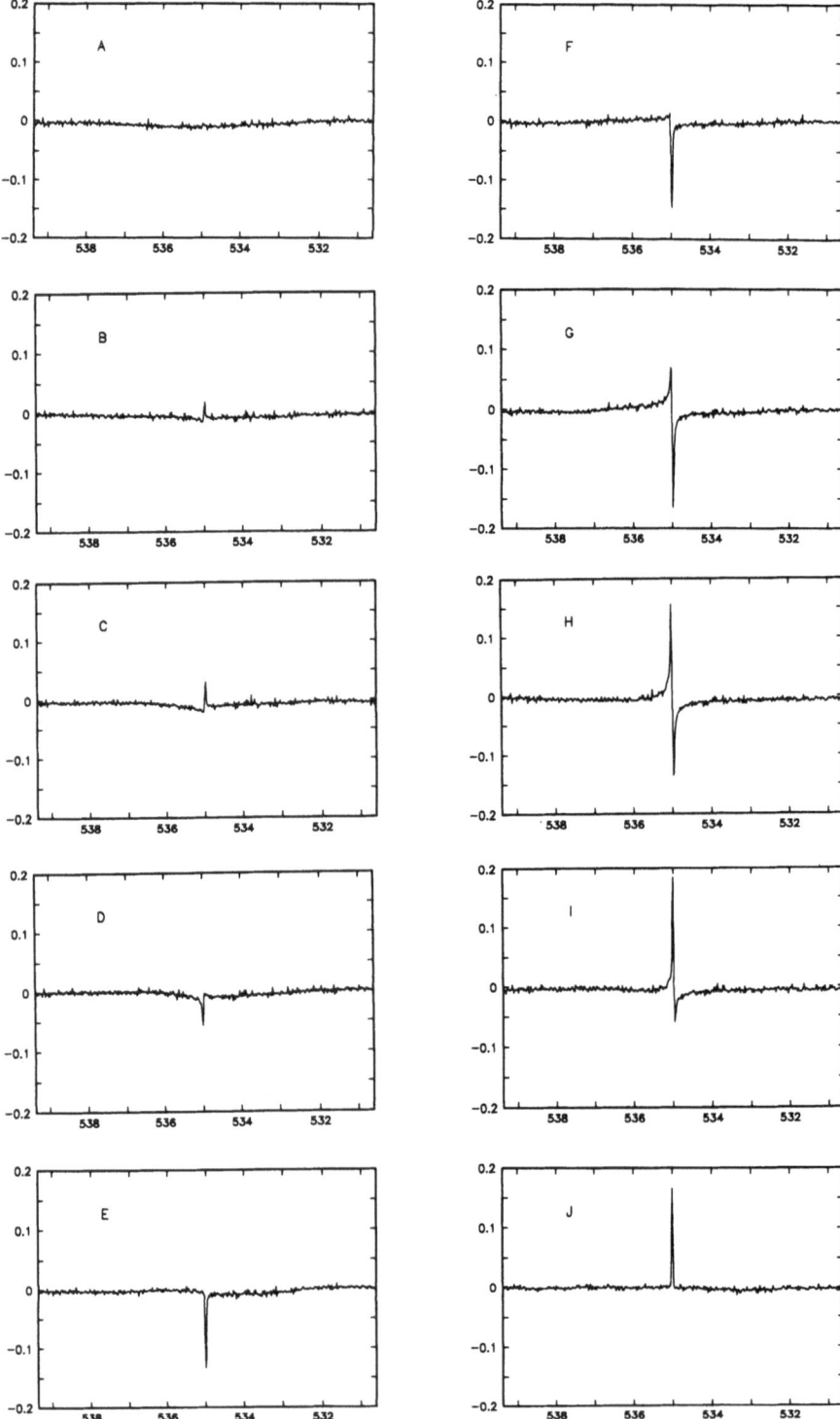

transient absorption spectra would be like. For long pump–probe delays, i.e. delays longer than a picosecond or so, we knew that we would be probing essentially free atoms, and that the absorption spectrum would simply be that of the particular (thallium) atomic resonance lines that happen to fall within the spectral region of the broadband probe pulse. As we dialed the pump–probe separation to be less and less, we thought that perhaps the absorption peak would shift to red wavelengths, or perhaps to blue wavelengths, since the molecular transition frequency being probed (also known as the difference frequency) was fully expected to vary with R, especially as we probed closer and closer to the molecular configuration.

However, when we performed the TlI experiments, something strikingly different was observed, as can be seen in Fig. 1.3. Here are shown some relatively recent transient absorption spectra taken in the vicinity of the 535-nm Tl resonance line, for various values of the pump–probe separation, which increase in order of alphabetical designation. (The actual values of the pump–probe separation can be seen in Fig. 1.4.) In this experiment, the photolysis was produced by a 160-fs, 308-nm laser pulse. Note, firstly, that at long times the absorption profile is just what would be expected. One sees simply absorption

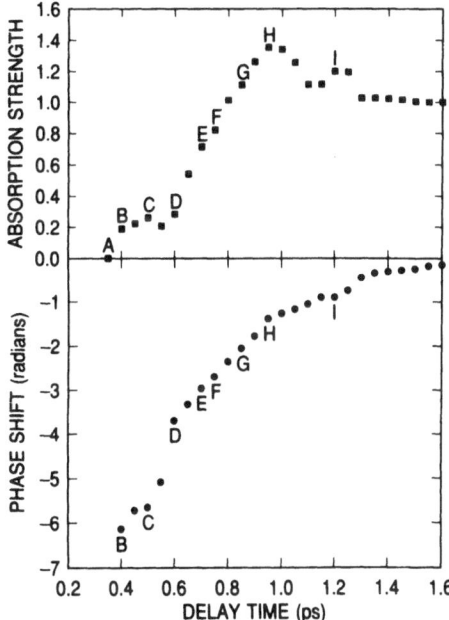

Fig. 1.4. Absorption strength and measured accumulated phase shift for experimental run corresponding to Fig. 1.3

Fig. 1.3. Transient absorption spectra taken in vicinity of 535-nm Tl resonance line following photolysis of TlI vapor by 160-fs, 308-nm pump pulse. Pump–probe delays as indicated in Fig. 1.4. Vertical scale plots absorbance. Horizontal scale shows wavelength in nm

due to the free atom resonance line. As one starts to decrease the pump–probe delay time, however, the absorption profile starts to change, becoming first dispersive, then displaying clearly a negative absorption profile, then dispersive with a change of sign, and, finally, almost approaching a positive absorption profile again, before the signal finally fades. The second feature to note is that the transient features just described are all centered around the free atom spectral line position. There is no sign of any early red (or blue) shift that one might have expected. These two general features were observed for a number of Tl resonance lines in the 308-nm (and 248-nm) photolysis of TlI and TlCl, and also in the case of Tl-cyclopentadienyl, the latter being a polyatomic gas-phase organometallic molecule in which a Tl atom sits just above the plane of the five-membered ring in the parent molecule, but flies apart from it when the molecule absorbs a 308-nm (or 248-nm) photon.

1.2 "Impact" and "Statistical" Transient Absorption Spectra

From the efforts of our group, a fairly convincing explanation has evolved for the appearance of these observed transient absorption spectra. The original explanation [1.2, 3] began with the observation that the probe pulse induces a polarization of the medium, which radiates a phase-matched electric field in the forward direction (i.e. in the direction of the probe beam propagation). In a measurement of the transmitted spectral power density, spectral components of the probe field interfere with spectral components of the polarization field. The absorption spectrum will be determined by the relative phases of these two fields.

The polarization induced by the probe pulse was initially calculated using Schrodinger's equation for two-level atoms in the adiabatic approximation. (The adiabatic approximation allows for a continuously varying frequency shift to be incorporated into the Schrodinger equation.) With the use of Maxwell's equations, the field radiated by this polarization was then calculated. Evaluation of the interference of this field with the field of the probe pulse then enabled us to compute the transient absorption spectrum, when certain simplifying assumptions were made. While the computed spectra qualitatively seemed to account for the observed transient line shapes, the deduced functionality of the difference frequency, $\omega_M(R)$, was unrealistic.

More recent theoretical work [1.5], emphasizing a more global viewpoint, has resulted in a greatly increased understanding of the problem. It turns out that the classical model of time-resolved absorption spectra that our group had started to develop is very closely related to adiabatic descriptions of collision-broadened spectral lines [1.10, 11]. This relationship to collision broadening provides a number of useful concepts that are directly applicable to the time-resolved case. For example, time-resolved absorption spectra have an "impact region" centered near the transition frequency for separated atoms, and a "far-wing" region which includes spectral components that are shifted far from the atomic resonance frequency. The impact region is due to radiation emitted *after* the molecules have separated into atomic fragments, and the far-wing region is due to radiation emitted *during* the time of strong molecular interaction. One can broadly classify time-resolved absorption spectra into two categories: "impact spectra", where most of the probe-induced polarization is radiated after the dissociation process is completed, and "statistical spectra", where most of the probe-induced polarization is radiated near the time of excitation by the probe pulse. "Impact spectra" are sharp and atomic-like, always centered at ω_0, the transition frequency for separated atoms. "Statistical spectra" are broad and shifted far from ω_0. An important factor that determines whether spectra will be "impact" or "statistical" is the spread in phase shifts across the width of the wavepacket moving along the dissociative potential curve. This phase-shift spread is defined in the discussion below. In Table 1.1, we summarize the principal characteristics of the two types of time-resolved spectra.

Since the TlI spectra are obviously representative of "impact" spectra, we begin by discussing this spectral type. The actual shape of the absorption profile near ω_0 results from the fact that the phase of the probe-induced polarization is shifted with respect to that of the probe pulse itself by the effect of the "partial half-collision" that begins the moment the probe pulse is applied. This phase shift results from the fact that the transition frequency continually changes, from the moment the probe pulse is applied, until the atoms have fully separated. From a general Fourier integral expression for the absorption coefficient $\alpha(\Delta)$ contained in *Walkup* et al. [1.5] one can easily show that, for the ideal case of a delta-function probe, and with $\gamma T_c < 1$ and $|\Delta| T_c < 1$, one has

$$\alpha(\Delta) = \frac{8\pi^2}{\lambda} \frac{1}{V} \sum_k \frac{d_k(t^*) d_0}{\hbar} \left[\cos(\eta_k) L(\Delta) + \sin(\eta_k) D(\Delta) \right] . \tag{1.1}$$

Table 1.1. Principal characteristics of "statistical" and "impact" spectra

Statistical spectra	Impact spectra
Applies to absorption found in spectral region $|\Delta| T_c > 1$. (Here, $\Delta = \omega - \omega_0$, and T_c is the duration of the half-collision)	Applies to spectral region $|\Delta| T_c < 1$ (For $T_c = 1$ ps, $|\Delta| < 5$ cm^{-1})
Significant radiation by probe-induced polarization occurs during half-collision ... leads to "molecular-like" spectra	All radiation by probe–induced polarization occurs after half-collision is over ($\gamma T_c < 1$), where γ is the coherence decay rate for the probe-induced polarization ... leads to "atomic-like" spectra centered at ω_0
Local character: shape of the absorption spectrum at frequency ω due to molecules with internuclear separations, R, such that $\omega_M(R) = \omega$	Shape of absorption spectrum determined by accumulated phase shift of probe–induced polarization occurring during half-collision
Requires broad wave packets for full development	Requires narrow wave packets for full development

Here the sum is over all of the molecules in the volume V that are excited to state $|1\rangle$, t^* is the time that the delta-function probe pulse is applied to the molecules, relative to the pump pulse, and $d(t)$ is the dipole transition moment, viewed as a function of time. The functions $L(\Delta)$ and $D(\Delta)$ are Lorentzian and dispersion profiles:

$$L(\Delta) = \frac{\gamma}{\gamma^2 + \Delta^2} , \tag{1.2}$$

and

$$D(\Delta) = \frac{\Delta}{\gamma^2 + \Delta^2} . \tag{1.3}$$

The quantity η_k is the net phase shift for a "partial" collision:

$$\eta_k = \int_{t*}^{\infty} \delta\omega_k(t') \, dt' , \tag{1.4}$$

with

$$\delta\omega_k = \omega_M[R_k(t)] - \omega_0 \tag{1.5}$$

being the shift from the separated atom frequency for molecule k travelling on trajectory $R_k(t)$. Thus, for frequencies near the separated-atom resonance, the absorption spectrum is a weighted sum of Lorentzian and dispersion profiles, with weights that depend upon phase shifts η_k.

1.3 Extracting the Difference Potential

It should be clear from the above that, as the pump–probe separation is reduced, the amount of accumulated phase shift increases, and the transient absorption profile in the impact region should thus continually cycle between positive absorption, dispersion, and negative absorption, with the cycling becoming more and more rapid as one approaches the molecular configuration. Exactly this behavior, for roughly one such cycle, is seen in the TlI spectra. Before discussing what limits the amount of accumulated phase evolution in actual experiments, let us first indicate how this simple theory points to a potentially powerful way of obtaining the molecular transition frequency as a function of pump–probe delay, by analysis of the time–resolved absorption spectra.

By fitting the shape of the absorption spectrum in the impact region to a superposition of Lorentzian and dispersion functions, one can determine $\eta(t^*)$, the net phase shift accumulated from the time, t^*, of excitation by the probe pulse to $t = \infty$. The difference potential can then be determined by differentiation of the phase shift. (This is obvious from the expression for η_k, given above.)

If the relationship between internuclear separation and the time of excitation by the probe is known, then the difference potential, $\omega_M(R)$, can be determined vs. the internuclear separation, R.

The ability to extract the difference potential depends upon the width of the spatial distribution of molecules excited to state $|1\rangle$ by the pump pulse. If the dissociation process yields a narrow spatial distribution on state $|1\rangle$, then all of the molecules will contribute with essentially the same phase shift. However, if there is substantial variation of the phase shift across the width of the distribution, then there will be destructive interference of the phase-shifted radiated waves. Thus, the absorption spectrum near ω_0 will be very small in amplitude, and may not reflect the average phase shift. This, in fact, is why one only sees roughly one cycle of phase evolution in the TlI spectra in Fig. 1.3. In Fig. 1.4, both the transient signal strength and corresponding accumulated phase are plotted as a function of pump-probe delay for the complete run that forms the basis of Fig. 1.3. We note (again) that a phase accumulation of almost 2π evolves before the transient signal completely vanishes when the pump–probe delay is made less than 0.4 ps. If we globally fit the phase shift vs. delay curve, and then analytically differentiate the fitted function, we approximately recover an effective C_6 coefficient for the difference potential. For the TlI system, this procedure agrees with the difference between the C_6 coefficients for the analogous states of the Tl–Xe system, to within an estimated error of about 30%. With reference to Fig. 1.1, the center of the wavepacket moving on state $|1\rangle$ has already reached an R value of about 6 Å when the transient absorption signal first starts to appear. At this point, the probe transition frequency is redshifted from the separated-atom value by about 200 cm^{-1}. Only the van der Waals interaction region has thus far really been probed in real time.

What, then, can be done about probing the potential difference closer to the molecular configuration? Figure 1.5 shows some realistic estimates for the amount of accumulated phase shift that should be observable in TlI under varying pump conditions, with the abscissa being the temperature of the parent

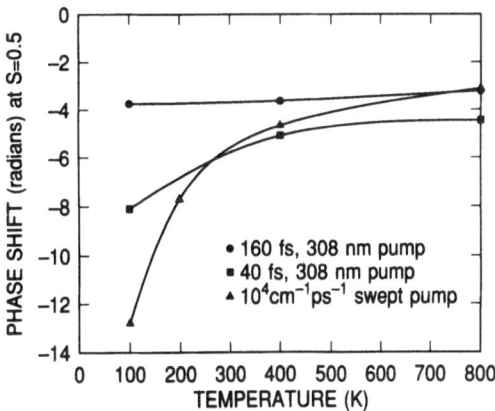

Fig. 1.5. Estimated maximum observable accumulated phase shift for the TlI system modeled in Fig. 1.1, as a function of pump pulse characteristics and initial temperature of parent gas molecules

gas. (Note that the maximum phase shift is here defined more conservatively than in the case of Fig. 1.4, being associated with the point at which the amplitude of the transient signal is half the asymptotic value, rather than with the point at which the signal disappears entirely.) The top curve applies to our experiments with TlI – it is seen that not much benefit would result from cooling of the gas. For a 40-fs pump pulse, twice as much phase evolution can be expected, if the gas is near $100°$ K. The most interesting looking curve – the one with the most accumulated phase evolution at low temperatures – corresponds to a frequency-swept pump pulse, whose sweep partially compensates the velocity dispersion of molecules moving on state $|1\rangle$ and leads to subsequent momentary pulse compression of the wavepacket moving on state $|1\rangle$. (This effect, if it could be demonstrated, would be somewhat similar to the pulse compression of a frequency-swept optical pulse propagating in a dispersive medium.)

Finally, computer simulations [1.5] for the model TlI system show that, at $0°$ K, for delta-function pump and probe pulses, the difference potential recovered via the impact approximation represents fairly accurately the actual difference potential seen at the center of the wave packet moving on state $|1\rangle$, all the way back to the parent molecule configuration. Thus, the importance of going to low temperatures, for the immediate purpose of minimizing wave-packet widths on state $|1\rangle$, and for the ultimate purpose of recovering difference potentials, is again indicated.

The question, then, is: how can very low temperature TlI gas molecules be produced? The answer is through the use of a pulsed planar molecular jet which has been created by expansion of a heated mixture of TlI vapor and high-pressure argon gas through a 10-cm-long, 20-μm-wide slit into a vacuum. A number of chemical physicists [1.12, 13] have utilized the tremendous cooling obtained in such jets to record single-pass, low-temperature, absorption spectra (with the probe beam transverse to the direction of the jet) of even very weakly absorbing molecules. The atomic transitions we usually probe in our experiments are very strong, so detecting the transient signals should not be impossibly difficult. This would be an interesting experiment to try, but it has to be done in collaboration with someone with access to a pulsed planar nozzle.

One advantage that the Caltech approach to transition-state spectroscopy, in which fluorescence is induced by a short optical probe pulse, has over the approach we have followed should here be pointed out. Since the fluorescence signal is not sensitive to the relative phases of two fields, there will not be an inherent cancellation of intensities when a broad wave-packet on state $|1\rangle$ is probed at early times in the dissociative half-collision, unlike the situation that occurs when the transmitted spectral power density is monitored, especially in the impact region. This should allow probing closer to the molecular configuration without the necessity of going to very low temperatures, for example. However, there appear also to be some severe limitations with this method when 100-fs probe pulses are employed, and the difference potential varies greatly with R, as it does close to the molecular configuration. Essentially, the problem is that for molecules moving on state $|1\rangle$, the time that they spend in resonance with the probe laser is much shorter than the probe laser pulsewidth, thus reducing the probability of excitation. One can, of course, shorten the probe laser pulsewidth, but then one tends to lose information about the difference potential at the time of excitation.

In the limit of a delta-function probe pulse, fluorescence monitoring would simply reflect the square of the transition dipole moment, averaged over the distribution function at the time of excitation by the probe pulse. Thus, the dependence of the transition dipole moment on internuclear separation can, in principle, be simply determined experimentally.

1.4 Examples of "Statistical" Spectra: Bi$_2$

Let us now discuss some theory and experiments which involve far-wing spectra. From the summary table given above, it appears that transient absorption spectra of this type are determined by the characteristics of radiation emitted during the half-collision. In particular, absorption at a frequency ω can be directly attributed to molecules with an internuclear separation R, such that the difference potential at R equals ω. Mathematically, the Fourier integral in the general expression for the absorption spectrum is approximated using the method of stationary phase [1.5]. It turns out that the resulting expression for the absorption coefficient as a function of frequency, $\alpha(\Delta)$, simplifies considerably in the limit of large phase variation, $\delta\eta$, across the distribution. Let us examine this limit for the ideal case of delta-function pump and probe pulses. The result is:

$$\alpha(\omega) \approx \frac{8\pi^3}{\hbar\lambda} n_1 d^2(R) \frac{\rho_1(R;t^*)}{\left|\dfrac{\partial\omega_{\mathrm{M}}(R)}{\partial R}\right|} \ . \tag{1.6}$$

In this expression, ω and R are related by $\omega_{\mathrm{M}}(R) = \omega$, and $\rho_1(R;t^*)$ is the normalized distribution of internuclear separations for molecules on state $|1\rangle$ at the time, t^*, of excitation by the probe pulse.

Note that this expression for $\alpha(\omega)$ is positive definite. In this limit, the spectrum is a scaled image in frequency space of the wave packet (actually, the probability density) in configuration space. This lineshape would be referred to as a statistical lineshape in the language of collision broadening, because the probability of absorption within $d\omega$ of a given frequency shift ω is simply proportional to the probability of finding molecules within dR of R, where, again, $\omega_{\mathrm{M}}(R) = \omega$:

$$\alpha(\omega)\, d\omega \propto d^2(R)\rho_1(R;t^*)\, dR \ . \tag{1.7}$$

We emphasize that the far-wing region is described by the statistical profile only in the limit of a large phase variation over the width of the distribution. In this same limit, the amplitude of the spectrum in the impact region tends to zero, due to destructive interference of the phase-shifted radiated waves; and thus the statistical profile describes the whole spectrum. If the time-resolved spectra show prominent derivative-like features in the impact region, then the far-wing region is not described by a statistical profile. Conversely, if the amplitude in the impact region vanishes, then the spectrum will be described by a statistical profile.

In some work performed almost two years ago [1.4], a photodissociative system was found in which the time-resolved absorption spectra were observed to be approximately statistical. This system was gaseous diatomic bismuth, photolyzed at 308-nm. In Bi_2 photolysis, the Bi–Bi interaction time is relatively long (3–4 ps). This is a result of three main factors: (1) Because Bi_2 is a homonuclear diatomic molecule, some of the excited state potential curves display, in addition to the usual van der Waals interaction, a long-range $(1/R^3)$ dipole–dipole interaction, not present in heteronuclear diatomics such as TlI; (2) The dissociative Bi_2 potential curves accessed by 308-nm pump light are very shallow, meaning that very little kinetic energy is released on photolysis; (3) Bi atoms are relatively heavy. Thus, for a given release of kinetic energy, the two Bi atoms move apart relatively slowly. The Bi_2 experiments were conducted at a relatively high temperature (1170° K). This fact, in conjunction with the fact of shallow potential curves, results in the formation of very broad wavepackets on the dissociative state $|1\rangle$ in the case of Bi_2. Thus, it is understandable why the Bi_2 transient absorption spectra are generally observed to display statistical lineshapes, and not "impact" lineshapes, such as are seen in the case of TlI.

In somewhat greater detail, Fig. 1.6 shows the electronic states of Bi_2, as represented in the previous scientific literature [1.14]. Of interest is the fact that a dissociative state M exists which would absorb at 308-nm and which would, according to this figure, produce at least one excited state, the $6p^3\ ^2D^0_{3/2}$. We found that photolysis of Bi_2 by 308-nm light produces, in fact, not only $^4S^0_{3/2}$ + $^2D^0_{3/2}$ atoms, but also $^4S^0_{3/2}$ + $^2D^0_{5/2}$ atoms, in roughly the same abundance. For this reason, we have added another potential curve M′ to the others shown in the figure. In reality, we know very little about the shapes of the potential curves M and M′ close to the molecular configuration, and it is even likely that in view of the apparent smoothness of the Bi_2 continuum absorption band (Fig. 1.7) through which the photodissociation is accessed, there is only one state with which the pump pulse interacts. The basis for the observed second exit channel could then well be an avoided crossing as shown in Fig. 1.8, as recently suggested by *Bowman* et al. [1.15].

At any rate, we observed this second exit channel, and found, moreover, that the appearance of $^2D^0_{5/2}$ atoms was delayed by 0.5 ps with respect to the appearance of $^2D^0_{3/2}$ atoms. We did not try to time-resolve the appearance of $^4S^0_{3/2}$ atoms, since there are always a large number of these, bismuth vapor being roughly half atomic and half molecular.

The appearance of transient absorption spectra in the vicinity of the 302.5-nm line, which, in the free atom, originates from the $^2D^0_{5/2}$ state, is shown

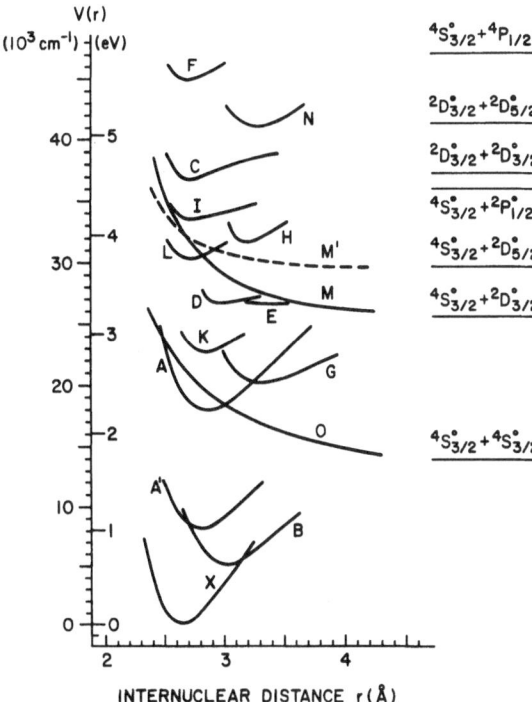

Fig. 1.6. Electronic states of Bi$_2$ molecules, as represented in [1.14]. The dashed potential curve M' has been added to indicate the existence of a second dissociative channel observed with 308-nm photolysis

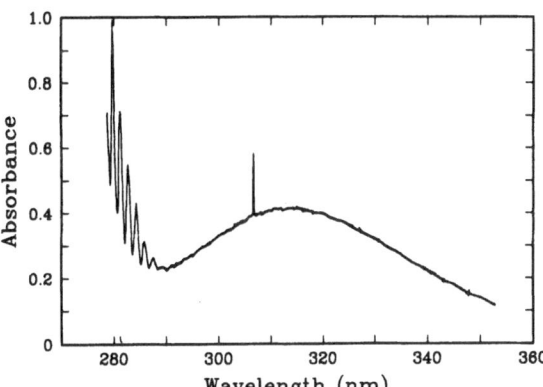

Fig. 1.7. Absorption spectrum of bismuth vapor measured near 308-nm. The absorption line at 307.7-nm is due to Bi atoms

in Fig. 1.9. Note that at early times the absorption is entirely concentrated on the red side of the free atom line position, and that, at the earliest time shown, the wing extends some 60 cm^{-1} beyond the free atom position. The wing gradually pulls in, and the absorption profile eventually becomes that of the

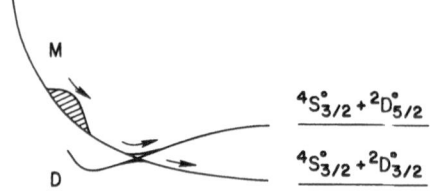

$$^4S_{3/2}^\circ + {}^2D_{5/2}^\circ$$

$$^4S_{3/2}^\circ + {}^2D_{3/2}^\circ$$

Fig. 1.8. Possible avoided-crossing basis for the second dissociative channel observed with 308-nm photolysis of Bi_2

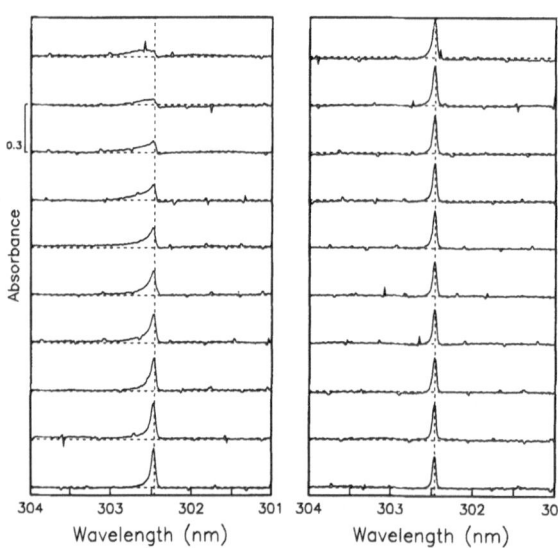

Fig. 1.9. Transient absorption spectra taken in the vicinity of the 302.5-nm Bi absorption line following photolysis of Bi_2 vapor with 160-fs, 308-nm pump pulses. Separation between pump and probe pulses increased by 100-fs between successive spectra

separated atom. There is no sign of the cycling between positive absorption, dispersion, and negative absorption profiles that we saw in the TlI spectra.

Another example of nearly statistical behavior is shown in Fig. 1.10. Here, again, the free-atom transition monitors the population in the upper doublet D level. This time the transient absorption is blueshifted with respect to the

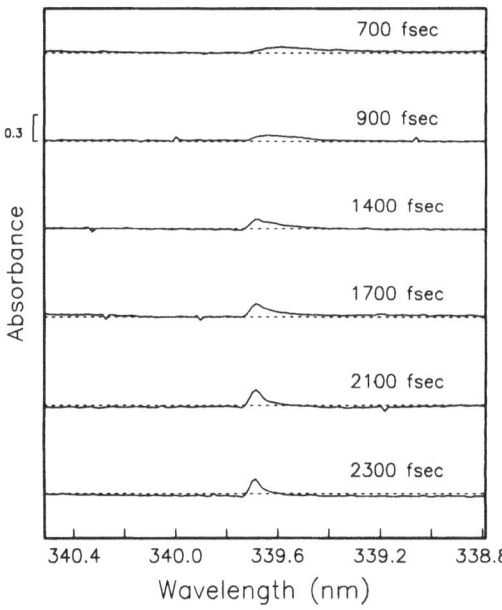

Fig. 1.10. Transient absorption spectra taken in the vicinity of the 339.7-nm Bi absorption line following photolysis of Bi_2 vapor with 160-fs, 308-nm pump pulses. Pump–probe delays indicated

separated-atom frequency, showing that the difference potential in the case of this probe transition is blueshifted when the probe pulse is applied. Again, the maximum extent of the blue wing shown here is about 50 wavenumbers.

An attempt to model qualitatively these transients is shown in Fig. 1.11 [1.5]. The left panel shows the distribution of internuclear separations calcu-

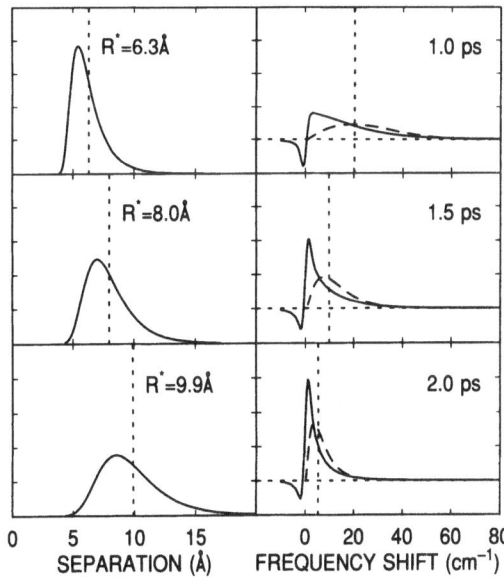

Fig. 1.11. Left panel: distribution of internuclear separations calculated for 308-nm photolysis of Bi_2 on the "slow" dissociation channel. The vertical dotted lines indicate the centroid of the distribution for delay times of 1.0, 1.5, and 2.0 ps. Right panel: corresponding calculated time-resolved absorption spectra, using the exact Fourier integral expression (solid curves) and the "statistical spectra" model, (1.6), (dashed curves). The vertical dotted lines indicate the frequency shift for molecules at the centroid of the distribution, and the horizontal dotted lines indicate zero absorption

lated for 308-nm photolysis of Bi_2 on the "slow" dissociation channel. The vertical dotted lines indicate the centroid of the distribution for delay times of 1.0, 1.5, and 2.0 ps. The right panel shows the corresponding time-resolved absorption spectra, using the exact Fourier integral expression (solid curves), and the statistical approximation (dashed curves). The vertical dotted lines indicate the frequency shift for molecules at the centroid of the distribution, and the horizontal dotted lines indicate zero absorption. The calculated spectra assume a difference potential that has the dipole–dipole form, varying as R^{-3}, with a coefficient C_3 appropriate for states coupled by dipole transitions with oscillator strengths of about 0.05, as is the case for the states involved in the Bi_2 system.

We note that the far-wing spectra calculated from the exact Fourier integral expression have a substantially different character from ones based upon the impact approximation. There is still a noticeable derivative-like feature near ω_0 in the calculated spectra. However, the dominant feature is a broad wing, which is always positive, and which gradually pulls in towards the separated-atom frequency as the average internuclear separation increases. The statistical spectra are seen not to be quantitatively accurate, and the reason is simply that there is not enough phase variation over the distribution. For example, at 1.0-ps delay, the phase variation is $\delta\eta \approx \delta\omega(R^*)\delta R/v \approx 4$ rad. (Here $\delta\omega(R^*)$ is the shift of the molecular transition frequency from the separated-atom limit for a molecule at position R^*, which is given by the center of the wavepacket on state $|1\rangle$ at the time of excitation by the probe pulse. The width of the wavepacket at

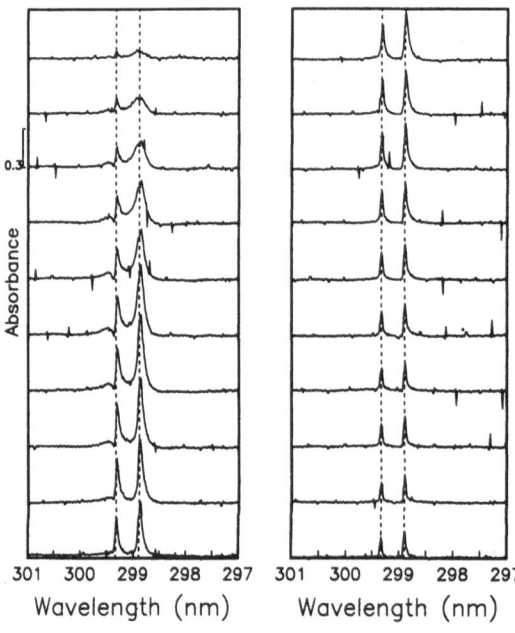

Fig. 1.12. Time-resolved absorption spectra taken in the vicinity of the 299.0- and 299.3-nm Bi lines, following photolysis of Bi_2 vapor with 160-fs, 308-nm pump pulses. Pump–probe delay increased by 100-fs between successive spectra

the time of excitation by the probe pulse is δR, and v is a characteristic relative velocity.) Calculations show that far-wing spectra would more closely approach the statistical shape for systems with $\delta \eta \geq 10$ rad. Since the spectra in both Figs. 1.9 and 1.10 do appear to have statistical lineshapes, we must tentatively conclude that the actual wavepacket distributions are even broader than assumed for the calculations of Fig. 1.11.

It was assumed in the above calculations that the transition dipole moment is independent of internuclear separation. However, for some of the transitions we have monitored in bismuth, this does not seem to be true at all. As evidence of this statement, consider Fig. 1.12, in which Bi_2 molecules dissociating in the "fast" channel are stroboscopically viewed through a particular probe transition. Spectroscopic features as surprising as the ones shown in this figure will undoubtedly continue to be found in future real-time absorption studies of molecular dissociation.

1.5 Conclusions

We have tried to present a picture of what is conceptually involved in our studies thus far. The major finding has been the observation that two widely different types of transient absorption spectra are to be seen in nature, the type we call "impact spectra", and the type we term "statistical spectra". Both types, when interpreted and analyzed properly, can provide information about difference potentials.

Acknowledgements. Our research has been supported, in part, by the U.S. Army Research Office.

References

1.1   R.L. Fork, B.I. Greene, C.V. Shank: Appl. Phys. Lett. **38**, 671 (1981)
1.2   J.A. Misewich, J.H. Glownia, J.E. Rothenberg, P.P. Sorokin: Chem. Phys. Lett. **150**, 374 (1988)
1.3   J.H. Glownia, J. Misewich, P.P. Sorokin: *The Supercontinuum Laser Source*, ed. by R.R. Alfano (Springer, New York 1989) pp. 337–376
1.4   J.H. Glownia, J.A. Misewich, P.P. Sorokin: J. Chem. Phys. **92**, 3335 (1990)
1.5   R.E. Walkup, J.A. Misewich, J.H. Glownia, P.P. Sorokin: J. Chem. Phys. **94**, 3389 (1991)
1.6   P. Davidovits, J.A. Bellisio: J. Chem. Phys. **50**, 3560 (1969)
1.7   E. Czuchaj, J. Sienkiewicz: Z. Naturforsch. **39a**, 513 (1984)
1.8   N.J.A. Van Veen, M.S. De Vries, T. Baller, A.E. De Vries: Chem. Phys. **55**, 371 (1981)
1.9   M. Dantus, M.J. Rosker, A.H. Zewail: J. Chem. Phys. **87**, 2395 (1987)
1.10 I.I. Sobelman, L.A. Vainshtein, E.A. Yukov: *Excitation of Atoms and Broadening of Spectral Lines* (Springer, Berlin, Heidelberg 1981)
1.11 N. Allard, J. Kielkopf: Rev. Mod. Phys. **54**, 1103 (1982)
1.12 A. Amirav, U. Even, J. Jortner: Chem. Phys. Lett. **83**, 1 (1981)

1.13   K. Veeken, J. Reuss: Appl. Phys. **B38,** 117 (1985)
1.14   G. Gerber, H.P. Broida: J. Chem. Phys. **64,** 3423 (1976)
1.15   R.M. Bowman, J.J. Gerdy, G. Roberts, A.H. Zewail: J. Phys. Chem. **95,** 4635 (1991)

Peter P. Sorokin   James H. Glownia

2. Dye Lasers and Laser Dyes in Physical Chemistry

F.P. Schäfer

With 22 Figures

A symposium celebrating 25 years of dye lasers is the right occasion to remember the fact that it was by the introduction of dye lasers that physical chemistry (and to some extent even organic chemistry) started to play an important rôle in the further development of lasers. Even after such a long time there are still many physicochemical aspects of dye lasers and laser dyes that deserve a closer look and that might even affect the prospects for dye lasers in the future.

Dye lasers are certainly the most versatile class of lasers because of the virtually unlimited variety of dye molecules that can be used as active medium in dye lasers. But they also exhibit certain limitations in their performance as well as some peculiarities which are inseparably linked to inherent properties of the active medium. So it seems appropriate to start with some remarks on the active medium, i.e. laser dyes and their environment.

The chemical constitution and the absorption and fluorescence spectra of the first two dyes ever used in dye lasers are quite typical and are therefore reproduced in Fig. 2.1. Chloro-aluminum-phthalocyanine contains a ring-shaped π-electron cloud; 3,3'-diethylthiatricarbocyanine iodide has a linearly extended one. In such dyes the position and strength of the longest wavelength absorptive and emissive transition can readily be calculated using *Kuhn*'s electron gas theory of organic dyes [2.1]. Incidentally, the cross-sections of these transitions are reasonably well approximated by the area covered by a projection of the π-electron cloud on the nuclear plane. Since the length of the linear or circular π-electron cloud is responsible for the wavelength of absorption and emission, the question immediately arises as to what are the minimum and maximum wavelengths that can be reached with dye molecules and how do they translate into the wavelength range of dye lasers.

2.1 The Wavelength Limits of Dye Lasers

While in common usage the word "dye" means a substance that absorbs in some part of the visibile spectrum, thus creating a color sensation, it has become standard practice in science to include all substances with the same absorption mechanism by a π-electron cloud under this generic term. So the question of finding the shortest-wavelength absorbing dyes could be answered by

Topics in Applied Physics, Vol. 70
Dye Lasers: 25 Years Ed.: Dr. Michael Stuke
© Springer-Verlag Berlin Heidelberg 1992

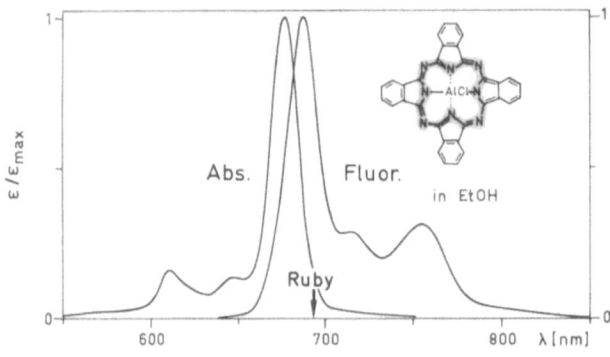

P.P. Sorokin and J.R. Lankard,
IBM J. Res. Developm. 10, 162 (1966)

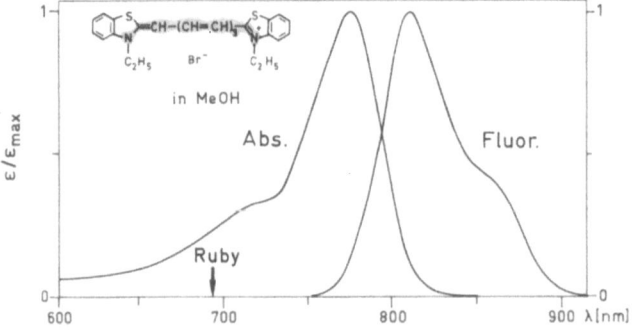

F.P. Schäfer, W. Schmidt and J. Volze,
Appl. Physics Letters 9, 306 (1966)

Fig. 2.1. Chemical formulae and spectra of the two first laser dyes

considering ethenes, possessing only two π-electrons and absorbing around 150 nm. Such high quantum energies, however, immediately destroy the molecules photochemically since all C–C and C–H bond energies are much weaker. So a particularly stable small molecule, namely, the guanidinium ion, shown with its absorption band in Fig. 2.2, could perhaps be more suitable, since its absorption maximum occurs at 182 nm. The first obstacle to achieving dye laser action with this dye is the extremely small fluorescence quantum yield. This can be overcome in many cases, as will be explained below, by excitation with ultrashort pulses in a travelling-wave pump arrangement. Even with this special arrangement all attempts to achieve laser emission (expected in the region around 200 nm) failed [2.2]. The reason for this, as was soon found out, was the two-step excitation into the dissociation and ionization continuum which evidently had a higher transition probability than the transition by stimulated emission to the ground state (cf. Fig. 2.2). Since it is highly probable that this

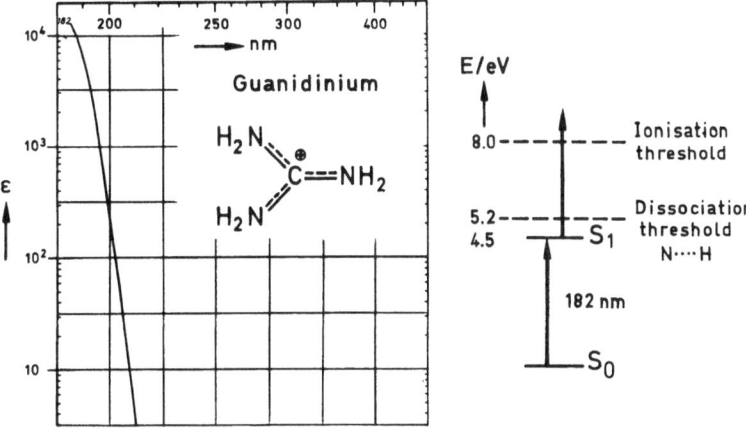

Fig. 2.2. Absorption spectrum and energy levels of the guanidinium ion

molecule changes upon excitation from a flat configuration into an umbrella-like shape which would facilitate subsequent dissociation as well as ionization, it would be worthwhile to try a supramolecular approach forcing the molecule to retain its planar shape upon electronic excitation. One such possibility would be to fix the guanidinium ion into a molecular cage as recently demonstrated by *Lehn* [2.3], where the guanidinium is fixed by six hydrogen-bonds (Fig. 2.3). Unfortunately, this possibility was not available at the time of the original laser experiments. At present the shortest dye laser wavelength remains at 308.5 nm [2.4].

The long-wavelength limit of dye lasers is essentially determined by the chemical stability of the dyes. The reason for this is the high reactivity of triplet states that can already be populated thermally at room temperature, when the absorption is in the near infrared, resulting in a low-lying triplet of less than

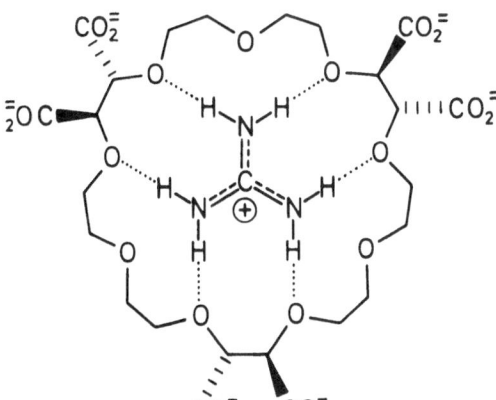

Fig. 2.3. Supramolecular fixation of a guanidinium ion

20 kcal/mol. It has been estimated that there is little hope of ever preparing a dye absorbing beyond 1.7 μm [2.5]. This limit seems to have been nearly reached with some new laser dyes synthesized by *Drexhage* and coworkers which have their peak absorption at up to 1.45 μm, thus enabling dye laser action to be extended to 1.85 μm [2.6].

One way around this stability limit would be to create an unstable long-wavelength absorbing dye photochemically from a short-wavelength absorbing stable precursor molecule. After laser action the unstable dye would quickly recombine to recover the precursor. An illuminating example of how this can be done is given by *Ishizaka* and *Kotani* [2.7], even though this work was done for a different purpose and the resulting dye laser wavelength was still at the red end of the visible. In this work an aromatic disulfide absorbing in the ultraviolet was dissociated by a photon from a XeCl-excimer laser at 308 nm to form two radicals with an absorption band peaking at 560 nm (Fig. 2.4). These highly

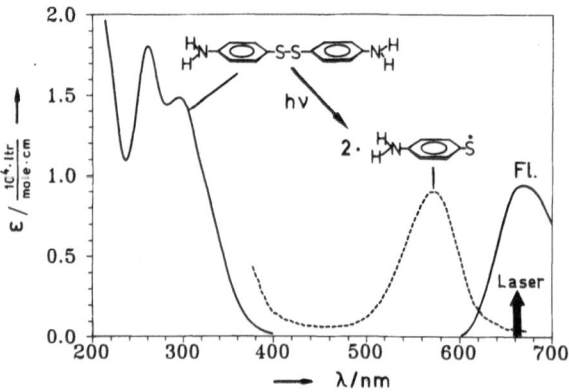

Fig. 2.4. Absorption spectrum of a disulfide (solid line) and of the corresponding thiyl-radical (dashed line) and fluorescence spectrum and wavelength of laser action of the thiyl-radical (adapted from [2.8])

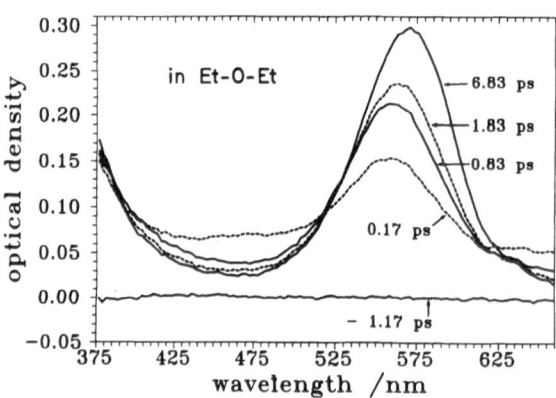

Fig. 2.5. Time-resolved absorption spectrum of the thiyl-radical of Fig. 2.4. Parameter: Time after maximum of the dissociating laser pulse at 308 nm. Adapted from [2.8]

fluorescent molecules were than pumped by the same XeCl laser pulse to give dye laser action at 670 nm, finally recombining to the disulfide in a dark-reaction after a few microseconds. This reaction was recently studied by *Ernsting* using a femtosecond-absorption spectrometer who could clearly show the initial formation of the geminal pair of radicals and the subsequent solvation by ether [2.8]. Figure 2.5, reproduced from Ernsting's work, demonstrates the rise of the radical absorption band and the incipient wavelength shift due to solvation.

2.2 Limitations by Radiationless Transitions

Radiationless transitions from the first excited state to the ground state or triplet state reduce the quantum yield q of fluorescence, defined as the ratio of the Einstein coefficient for spontaneous emission A and the sum of the rate K_{rl} of radiationless transitions and A: $q = A/(A + k_{rl})$. Since all laser action starts from spontaneous emission, it becomes increasingly difficult to operate a dye laser the stronger the quenching of the fluorescence by radiationless transitions for the dye used. For a long time it was thought that a quantum yield of about 10% was the limit for practicable dye laser operation. One way around this limitation would be to force the dye molecule to make a transition to the ground state by stimulated emission immediately after excitation, before radiationless transitions has had a chance to deactivate the molecule. Assuming a dye radiative lifetime $1/A \approx 10$ ns and a quantum yield of $q = 10^{-4}$, this would mean that stimulated emission on the average should occur in less than 1 ps. An ingenious solution of this problem was found independently by two groups [2.9, 10] and is shown in Fig. 2.6. A picosecond pulse from a suitable pump laser is sent through a beam-expanding cylinder-lens telescope to illuminate N lines of an optical grating under the blaze angle so that most of the energy is diffracted into the first order. From elementary theory of diffraction it is clear that the contribution to the diffracted beam from each line has a delay of one wavelength with respect to the neighboring line, so that the delay between the left and the

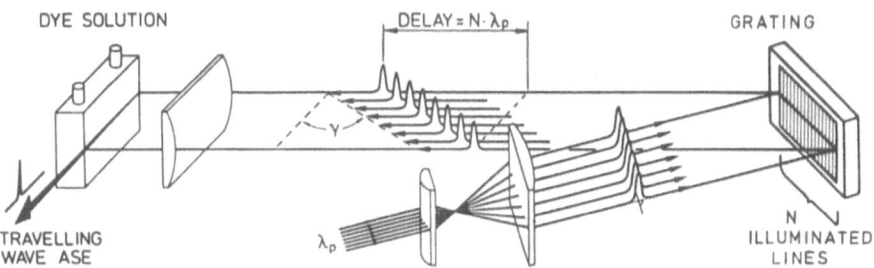

Fig. 2.6. Travelling-wave excitation of amplified spontaneous emission (ASE) in a dye cell. From [2.9]

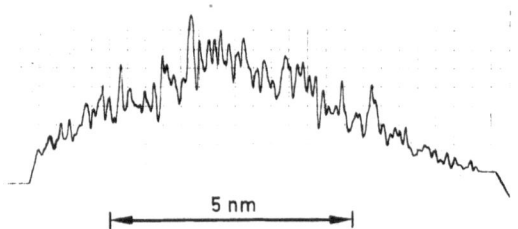

Fig. 2.7. Typical output spectrum of an ultrashort pulse generated with the experimental setup of Fig. 2.6. From [2.23]

5 nm

right border of the diffracted beam is N wavelengths. This means that the pulse front is tilted by an angle γ. Now this pulse is focused by a cylinder lens on the inner surface of a dye cell containing a weakly fluorescent dye solution with a refractive index n at the wavelength of the fluorescence maximum. It is immediately obvious that the pulse hitting first the near end of the cell will travel towards the far end across the cell with exactly the same speed as the fluorescence light emitted in the same direction, provided that the tilt of the pulse front is correctly chosen so that $\tan \gamma = n$. If the intensity of the pump pulse is sufficiently high, a few fluorescence photons will start at the near end in the direction towards the far end of the cell always seeing dye molecules that were excited into the lasing level less than 1 ps ago, so that stimulated emission has a higher probability than radiationless transitions, the fluorescence photon flux is strongly amplified, and a strong output beam of amplified spontaneous emission exits from the far end of the cell.

One disadvantage of this method is the broad spectral bandwidth of irregular shape (Fig. 2.7) with strong fluctuations from pulse to pulse [2.23]. A simple solution is shown in Fig. 2.8, where a narrow spectral band from the output of dye cell DC1 is selected by a grating G2, a lens L1, and a slit S. This spectrally narrowed pulse is then amplified in a second dye cell DC2 which is pumped by a part of the pump pulse that is sent over a delay path via mirror M1, adjusted so that the travelling-wave pump and the dye laser pulse are again in exact synchrony, as described above. By rotating the grating G2 one can now select a spectral region with a pulse duration limited by the bandwidth over a wide spectral range [2.23]. In this way it was possible to obtain the longest-wavelength dye laser pulses mentioned above [2.6] with the dye shown with its spectra in Fig. 2.9, reaching a pump energy conversion of 1%, even though the fluorescence quantum yield was only $q \approx 10^{-4}$. This method can probably be extended to even lower quantum yields using powerful femtosecond pump pulses and is only limited by nonlinear effects like multiphoton absorption etc. at very high intensities.

A nice visual observation of the suppression not only of radiationless transitions but also of the spontaneous emission by strong stimulated emission is possible if one fills the two dye cells in Fig. 2.8 with a strongly fluorescing dye like rhodamine 6G. If the output beam from DC1 is blocked and the cylinder lens in front of DC2 is moved a few millimeters from its exactly focused position,

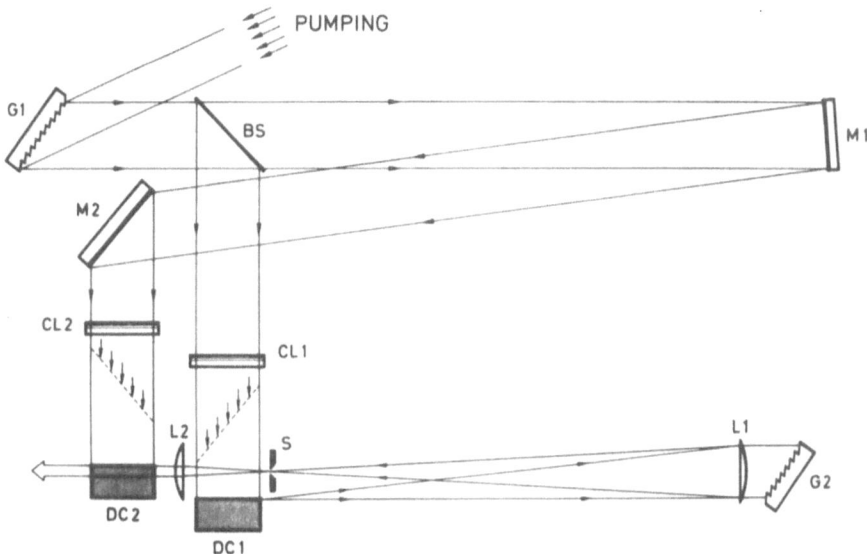

Fig. 2.8. Extension of the travelling-wave pumping arrangement of Fig. 2.6 for spectral selection and amplification of the ultrashort pulse output. Symbols explained in the text. From [2.23]

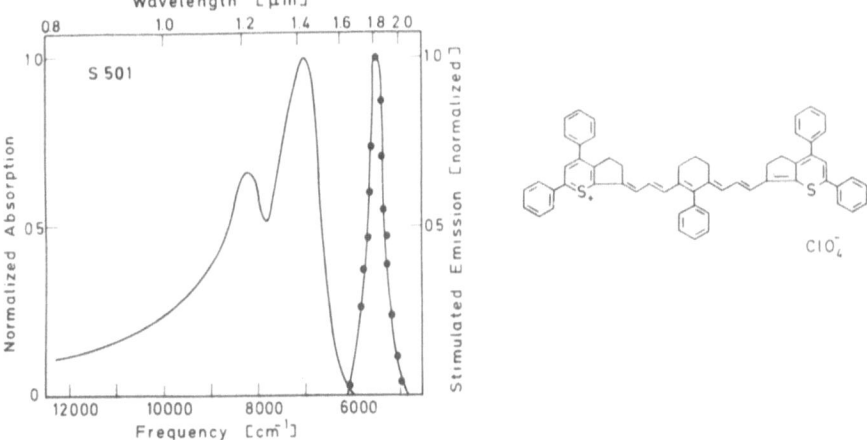

Fig. 2.9. Structural formula, absorption spectrum, and tuning curve of dye laser emission of the longest wavelength laser dye. From [2.6]

a brightly fluorescent band is seen on DC2. If the stimulating beam from DC1 is now admitted, a black trace on this bright stripe indicates where the stimulating beam has deactivated practically all excited dye molecules in passing before they could radiate spontaneously.

2.3 The Smallest and Largest Dye Lasers

Because of the very high gain of dye lasers, one can build useful devices of microscopic dimensions. On the other hand, the cheapness of dye solutions also allows one to build huge dye lasers of dimensions that are hardly possible with any other class of lasers. Two examples will illustrate this statement.

The first dye laser that could be classed as very small in at least one dimension is shown in Fig. 2.10a. It consisted of two dichroic mirrors highly transmitting for the ruby laser pump radiation and highly reflective for the dye laser radiation in the near infrared. The mirrors were separated and held parallel by spacers ranging in thickness from 100 to about 5 μm, thus forming a

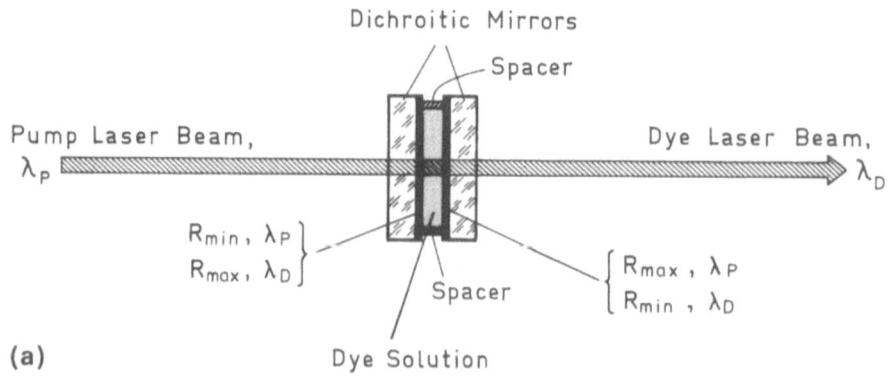

(a)

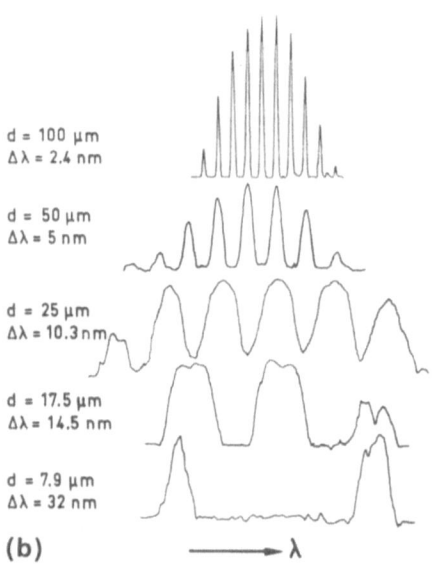

(b)

Fig. 2.10. (a) Cross section through a short-cavity dye laser; (b) spectra of the short-cavity dye laser output, d: spacer thickness, $\Delta\lambda$: spectral distance between the longitudinal modes. From [2.11]

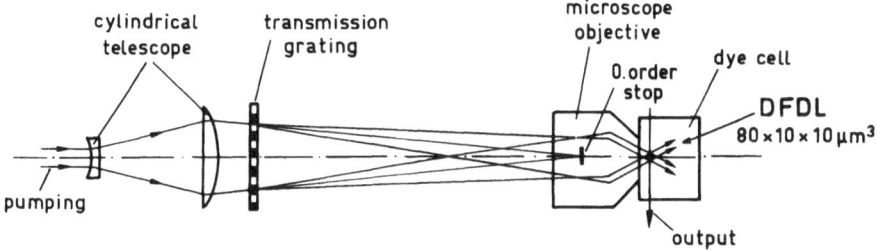

Fig. 2.11. Tunable distributed-feedback dye laser (DFDL). From [2.12]

resonator that could be filled with dye solution [2.11]. The spectra for different spacers are shown in Fig. 2.10b. One can see the longitudinal modes decreasing in number with resonator length until a single longitudinal mode remains for a cavity length of about 5 μm (not shown here) and the dye laser wavelength could then be tuned nicely by compressing the spacer mechanically.

A much more refined method of tuning a dye laser of extremely small volume was reported in [2.12] and is shown in Fig. 2.11. It makes use of the distributed-feedback principle that was first applied to dye lasers by *Kogelnik* and *Shank* [2.13] in 1971 and shown by *Bor* [2.14] to be very useful for the generation of ultrashort pulses. The distributed-feedback structure is created by imaging a coarse transmission grating into a dye solution using a microscope objective for demagnification. The transmission grating was illuminated by a dye laser pump pulse of 3–8 ps duration. Only the plus and minus first orders were used, while the undiffracted pump light was blocked by a stop in front of the microscope objective in order to obtain an interference pattern of highest visibility. With proper illumination of the grating the distributed-feedback structure had a volume of $80 \times 10 \times 10$ μm^3 and contained only 10^9–10^{10} dye molecules, depending on the concentration of the dye solution. When pumped up to twice threshold energy, the output consisted of a single pulse of 0.5 ps with a few kilowatts peak power. The wavelength could be changed continuously by varying the distance between the grating and the microscope objective and hence the demagnification factor.

Such a distributed-feedback dye laser (DFDL) was tuned to 497 nm, amplified and frequency-doubled to 248.7 nm and used as input signal for a KrF-excimer laser to produce powerful ultraviolet pulses. Another interesting application for this DFDL was reported recently [2.15] and is shown in Fig. 2.12. In this arrangement the output pulses were first compressed in a grating compressor to 200 fs, which was found possible because of their considerable chirp, and then focused into a fused silica sample for continuum production. The white pulse was then double-passed through an amplifier dye cell, which, of course, only amplified that part of the white pulse spectrum within the gain bandwidth of the dye used in the amplifier cell. This pulse also exhibited a considerable chirp and was compressed in a two-prism double-pass pulse compressor that at the same time permitted wavelength fine-tuning by shifting a

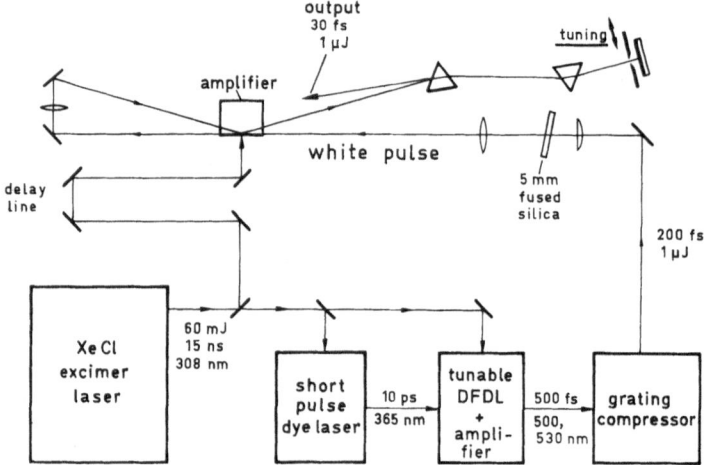

Fig. 2.12. Scheme for producing 30 fs laser pulses tunable over the whole visible spectrum. From [2.15]

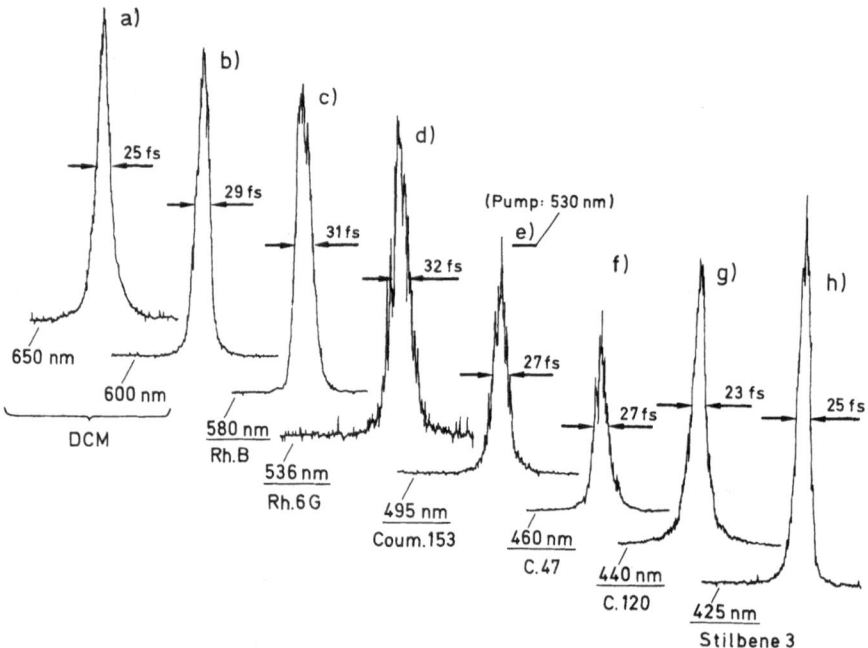

Fig. 2.13. Autocorrelation traces of output pulses from the setup of Fig. 2.12 with different dye solutions in the amplifier cell. From [2.15]

slit in front of the back-reflecting mirror in the way indicated in the figure. In this way, 30-fs pulses could be obtained over the whole visible range as shown by the autocorrelation traces of Fig. 2.13.

In contrast to these tiny DFDLs with an active volume of only 8 p, one might wonder what would constitute the upper limit for the active volume of a dye laser. Large flashlamp-pumped dye lasers can have an active volume of about a liter. Dyes with their short excited-state lifetime cannot store pump energy for longer than a few nanoseconds, in contrast to solid-state lasers with storage times of up to milliseconds. This means that flashlamps for dye lasers should be short-risetime, short-pulse devices and cannot have arbitrarily large dimensions. On the other hand pumping large active volumes with a multitude of smaller fast flashlamps would be cumbersome and costly. However, it was pointed out as early as 1969 that one could use explosively driven shockwaves in argon to pump very large active volumes with intense microsecond light pulses [2.16]. One possibility for realizing such a truly single-shot dye laser is shown in cross section in Fig. 2.14. The dye solution is contained in a double-walled cylindrical (plastic) cell and is surrounded by a cylinder of highly explosive material with a somewhat larger diameter and the space between the explosive mantle and the dye cell being filled with argon. Upon simultaneous ignition of the explosive mantle a shockwave is launched towards the center, radiating with a blackbody temperature of up to 20 000 K for a few microseconds. The output could for example be focused by parabolic mirrors of the silvered-foil type used for space applications on a target for plasma and/or X-ray generation. One kilogram of plastic explosive generates about 5 MJ mechanical energy, of which up to 10% is converted into light output, of which in turn up to 1% is converted into dye laser output, resulting in a microsecond pulse of up to 5 kJ energy, i.e. several GW peak power. There have been rumors that a similar design was tested successfully not too long ago.

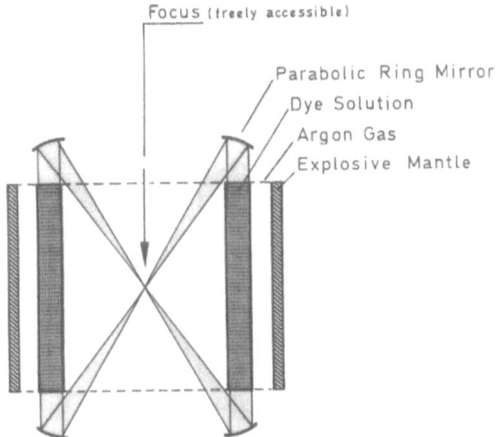

Focus (freely accessible)

Parabolic Ring Mirror
Dye Solution
Argon Gas
Explosive Mantle

Fig. 2.14. Cross section through a dye laser pumped by an argon bomb. From [2.16]

2.4 Increasing the Efficiency of Dye Lasers

While laser-pumped dye lasers often convert more than 50% of the pump energy
into dye laser output energy, flashlamp-pumped dye lasers rarely reach even 1%.
The reason for this becomes apparent when one looks at Fig. 2.15, where the
emission spectrum of a typical xenon-flashlamp is compared to the absorption
spectrum of the typical laser dye rhodamine 6G. Because of the narrow spectral
width of the dye absorption band, only a small fraction of the flashlamp output
is absorbed. This fact was recognized quite early and a remedy was sought in dye
mixtures where a short-wavelength absorbing dye whose fluorescence band
overlapped with the absorption band of the laser dye was intended to transfer its
absorbed energy to the laser dye by absorption of its fluorescence energy. This
scheme worked in only a very few exceptional cases; normally it was counter-
productive up to the complete quenching of the dye laser emission. The cause of
this unexpected behavior was soon found to be the triplet–triplet absorption of
the absorber dye that very often overlapped with the fluorescence band of the
laser dye, thus producing great losses at the dye laser wavelength.

The best way to avoid this difficulty would be to make the energy transfer
from the first excited singlet state of the absorber dye to the laser dye much faster
than the triplet-producing intersystem crossing in the absorber (donor) dye. This
can be achieved by making the average distance between the donor and the
acceptor (laser dye) so small that radiationless energy transfer of the *Förster*-
type [2.17] occurs efficiently, meaning distances smaller than about 30 Å, while
still separating the two π-electron clouds, thus keeping the spectral signatures of
the two dyes unchanged. Even though the strict validity of Förster theory is
questionable in the case where the distance between donor and acceptor is

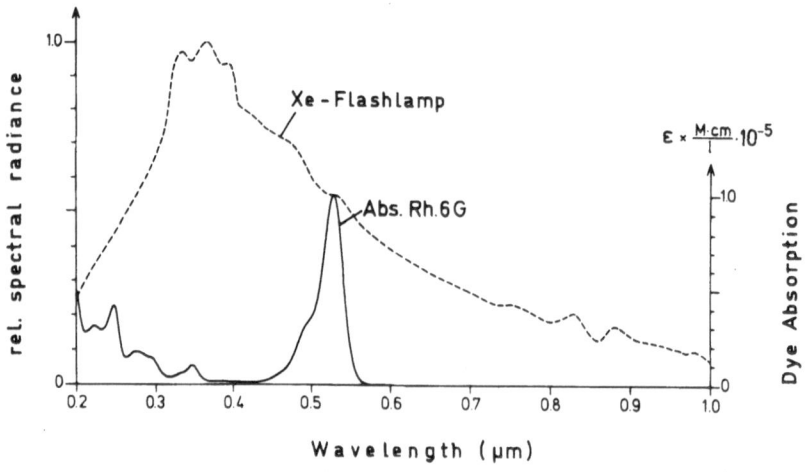

Fig. 2.15. Emission spectrum of a xenon flashlamp (dashed line) and absorption spectrum of
rhodamine 6G (solid line)

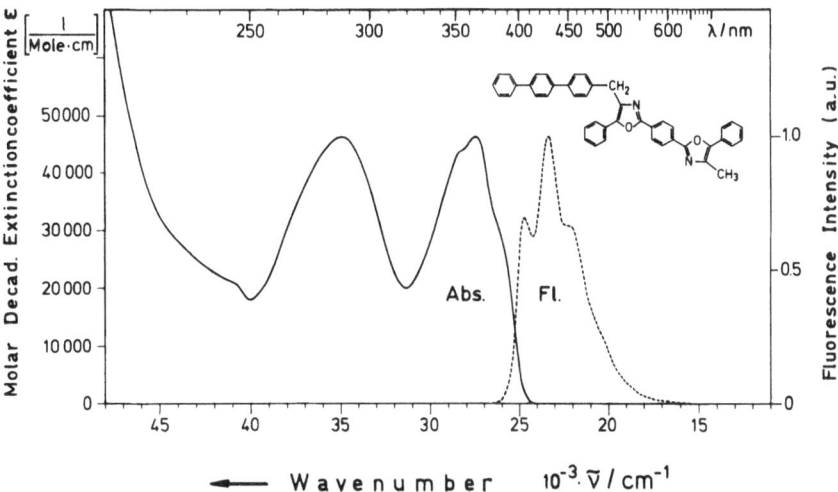

Fig. 2.16. Structural formula and absorption and fluorescence spectrum of the first bifluorophoric laser dye. From [2.18]

smaller than the molecular dimensions, one can nevertheless estimate that the time for the energy transfer is much shorter than a picosecond, thus making the energy transfer much faster than all radiationless transitions, leaving the donor molecule no time to make a transition to the triplet state.

The realization of this idea [2.18], was first achieved by linking p-terphenyl (donor) with the laser dye dimethyl-POPOP (acceptor), as shown in Fig. 2.16. When a solution of this bifluorophoric dye was compared in a flashlamp-pumped dye laser with dimethyl-POPOP proper and an equimolar mixture of dimethyl-POPOP and p-terphenyl by measuring the laser output as function of the electrical input energy to the flashlamp, it showed an increase in output power that agreed well with the expected increase due to the increased absorption by the p-terphenyl moiety, whereas the mixture exhibited a more than threefold increase of the threshold of laser action and a very low slope efficiency due to triplet absorption by p-terphenyl (Fig. 2.17).

A measurement of the dye laser pulse duration revealed that it was exactly the same, namely 60 ns, for the bifluorophoric dye and dimethyl-POPOP proper, while the pump flash was about 1 μs long. An obvious conclusion is that after 60 ns the dimethyl-POPOP triplet population was already so high that its absorption, overlapping the fluorescence band, would quench the dye laser emission.

In this situation an idea that suggests itself is the extension of the principle of intramolecular energy transfer also to the triplet states. In fact, the most elegant solution would be to use, in a bifluorophoric molecule, a singlet-energy donor moiety with a low-lying triplet state, so that one could use this moiety as

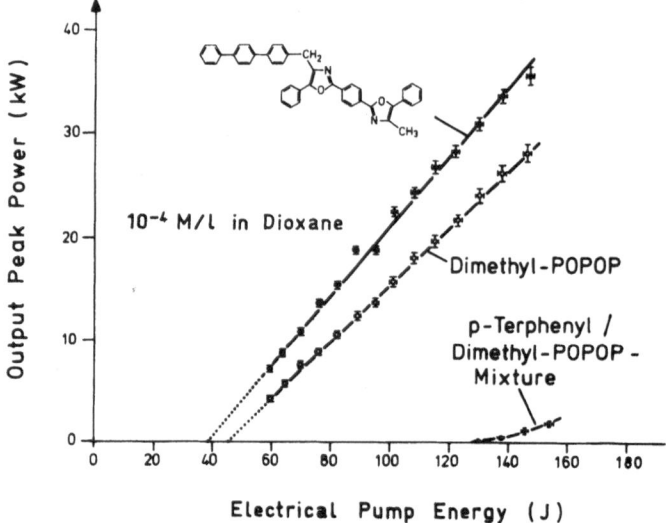

Fig. 2.17. Output peak power vs. electrical pump energy of a flashlamp-pumped dye laser using the first bifluorophoric dye as compared to dimethyl-POPOP and a mixture of *p*-terphenyl with dimethyl-POPOP. From [2.18]

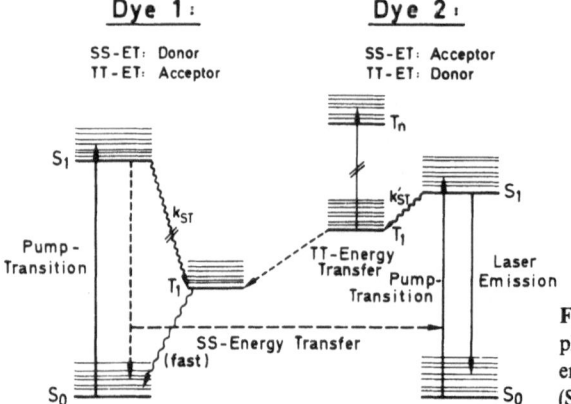

Fig. 2.18. Pump-cycle in bifluorophoric dyes with intramolecular energy transfer between singlet states (SS-ET) and triplet states (TT-ET)

acceptor in an intramolecular TT-energy transfer from the lasing moiety (TT-energy transfer donor) to the other moiety, as shown in the schematic energy-level diagram of Fig. 2.18. A search for a suitable molecule has led to stilbene, which absorbs at shorter wavelength than dimethyl-POPOP (DMP) while its fluorescence shows a perfect overlap with the absorption of DMP; its triplet level lies below that of DMP and has a very short lifetime. The results of laser tests as described above were in complete agreement with these ideas [2.19] and are shown in Fig. 2.19. When a molecule with one CH_2-group between stilbene and DMP was compared to the above-mentioned molecule with *p*-terphenyl,

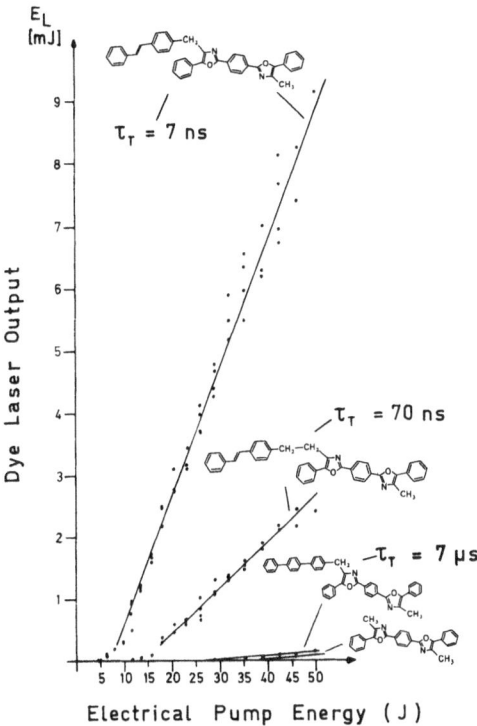

Fig. 2.19. Dye laser output energy vs. electrical pump energy for two bifluorophoric dyes using SS-ET + TT-ET in comparison to the first bifluorophoric dye using only SS-ET and to dimethyl-POPOP. τ_T: measured values of triplet lifetime of the dimethyl-POPOP moiety of the three bifluorophoric dyes. From [2.19, 20]

the output energy was found to be 60 times higher; when the link consisted of two CH_2-groups, it was still 18 times higher. In a separate experiment [2.20] the triplet lifetimes of these molecules were measured and found to be 7 μs for DMP, 70 ns for the molecule with two CH_2-groups, and 7 ns for the molecule with one CH_2-group.

2.5 Photodegradation of Laser Dyes

Solutions of the above-mentioned three dyes were also tested for photo-degradation by irradiation with the filtered light of a mercury lamp at 366 nm [2.21]. The decrease of the long-wavelength absorption by photodecomposition is shown in Fig. 2.20. It clearly demonstrates the strong decrease of the photodecomposition rate with decreasing lifetime of the highly reactive triplet state of the DMP-moiety in the bifluorophoric molecules. Nevertheless, even the

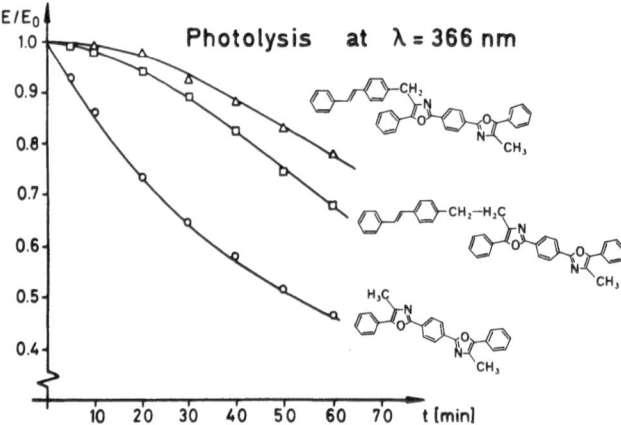

Fig. 2.20. Decrease of extinction vs. time by photolysis of two bifluorophoric dyes in comparison to dimethyl-POPOP. From [2.21]

molecule with only 7 ns triplet lifetime of the DMP-moiety exhibited a non-negligible photodecomposition rate. It was thought that this might be caused by highly reactive singlet oxygen molecules that are created by spin exchange between a triplet of a dye moiety D (stilbene or DMP) and a ground state (triplet) oxygen molecule dissolved in the solution: $^3D + {}^3O_2 \rightarrow {}^1D + {}^1O_2$. Since it is well known that some substances, e.g. tertiary amines, cause singlet oxygen to revert to the ground state in collisions, 5×10^{-2} M DABCO (1,4-diazabicyclo[2.2.2]octan) was added to a solution of DMP in order to test this hypothesis [2.22]. It was seen to reduce the photodecomposition rate from 2.2×10^{-4} without DABCO to 0.5×10^{-4} with DABCO added, so that the efficiency of DABCO in quenching those singlet oxygen molecules that otherwise would have destroyed a DMP molecule was 77% (difference of rates with and without divided by the rate without DABCO).

This suggested that one again try an intramolecular approach by introducing one or two chemically linked tertiary amino-groups into the DMP molecule in such a way that the absorption spectrum remained essentially unchanged, as shown in Fig. 2.21. In both cases the photodecomposition *increased* instead of decreased with the number of diethylamino-groups. This was probably caused by the photochemically more sensitive diethylamino-groups, which were used for simplicity only since the introduction of two rigid DABCO-groups would have taken much more time than was available for this project. Notwithstanding this complication, the effective protection of DMP by the diethylamino-groups could be seen by the reduction of the effect of externally added DABCO in comparison to that in DMP proper. It was found to be 35% with one and only 24% with two added diethylamino-groups. This clearly indicates that the singlet oxygen molecules were quenched significantly faster intramolecularly than by the extremely high concentration of externally added DABCO.

Intramolecular $^1O_2$-Quenching with Tertiary Amines

$$Q = \frac{\text{Number of molecules destroyed}}{\text{Number of photons absorbed (308 nm)}}$$

Q': with $5 \cdot 10^{-2}$ M <u>DABCO</u> added

compound	Q	Q'	DABCO EFFICIENCY $\frac{Q-Q'}{Q}$
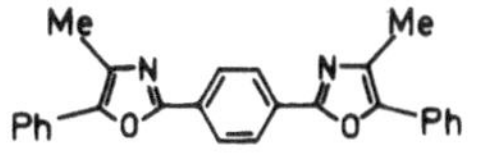 Me, Me (DMP) Ph, Ph	$2.2 \cdot 10^{-4}$	$0.5 \cdot 10^{-4}$	77 %
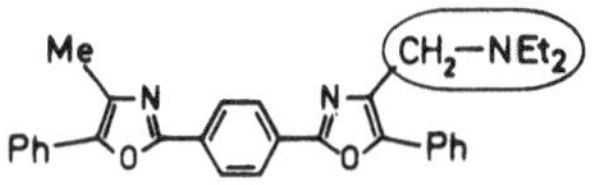 Me, CH_2-NEt_2 Ph, Ph	$2.6 \cdot 10^{-4}$	$1.7 \cdot 10^{-4}$	35 %
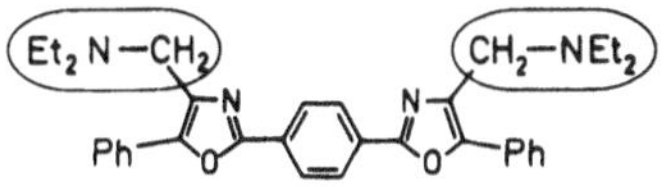 Et_2N-CH_2, CH_2-NEt_2 Ph, Ph	$3.4 \cdot 10^{-4}$	$2.6 \cdot 10^{-4}$	24 %

Fig. 2.21. Experimental data for intramolecular singlet-oxygen quenching with tertiary amines. From [2.22]

The Ideal Laser-Dye Molecule

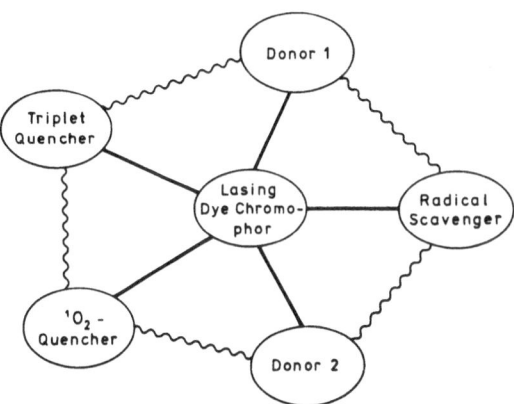

Fig. 2.22. The ideal laser dye molecule

Unfortunately, this interesting exercise in supramolecular chemistry had to be discontinued for formal reasons, otherwise it might eventually have led to the final goal of a highly efficient and stable class of laser dyes as shown schematically in Fig. 2.22.

References

2.1   H. Kuhn: In *Progress in the Chemistry of Organic Natural Products*, ed. by D.L. Zechmeister Vol. 16 (Springer, Wien 1959) p. 17

2.2   F.-G. Zhang, S. Szatmári, F.P. Schäfer: unpublished

2.3   J.M. Lehn, P. Vierling, R.C. Hayward: J. Chem. Soc. Chem. Commun. 296 (1979)

2.4   F.-G. Zhang, F.P. Schäfer: Appl. Phys. B **26**, 211 (1981)

2.5   F.P. Schäfer: Principles of dye laser operation. In *Dye Lasers*. 3rd. enlarged and revised edition, ed. by F.P. Schäfer, Topics Appl. Phys. Vol. 1 (Springer, Berlin, Heidelberg 1990) p. 7

2.6   H.J. Polland, T. Elsaesser, A. Seilmeier, W. Kaiser, M. Kussler, N.J. Marx, B. Sens, K.H. Drexhage: Appl. Phys. B **32**, 53 (1983)

2.7   Shin-ichi Ishizaka: Appl. Phys. B **48**, 111 (1989)

2.8   N.P. Ernsting: Chem. Phys. Lett. **166**, 221 (1990)

2.9   Zs. Bor, S. Szatmári, A. Müller: Appl. Phys. B **32**, 101 (1983)

2.10 F. Wondrazek, A. Seilmeier, W. Kaiser: Appl. Phys. B **32**, 39 (1983)

2.11 F.P. Schäfer: Principles of dye laser operation. In *Dye Lasers*. 3rd. enlarged and revised edition, ed. by F.P. Schäfer, Topics Appl. Phys. Vol. 1 (Springer, Berlin, Heidelberg 1990) p. 75

2.12 S. Szatmári, F.P. Schäfer: Appl. Phys. B **46**, 305 (1988)

2.13 H. Kogelnik, C.V. Shank: Appl. Phys. Lett. **18**, 152 (1971)

2.14 Zs. Bor: IEEE J. Quant. Electron. **QE-16**, 517 (1980)

2.15 P. Simon, S. Szatmári, F.P. Schäfer: Opt. Lett. **16**, 1569 (1991)

2.16 F.P. Schäfer: Principles of dye laser operation. In *Dye Lasers*. 3rd. enlarged and revised edition, ed. by F.P. Schäfer. Topics Appl. Phys. Vol. 1 (Springer, Berlin, Heidelberg 1990) p. 89

2.17 Th. Förster: Discuss. Faraday Soc. **27**, 7 (1959)

2.18 F.P. Schäfer, Zs. Bor, W. Lüttke, B. Liphardt: Chem. Phys. Lett. **56**, 455 (1978)

2.19 Bo. Liphardt, Be. Liphardt, W. Lüttke: Opt. Commun. **38**, 207 (1981)

2.20 F.P. Schäfer, F.-G. Zhang, J. Jethwa: Appl. Phys. B **28**, 37 (1982)

2.21 Bo. Liphardt, Be. Liphardt: Private communication

2.22 Be. Liphardt: Private communication

2.23 S. Szatmári, F.P. Schäfer: Opt. Commun..**49**, 281 (1984)

Fritz P. Schäfer

3. Single-Atom Experiments and the Test of Fundamental Quantum Physics

H. Walther

With 10 Figures

This chapter will review work on the one-atom maser. The superconducting maser cavity made of niobium and cooled to 0.5 K to exclude thermal photons has a quality factor of 3×10^{10}. Velocity-selected Rydberg atoms pump the maser. The field inside the cavity is investigated by probing the atoms leaving the cavity. The dynamics of the Rabi nutation has been investigated. Furthermore the statistics of the photons in the cavity was determined to be sub-Poissionian corresponding to photon-number fluctuations of 70% below the vacuum-state limit. The one-atom maser is the only maser system which leads to nonclassical radiation even when pumping processes with Poissonian statistics are used. Some applications of the one-atom maser to study the quantum measurement process are also discussed.

3.1 Overview

Modern methods of dye-laser spectroscopy allow us to study single atoms or ions in an unperturbed environment. This has opened up interesting new experiments, among them the detailed study of radiation–atom coupling. In the following, one group of experiments of this type is reviewed: the single-atom maser.

The simplest and most fundamental system for studying radiation–matter coupling is a single two-level atom interacting with a single mode of an electromagnetic field in a cavity. It received a great deal of attention shortly after the maser was invented, but at that time, the problem was of purely academic interest as the matrix elements describing the radiation–atom interaction are so small and the field of a single photon is not sufficient to lead to an atom–field evolution time shorter than other characteristic times of the system, such as the excited state lifetime, the time of flight of the atom through the cavity and the cavity mode damping time. It was therefore not possible to test experimentally the fundamental theories of radiation–matter interaction which predicted amongst other effects:

a) a modification of the spontaneous emission rate of a single atom in a resonant cavity,

b) oscillatory energy exchange between a single atom and the cavity mode, and

Topics in Applied Physics, Vol. 70
Dye Lasers: 25 Years Ed.: Dr. Michael Stuke
© Springer-Verlag Berlin Heidelberg 1992

c) the disappearance and quantum revival of optical (Rabi) nutation induced in a single atom by a resonant field.

The situation has changed drastically in the last few years with the introduction of frequency-tunable lasers that can excite large populations of highly excited atomic states, characterized by a high principal quantum number n of the valence electron. These states are generally called Rydberg states since their energy levels can be described by the simple Rydberg formula. Such excited atoms are very suitable for observing quantum effects in radiation–atom coupling for three reasons. First, the states are very strongly coupled to the radiation field (the induced transition rates between neighbouring levels scale as n^4); second, transitions are in the millimetre wave region, so that low-order mode cavities can be made large enough to allow rather long interaction times; finally, Rydberg states have relatively long lifetimes with respect to spontaneous decay [3.1].

The strong coupling of Rydberg states to radiation resonant with transitions to neighbouring levels can be understood in terms of the correspondence principle: with increasing n the classical evolution frequency of the highly excited electron becomes identical with the transition frequency to the neighbouring level; the atom therefore corresponds to a large dipole oscillating with the resonance frequency. (The dipole moment is very large since the atomic radius scales with n^2).

In order to understand the modification of the spontaneous emission rate in an external cavity, we have to remember that in quantum electrodynamics this rate is determined by the density of modes of the electromagnetic field at the atomic transition frequency ω_0, which in turn depends on the square of the frequency. If the atom is not in free space, but in a resonant cavity, the continuum of modes is changed into a spectrum of discrete modes of which one may be in resonance with the atom. The spontaneous decay rate of the atom in the cavity γ_c will then be enhanced in relation to that in free space γ_f by a factor given by the ratio of the corresponding mode densities:

$$\frac{\gamma_c}{\gamma_f} = \frac{\rho_c(\omega_0)}{\rho_f(\omega_0)} = \frac{2\pi Q}{V_c \omega_0^3} = \frac{Q \lambda_0^3}{4\pi^2 V_c} \, ,$$

where V_c is the volume of the cavity and Q is the quality factor of the cavity which expresses the sharpness of the mode. For low-order cavities in the microwave region $V_c = \lambda_0^3$, which means that the spontaneous emission rate is increased roughly by a factor of Q. However, if the cavity is detuned, the decay rate will decrease. In this case, the atom cannot emit a photon, since the cavity is not able to accept it, and therefore the energy has to stay with the atom.

Recently, a number of experiments have been made with Rydberg atoms to demonstrate this enhancement and inhibition of spontaneous decay in external cavities or cavity-like structures (for the most recent experiments see [3.2]).

More subtle effects due to the change of the mode density can also be expected: radiation corrections such as the Lamb shift and the anomalous magnetic dipole moment of the electron are modified with respect to the free space value [3.3], although changes are of the same order as present experimental accuracy. Roughly speaking, one can say that such effects are determined by a change of virtual transitions and not by real transitions as in the case of spontaneous decay.

In the following, attention is focussed on the one-atom maser in which the idealized case of a two-level atom interacting with a single mode of a radiation field is realized. The theory of this system was treated by *Jaynes* and *Cummings* [3.4] many years ago and we shall concentrate on the dynamics of the atom–field interaction predicted by this theory. Some of the features are explicitly a consequence of the quantum nature of the electromagnetic field: the statistical and discrete nature of the photon field leads to new dynamic characteristics such as collapse and revivals in the Rabi nutation.

3.2 The One-Atom Maser

The detailed experimental setup of the one-atom maser [3.5] is shown in Fig. 3.1. A highly collimated beam of rubidium atoms passes through a Fizeau velocity selector. Before entering the superconducting cavity, the atoms are excited into the upper maser level $63p_{3/2}$ by the frequency-doubled light of a cw ring dye laser. The laser frequency is stabilized onto the atomic transition $5s_{1/2} \rightarrow 63p_{3/2}$ which has a width determined by the laser linewidth and the transit time broadening corresponding to a total of a few MHz. In this way, it is possible to prepare a very stable beam of excited atoms. The ultraviolet light is linearly polarized parallel to the electric field of the cavity. Therefore only $\Delta m = 0$ transitions are excited by both the laser beam and the microwave field.

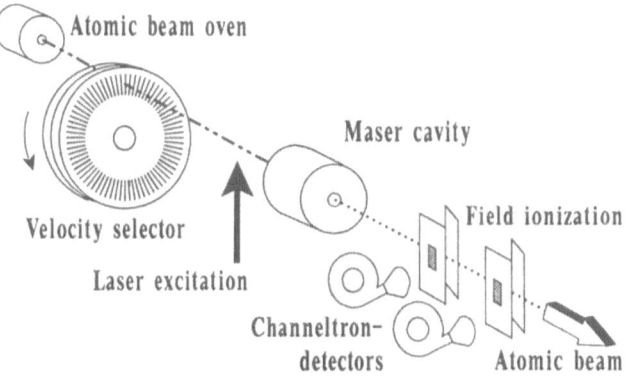

Fig. 3.1. Scheme of the experimental set-up. To suppress black-body-induced transitions to neighbouring states, the Rydberg atoms are excited inside a liquid-helium-cooled environment

The superconducting niobium maser cavity is cooled down to a temperature of 0.5 K by means of a ^{3}He cryostat. At such a low temperature the number of thermal photons is reduced to about 0.15 at a frequency of 21.5 GHz. The cryostat is carefully designed to prevent room temperature microwave photons from leaking into the cavity. This would considerably increase the temperature of the radiation field above the temperature of the cavity walls. The quality factor of the cavity is 3×10^{10} corresponding to a photon storage time of about 0.2 s. The cavity is carefully shielded against magnetic fields by several layers of cryoperm. In addition, three pairs of Helmholtz-coils are used to compensate the earth's magnetic field to a value of several mG in a volume of $10 \times 4 \times 4$ cm^3. This is necessary in order to achieve the high quality factor and to prevent the different magnetic substates of the maser levels from mixing during the atom–field interaction time. Two maser transitions from the $63p_{3/2}$ level to the $61d_{3/2}$ and to the $61d_{5/2}$ level are studied.

The Rydberg atoms in the upper and lower maser levels are detected in two separate field ionization detectors. The field strength is adjusted so as to ensure that in the first detector the atoms in the upper level are ionized, but not those in the lower level.

To demonstrate maser operation, the cavity is tuned over the $63p_{3/2}$–$61d_{3/2}$ transition and the flux of atoms in the excited state is recorded simultaneously. Transitions from the initially prepared $63p_{3/2}$ state to the $61d_{3/2}$ level (21.50658 GHz) are detected by a reduction of the electron count rate.

In the case of measurements at a cavity temperature of 0.5 K, shown in Fig. 3.2, a reduction of the $63p_{3/2}$ signal can be clearly seen for atomic fluxes as small

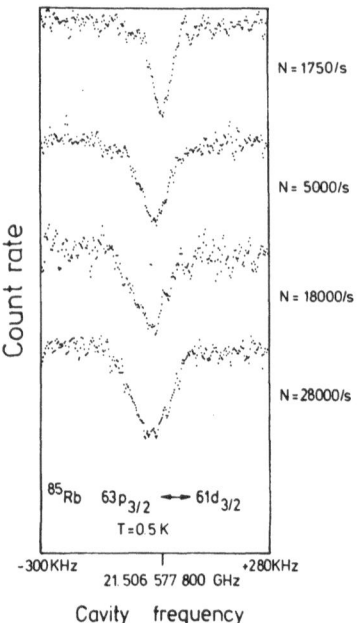

Fig. 3.2. Maser operation of the one-atom maser manifests itself in a decrease of the number of atoms in the excited state. The flux of excited atoms N governs the pump intensity. Power broadening of the resonance line demonstrates the multiple exchange of a photon between the cavity field and the atom passing through the resonator. The measurement was performed with the transition $63p_{3/2} \to 61d_{3/2}$ having a single photon Rabi frequency of 10 kHz. The Q value of the cavity for this measurement was 4×10^9 which is lower than the one used for the measurements of the $63p_{3/2} \to 61d_{5/2}$ transition being 3×10^{10}. The temperature of the cavity was 0.5 K. For the transition $63p_{3/2} \to 61d_{5/2}$ used also for many of the measurements described in this paper the single photon Rabi frequency is 40 kHz

as 1750 atoms/s. An increase in flux causes power broadening and a small shift. This shift is attributed to the ac Stark effect, caused predominantly by virtual transitions to neighbouring Rydberg levels. Over the range from 1750 to 28 000 atoms/s, the field ionization signal at resonance is independent of the particle flux, indicating that the transition is saturated. This, and the observed power broadening show that there is a multiple exchange of photons between Rydberg atoms and the cavity field.

For an average transit time of Rydberg atoms through the cavity of 50 μs and a flux of 1750 atoms/s we find that approximately 0.09 Rydberg atoms are in the cavity on the average. According to Poisson statistics this implies that more than 90% of the events are due to single atoms. This clearly demonstrates that single atoms are able to maintain a continuous oscillation of the cavity with a mean number of photons between unity and several hundreds.

3.3 Dynamics of a Single Atom

With very low atomic-beam fluxes, the cavity of the single-atom maser contains essentially thermal photons only, whose number varies randomly obeying Bose–Einstein statistics. At high fluxes, the atoms deposit energy in the cavity and the maser reaches the threshold so that the number of photons stored in the cavity increases and their statistics changes.

For a coherent field the probability distribution is Poissonian which results in a dephasing of the Rabi oscillations, and therefore the envelope of $P_e(t)$, the probability of finding the atom in the upper maser level, collapses. After the collapse, $P_e(t)$ starts oscillating again in a very complex way. Such changes recur periodically, the time interval being proportional to the square root of the photon number stored in the cavity. Both collapse and revival in the coherent state are pure quantum features without any classical counterpart [3.6].

Collapse and revival also occur in the case of a thermal Bose–Einstein field where the spread in the photon number is far larger than for a coherent state, and the collapse time is much shorter. In addition, revivals overlap completely and interfere, producing a very irregular time evolution. On the other hand, a classical thermal field represented by an exponential distribution of the intensity shows collapse, but no revival at all. From this it follows that revivals are pure quantum features of the thermal radiation field, whereas the collapse is less clear-cut as a quantum effect. [3.7].

The above-mentioned effects have been demonstrated experimentally [3.8], using the Fizeau velocity selector to vary the interaction time (Fig. 3.1). Figure 3.3 shows a series of measurements obtained with the single-atom maser, where $P_e(t)$ is plotted against interaction time for increasing atomic flux N. The strong variation of $P_e(t)$ for interaction times between 50 and 80 μs disappears for larger N and a revival shows up for $N = 3000$ s^{-1} for interaction times larger than 140 μs. The average photon number in the cavity varies between 2.5 and 5,

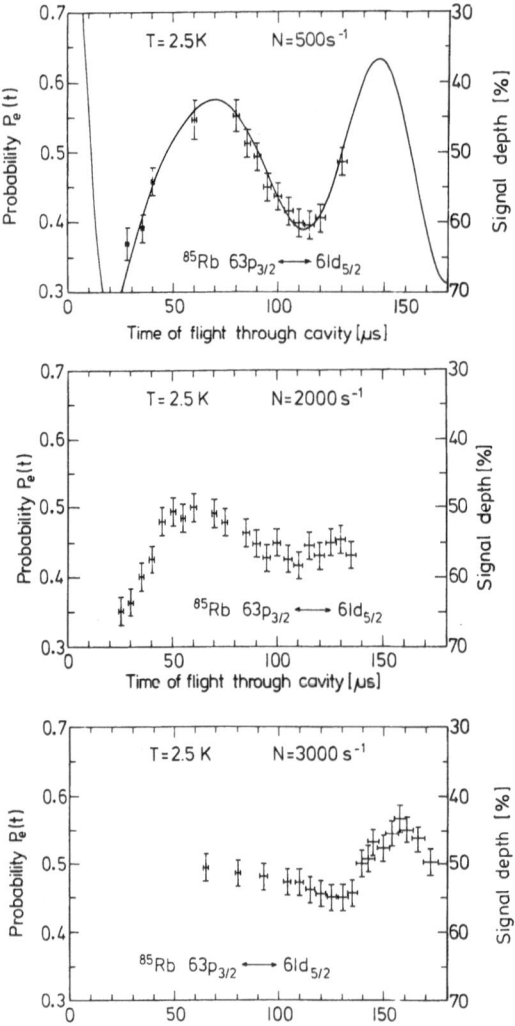

Fig. 3.3. Quantum collapse and revival in the one-atom maser. Plotted is the probability $P_e(t)$ of finding the atom in the upper maser level for different fluxes N of the atomic beam [3.8]. The measurements were performed with a cavity at 2.5 K

about 2 photons being due to the black-body field in the cavity corresponding to a temperature of 2.5 K.

3.4 A New Source of Nonclassical Light

One of the most interesting questions in connection with the one-atom maser is the photon statistics of the electromagnetic field generated in the superconducting cavity. This problem will be discussed in this section.

Electromagnetic radiation can show nonclassical properties [3.9], that is properties that cannot be explained by classical probability theory. Loosely speaking we need to invoke "negative probabilities" to get deeper insight into these features. We know of essentially three phenomena which demonstrate the nonclassical character of light: photon antibunching [3.10], sub-Poissonian photon statistics [3.11] and squeezing [3.12]. Methods of nonlinear optics are most often employed to generate nonclassical radiation. However, the fluorescence light from a single atom caught in a trap also exhibits nonclassical features [3.13, 14].

Another nonclassical light generator is the one-atom maser. We recall that the Fizeau velocity selector preselects the velocity of the atoms: Hence the interaction time is well-defined, leading to conditions usually not achievable in standard masers [3.15–20]. This has a very important consequence when the intensity of the maser field grows as more and more atoms give their excitation energy to the field: Even in the absence of dissipation, this increase in photon number is stopped when the increasing Rabi-frequency leads to a situation where the atoms reabsorb the photon and leave the cavity in the upper state. For any photon number, this can be achieved by appropriately adjusting the velocity of the atoms. In this case the maser field is not changed any more and the number distribution of the photons in the cavity is sub-Poissonian [3.15, 16], that is narrower than a Poisson distribution. Even a number state, i.e. a state of well-defined photon number, can be generated [3.17, 18] using a cavity with a high enough quality factor. If there are no thermal photons in the cavity – a condition achievable by cooling the resonator to an extremely low temperature – very interesting features such as trapping states show up [3.19]. In addition, steady state macroscopic quantum superpositions can be generated in the field of the one-atom maser pumped by two-level atoms injected in a coherent superposition of their upper and lower states [3.20].

Unfortunately the measurement of the nonclassical photon statistics in the cavity is not that straightforward. The measurement process of the field involves the coupling to a measuring device whereby losses inevitably lead to a destruction of the nonclassical properties. The ultimate technique to obtain information about the field employs the Rydberg atoms themselves: Measurement of the photon statistics via the dynamical behaviour of the atoms in the radiation field, that is via the collapse and the revivals of the Rabi oscillations, is one possibility. However, since the photon statistics depend on the interaction time which has to be changed when collapse and revivals are measured it is much better to probe the population of the atoms in the upper and lower maser levels when they leave the cavity. In this case, the interaction time is kept constant. Moreover, this measurement is relatively easy since electric fields can be used to perform a selective ionization of the atoms. The detection sensitivity is sufficient for the atomic statistics to be investigated. This technique maps the photon statistics of the field inside the cavity via the atomic statistics.

In this way, the number of maser photons can be inferred from the number of atoms detected in the lower level. In addition, the variance of the photon number distribution can be deduced from the number of fluctuations of the lower-level atoms [3.21]. In the experiment, we are therefore mainly interested in the atoms in the lower maser level. Experiments carried out along these lines are described in the next section (see also [3.22]).

3.5 Experimental Results – A Nonclassical Beam of Atoms

Under steady-state conditions, the photon statistics of the field are essentially determined by the dimensionless parameter $\theta = (N_{ex} + 1)^{1/2}\Omega t_{int}$, which can be understood as a pump parameter for the one-atom maser [3.15]. Here N_{ex} is the average number of atoms that enter the cavity during the lifetime of the field, t_{int} the time of flight of the atoms through the cavity and Ω the atom-field coupling constant (one-photon Rabi frequency). The one-atom maser threshold is reached for $\theta = 1$. At this value and also at $\theta = 2\pi$ and integer multiples thereof, the photon statistics are super-Poissonian. At these points, the maser field undergoes first-order phase transitions [3.15]. In the regions between those points sub-Poissonian statistics are expected. The experimental investigation of the photon number fluctuation is the subject of the following discussion.

In the experiments [3.22], the number N of atoms in the lower maser level is counted for a fixed time interval T roughly equal to the photon storage time T_{cav} of the cavity. By repeating this measurement many times the probability distribution $P(N)$ of finding N atoms in the lower level is obtained. The normalized variance [3.23] $Q_a = (\langle N^2 \rangle - \langle N \rangle^2 - \langle N \rangle)/\langle N \rangle$ is evaluated and is used to characterize the deviation from Poissonian statistics. A negative (positive) Q_a value indicates sub-Poissonian (super-Poissonian) statistics, while $Q_a = 0$ corresponds to a Poisson distribution with $\langle N^2 \rangle - \langle N \rangle^2 = \langle N \rangle$. The atomic Q_a is related to the normalized variance Q_f of the photon number by the formula

$$Q_a = \varepsilon P Q_f (2 + Q_f) , \tag{3.1}$$

which was derived in [3.21] with P denoting the probability of finding an atom in the lower maser level. It follows from formula (3.1) that the nonclassical photon statistics can be observed via sub-Poissonian atomic statistics. The detection efficiency ε for the Rydberg atoms reduces the sub-Poissonian character of the experimental result. The detection efficiency was 10% in our experiment; this includes the natural decay of the Rydberg states between the cavity and the field ionization detection [3.22].

We start the discussion of the experimental results by describing measurements in the build-up period of the maser field. For this purpose, the mean number of atoms passing through the cavity during a fixed sampling time interval shorter than the cavity decay photon storage time is varied. For each flux, the normalized variance Q_a of the probability distribution of atoms detected in both the upper and the lower maser level is determined. Experimental

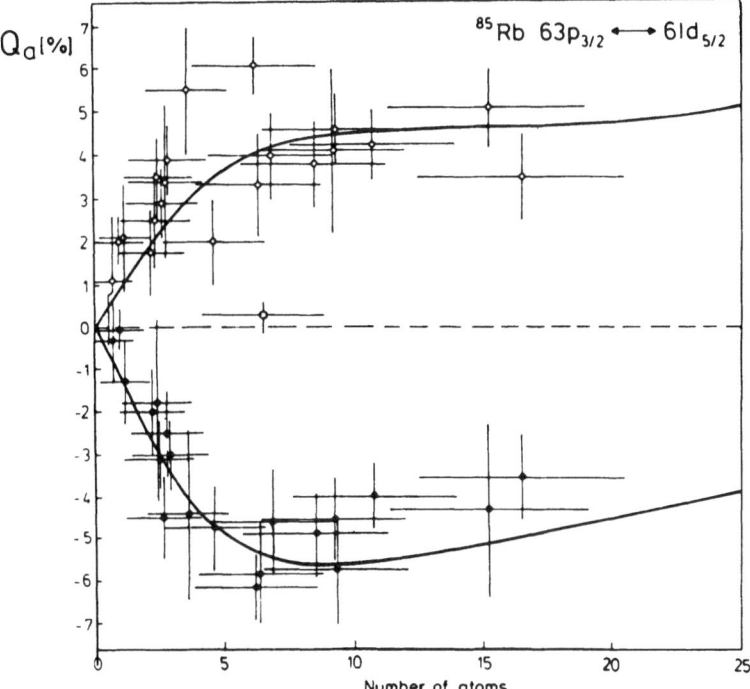

Fig. 3.4. Variance Q_a of the atoms in the lower (open circles) and upper level (full circles) as a function of the total number of atoms crossing the cavity

results are plotted in Fig. 3.4. Open circles (full circles) represent the normalized variance of atoms in the lower (upper) maser level leading to a sub-Poissonian (super-Poissonian) atom statistics. About 20 000 experiments are averaged for each data point to keep the uncertainty of Q_a below 1%. For a low temperature of the atomic beam oven, the horizontal error bars are determined by the Poisson statistics of the total flux of atoms. These statistics are measured with the cavity out of resonance so that all atoms leave the cavity in the upper level. The result is given by a Poisson distribution with $Q_a = 0$.

The two solid lines are calculated by a numerical simulation of the maser process which explicitly takes into account the measurement of atoms and the corresponding change of the cavity photon statistics. Details of this procedure are given in [3.21]. For the comparison with the results of Fig. 3.4 the method was extended to the transient regime of an increasing maser field starting from the thermal 0.5 K field. The detection efficiency of atoms in the lower and upper maser levels amounts to 10% and 7%, respectively. In the simulation, the Poisson statistics of the flux, the temperature of the cavity field (0.5 K), and the damping of the maser field in the time interval between adjacent atoms are also considered. The agreement between the numerical simulation and the experimental results is good.

The sub-Poissonian statistics of atoms in the lower level proves the non-classical character of the maser field. As is expected from simple statistical arguments, the probability distribution of atoms in the upper maser level is always super-Poissonian. With the single photon Rabi frequency given by $\Omega = 44$ KHz, $n = 2$ photons are able to induce a 2π Rabi nutation of the atom during the atom–field interaction time of $t_{\text{int}} = 40$ μs: $2\Omega(n + 1)^{1/2}t_{\text{int}} = 6.10$. This number is slightly smaller than 2π to reduce the influence of the velocity distribution on the maser photon statistics. The simulation shows that, averaged over many experiments, the probability distribution of finding n maser photons after about 8 atoms have crossed the cavity has a mean of $\langle n \rangle = 2$ and a variance of $\langle n^2 \rangle - \langle n \rangle^2 = 0.2\langle n \rangle$. In the experiment, this value corresponds to $Q_a = -6 \times 10^{-2}$. Figure 3.4 shows that for a higher flux of atoms (more than 20), the normalized variance of the probability distribution for atoms in the lower level increases slowly indicating that the 2π trapping condition is not exactly fulfilled.

Now we continue the discussion with our results on the maser in steady state. Experimental results for the transition $63p_{3/2} \leftrightarrow 61d_{3/2}$ are shown in Fig. 3.5. The measured normalized variance Q_a is plotted as a function of the flux of atoms. The atom–field interaction time is fixed at $t_{\text{int}} = 50$ μs. The atom–field coupling constant Ω is rather small for this transition, $\Omega = 10$ kHz. A relatively high flux of atoms $N_{\text{ex}} > 10$ is therefore needed to drive the one-atom maser above threshold. The large positive Q_a observed in the experiment proves the large intensity fluctuations at the onset of maser oscillation at $\theta = 1$. The solid

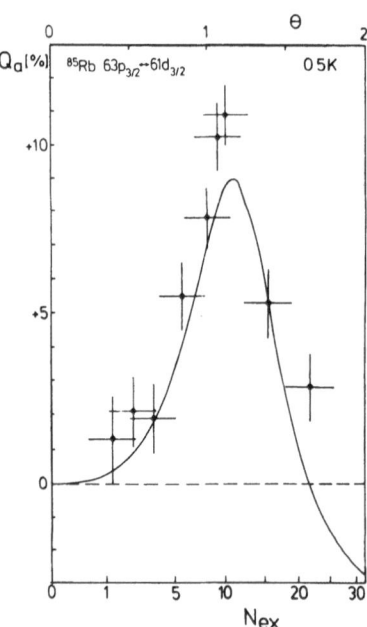

Fig. 3.5. Variance Q_a of the atoms in the lower maser level as a function of flux N_{ex} near the onset of maser oscillation for the $63p_{3/2} \rightarrow 61d_{3/2}$ transition [3.22]

line is plotted according to equation (1) using the theoretical predictions for Q_f of the photon statistics [3.15, 16]. The error in the signal follows from the statistics of the counting distribution $P(N)$. About 2×10^4 measurement intervals are needed to keep the error of Q_a below 1%. The statistics of the atomic beam is measured with a detuned cavity. The result is a Poisson distribution. The error bars of the flux follow from this measurement. The agreement between theory and experiment is good. The nonclassical photon statistics of the one-atom maser are observed at a higher flux of atoms or a larger atom–field coupling constant, therefore the $63p_{3/2} \leftrightarrow 61d_{5/2}$ maser transition with $\Omega = 44$ kHz was also studied. Experimental results for this transition are shown in Fig. 3.6. Fast atoms with an atom–cavity interaction time of $t_{int} = 35 \ \mu s$ are used. A very low flux of atoms of $N_{ex} > 1$ is already sufficient to generate a nonclassical maser field. This is the case since the vacuum field initiates a transition of the atom to the lower maser level, thus driving the maser above threshold. The sub-Poissonian statistics can be understood from Fig. 3.7, where the probability of finding the atom in the upper level is plotted as a function of the atomic flux. The oscillation observed is closely related to the Rabi nutation induced by the maser field. The solid curve was calculated according to the one-atom maser theory with a velocity dispersion of 4%. A higher flux generally leads to a higher photon number, but for $N_{ex} < 10$ the probability of finding the atom in the lower level decreases. An increase in the photon number is therefore counterbalanced by the fact that the probability of photon emission in the cavity

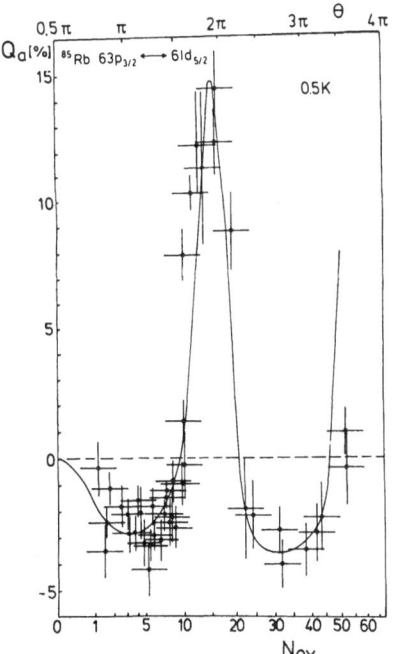

Fig. 3.6. Same as Fig. 3.3, but above threshold for the $63p_{3/2} \rightarrow 61d_{5/2}$ transition [3.22]

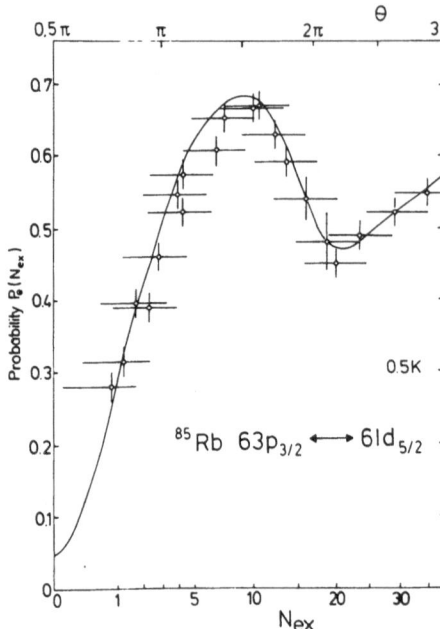

Fig. 3.7. Probability $P_e(N_{ex})$ of finding the atom in the upper maser level $63p_{3/2}$ for the $63p_{3/2} \rightarrow 61d_{5/2}$ transition as a function of the atomic flux

is reduced. This negative feedback leads to a stabilization of the photon number [3.21]. The feedback changes sign at a flux $N_{ex} \approx 10$, where the second maser phase transition is observed at $\theta = 2\pi$. This is again characterized by large fluctuations of the photon number. Here the probability of finding an atom in the lower level increases with increasing flux. For even higher fluxes, the state of the field is again highly nonclassical. The solid line in Fig. 3.6 represents the result of the one-atom maser theory using (3.1) to calculate Q_a. The agreement with the experiment is very good.

The sub-Poissonian statistics of atoms near $N_{ex} = 30$, $Q_a = -4\%$ and $P = 0.45$ (Fig. 3.6) are generated by a photon field with a variance $\langle n^2 \rangle - \langle n \rangle^2 = 0.3 \langle n \rangle$, which is 70% below the shot noise level. Again, this result agrees with the prediction of the theory [3.15, 16]. The mean number of photons in the cavity is about 2 and 13 in the regions $N_{ex} \approx 3$ and $N_{ex} \approx 30$, respectively. Near $N_{ex} \approx 15$, the photon number changes abruptly between these two values. The next maser phase transition with a super-Poissonian photon number distribution occurs above $N_{ex} \approx 50$.

Sub-Poissonian statistics are closely related to the phenomenon of antibunching for which the probability of detecting a next event shows a minimum immediately after a triggering event. The duration of the time interval with reduced probability is of the order of the coherence time of the radiation field. In our case this time is determined by the storage time of the photons. The Q_a value therefore depends on the measuring interval T. Experimental results for a flux $N_{ex} \approx 30$ are shown in Fig. 3.8. The measured Q_a value approaches a time-independent value for $T > T_{cav}$. For very short sampling intervals, the statistics of atoms in the lower level show a Poisson distribution. This means that the cavity cannot stabilize the flux of atoms in the lower level on a time scale which is short in relation to the intrinsic cavity damping time.

We want to emphasize that the reason for the sub-Poissonian atomic statistics is the following: A changing flux of atoms changes the Rabi frequency via the stored photon number in the cavity. By adjusting the interaction time, the phase of the Rabi-nutation cycle can be chosen so that the probability for the atoms leaving the cavity in the upper maser level increases when the flux and therefore the photon number is enlarged or vice versa. We observe sub-Poissonian atomic statistics in the case where the number of atoms in the lower state is decreasing with increasing flux and photon number in the cavity. The same argument can be applied to understand the nonclassical photon statistics of the maser field: Any deviation of the number of light quanta from its mean value is counterbalanced by a correspondingly changed probability of photon emissions of the atoms. This effect leads to a natural stabilization of the maser intensity by a feedback loop incorporated into the dynamics of the coupled atom–field system. This feedback mechanism is also demonstrated when the anticorrelation of atoms leaving the cavity in the lower state is investigated. Measurements of these "antibunching" phenomena for atoms are described in the following.

For steady state conditions, experimental results are displayed in Fig. 3.9. Plotted is the probability $g^{(2)}(t)$ of finding an atom in the lower maser level $61d_{5/2}$ at time t after a first one has been detected at $t = 0$. The probability $g^{(2)}(t)$ was calculated from the actual count rate by normalizing with the average number of atoms determined in a large time interval. Time is given in units of the

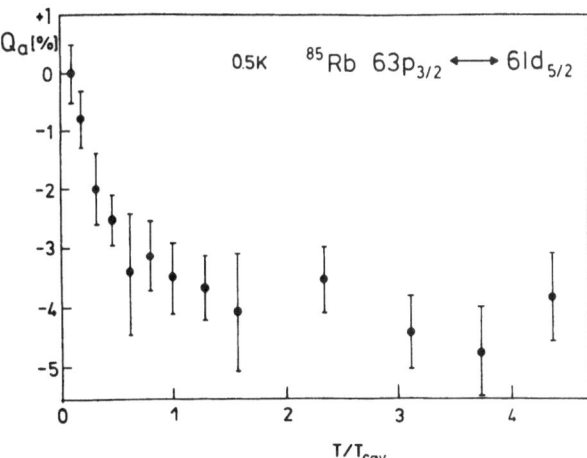

Fig. 3.8. Variance Q_a of the atoms in the lower maser level as a function of the measurement time interval T for a flux $N_{ex} \approx 30$ [3.22]

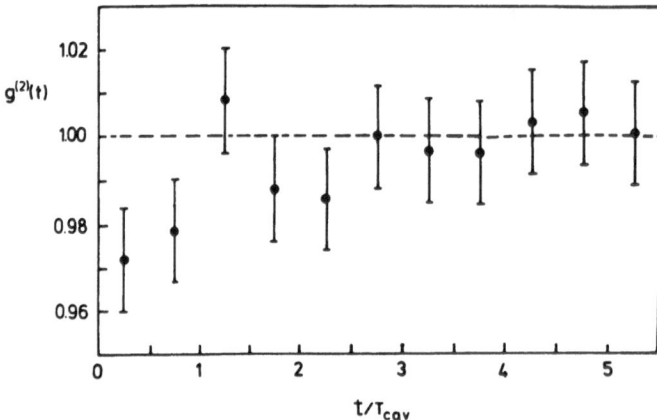

$g^{(2)}(t)$

t/τ_{cav}

Fig. 3.9. Anticorrelation of atoms in the lower maser state

photon storage time. The error bar of each data point is determined by the number of about 7000 lower level atoms counted. This corresponds to 2×10^6 atoms that have crossed the cavity during the total measurement time. The detection efficiency is near $\varepsilon = 10\%$. The measurements were taken for $N_{ex} = 30$. The time of flight of the atoms through the cavity was $t_{int} = 37\mu s$, leading to $\theta = 9$.

The fact that anticorrelation is observed shows that the atoms in the lower state are more equally spaced than expected for Poissonian distribution. It means when two atoms enter the cavity close to each other the second one performs a transition to the lower state with a reduced probability.

The experimental results presented here clearly show the sub-Poissonian photon statistics of the one-atom maser field. An increase in the flux of atoms leads to the predicted second maser phase transition. In addition, the maser experiment leads to an atomic beam with atoms in the lower maser level showing number fluctuations which are up to 40% below those of the Poissonian distribution usually found in atomic beams. This is interesting, because atoms found in the lower level have emitted a photon to compensate for cavity losses inevitably present in the maser under steady-state conditions. This is a purely dissipative phenomenon giving rise to fluctuations, nevertheless the atoms still obey sub-Poissonian statistics.

3.6 A New Probe of Complementarity in Quantum Mechanics

The preceding section discussed how to generate a nonclassical field inside the maser cavity. But this field is extremely fragile because any attenuation causes a considerable broadening of the photon number distribution. Therefore it is difficult to couple the field out of the cavity while preserving its nonclassical character. But what is the use of such a field? In the present section we want to propose a new series of experiments performed inside the maser cavity to test the "wave–particle" duality of nature, or better said "complementarity" in quantum mechanics.

Complementarity [3.24] lies at the heart of quantum mechanics: Matter sometimes displays wave-like properties manifesting themselves in interference phenomena, and at other times it displays particle-like behaviour thus providing "which-path" information. No other experiment illustrates this wave–particle duality in a more striking way than the classic Young's double-slit experiment [3.25, 26]. Here we find it impossible to tell which slit light went through while observing an interference pattern. In other words, any attempt to gain which-path information disturbs the light so as to wash out the interference fringes. This point has been emphasized by Bohr in his rebuttal to Einstein's ingenious proposal of using recoiling slits [3.26] to obtain "which-path" information while still observing interference. The physical positions of the recoiling slits, Bohr argues, are only known to within the uncertainty principle. This error contributes a random phase shift to the light beams which destroys the interference pattern.

Such random-phase arguments illustrating in a vivid way how the "which-path" information destroys the coherent-wave-like interference aspects of a given experimental set-up, are appealing. Unfortunately, they are incomplete: In principle, and in practice, it is possible to design experiments which provide "which-path" information via detectors which do not disturb the system in any noticeable way. Such "Welcher Weg" (German for "which-path") detectors have been considered recently within the context of studies involving spin coherence [3.27]. In the present section we describe a proposed quantum optical experiment [3.28] which shows that the loss of coherence occasioned by "which-path" information, that is, by the presence of a "Welcher Weg" detector, is due to the establishing of quantum correlations. It is in no way associated with large random-phase factors as in Einstein's recoiling slits.

The details of this application of the micromaser are discussed in [3.29]. Here only the essential features are given. We consider an atomic interferometer where the two particle beams pass through two maser cavities before they reach the two slits of the Young's interferometer. The interference pattern observed is then also determined by the state of the maser cavity. The interference term is given by

$$\langle \phi_1^{(f)}, \phi_2^{(i)} | \phi_1^{(i)}, \phi_2^{(f)} \rangle \, , \tag{3.2}$$

where $|\phi_j^{(i)}\rangle$ and $|\phi_j^{(f)}\rangle$ denote the initial and final states of the maser cavity.

Let us prepare, for example, both one-atom masers in coherent states $|\phi_j^{(i)}\rangle = |\alpha_j\rangle$ of large average photon number $\langle m \rangle = |a_j|^2 \gg 1$. The Poissonian photon number distribution of such a coherent state is very broad, $\Delta m \approx \alpha \gg 1$. Hence the two fields are not changed much by the addition of a single photon associated with the two corresponding transitions. We may therefore write

$$|\phi_j^{(f)}\rangle \cong |\alpha_j\rangle \,,$$

which to a very good approximation yields

$$\langle \phi_1^{(f)}, \phi_2^{(i)}|\phi_1^{(i)}, \phi_2^{(f)}\rangle \cong \langle \alpha_1, \alpha_2|\alpha_1, \alpha_2\rangle = 1 \,.$$

Thus there is an interference cross term (3.2) different from zero.

However, when we prepare both maser fields in number states $|n_j\rangle$ [3.17–19] the situation is quite different. After the transitions of the atom to the d-states, that is after emitting two photons, one in each cavity, the final states in the two cavities read

$$|\phi_j^{(f)}\rangle = |n_j + 1\rangle$$

and hence

$$\langle \phi_1^{(f)}, \phi_2^{(i)}|\phi_1^{(i)}, \phi_2^{(f)}\rangle = \langle n_1 + 1, n_2|n_1, n_2 + 1\rangle = 0 \,,$$

that is the coherence cross term vanishes and interferences are observed.

At first sight this result might seem a bit surprising when we recall that in the case of a coherent state the transitions did not destroy the coherent cross term, i.e. did not affect the temporal interference fringes. However, in the example of number states we can, by simply "looking" at the one-atom maser state, tell which "path" the atom took.

It should be pointed out that the beats disappear not only for a number state. For example, a thermal field leads to the same result. In this regard, we note that it is not enough to have an indeterminate photon number to ensure quantum beats. The state $|\phi_j^{(f)}\rangle$ goes as $a_j^+|\phi_j^{(i)}\rangle$, where a_j^+ is the creation operator for the jth maser. Hence the inner product

$$\langle \phi_j^{(i)}|\phi_j^{(f)}\rangle \rightarrow \langle \phi_j^{(i)}|a_j^+|\phi_j^{(i)}\rangle \,,$$

and in terms of a more general density matrix formalism we have

$$\langle \phi^{(i)}|\phi^{(f)}\rangle \rightarrow \sum_n \sqrt{n + 1}\,\rho_{n,n+1}^{(i)} \,.$$

Thus we see that an off-diagonal density matrix is needed for the production of beats. For example, a thermal field having indeterminate photon number would not lead to quantum beats since the photon number distribution is diagonal in this case.

The atomic interference experiment in connection with one-atom maser cavities is a rather complicated scheme for a "Welcher Weg" detector. There is a much simpler possibility which we will discuss briefly in the following. This is based on the logic of the famous "Ramsey fringe" experiment. In this experiment two microwave fields are applied to the atoms one after the other. The interference occurs since the transition from an upper state to a lower state may either occur in the first or in the second interaction region. In order to calculate the transition probability we must sum the two amplitudes and then square,

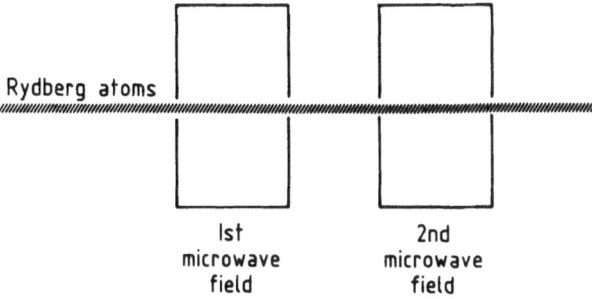

Fig. 3.10. Scheme of the Ramsey experiment

thus leading to an interference term. We will show here only the principle of this new experiment. In the set-up we discuss here, the two Ramsey fields are two one-atom maser cavities (Fig. 3.10). The atoms enter the first cavity in the upper state and are weakly driven into a lower state $|b\rangle$. That is, each microwave cavity induces a small transition amplitude governed by $m\tau$, where m is the atom–field coupling constant and τ_j is the time of flight across the cavity j; T is the time of flight between the cavities.

Now if the quantum state of the initial (final) field in the jth cavity is given by $\phi_j^{(i)}(\phi_j^{(f)})$ then the state of the atom + maser 1 + maser 2 at the various relevant times is given in terms of the coupling constant m_j and interaction times τ_j, initial $|\phi_j^{(i)}\rangle$ and final states $|\phi_j^{(f)}\rangle$ of the jth maser by

$$|\psi(0)\rangle = |a, \phi_1^{(i)}, \phi_2^{(i)}\rangle$$

$$|\psi(\tau_1)\rangle \cong |a, \phi_1^{(i)}, \phi_2^{(i)}\rangle - im_1\tau_1|b, \phi_1^{(f)}, \phi_2^{(i)}\rangle$$

$$|\psi(\tau_1 + T)\rangle \cong |a, \phi_1^{(i)}, \phi_2^{(i)}\rangle - im_1\tau_1|b, \phi_1^{(f)}, \phi_2^{(i)}\rangle e^{-i\Delta\omega T}$$

$$|\psi(\tau_1 + T + \tau_2)\rangle \cong |a, \phi_1^{(i)}, \phi_2^{(i)}\rangle - im_1\tau_1|b, \phi_1^{(f)}, \phi_2^{(i)}\rangle e^{-i\Delta\omega T}$$
$$- im_2\tau_2|b, \phi_1^{(i)}, \phi_2^{(f)}\rangle ,$$

where $\Delta\omega$ is the atom–cavity detuning and $T \gg \tau_j$ the time of flight between the two cavities. If we ask for P_b, the probability that the atom exits cavity 2 in the lower state $|b\rangle$, this is given by

$$P = \left(\langle\phi_1^{(f)}, \phi_2^{(i)}|m_1^*\tau_1 e^{i\Delta\omega T} + \langle\phi_1^{(i)}, \phi_2^{(f)}|m_2^*\tau_2\right) \times$$

$$\times \left(|\phi_1^{(f)}, \phi_2^{(i)}\rangle m_1\tau_1 e^{-i\Delta\omega T} + |\phi_1^{(i)}, \phi_2^{(f)}\rangle m_2\tau_2\right)$$

$$= m_1^* m_1\tau_1^2 + m_2^* m_2\tau_2^2 + (m_1^* m_2\tau_1\tau_2 e^{i\Delta\omega T}\langle\phi_1^{(f)}, \phi_2^{(i)}|\phi_1^{(i)}, \phi_2^{(f)}\rangle + \text{c.c.}) .$$

Now in the usual Ramsey experiment $|\phi_j^{(i)}\rangle = |\phi_j^{(f)}\rangle = |\alpha_j\rangle$, where $|\alpha_j\rangle$ is the coherent state in the jth maser which is not changed by the addition of a single photon. Thus the "fringes" appear behaving as $\exp(i\Delta\omega T)$. However, consider the situation in which $|\phi_j^{(i)}\rangle$ is a number state, e.g. the state $|0_j^{(i)}\rangle$, having no photons in the jth cavity initially; now we have

$$P = m_1^* m_1 \tau_1^2 + m_2^* m_2 \tau_2^2 + (m_1^* m_2 \tau_1 \tau_2\, e^{i\Delta\omega T}\langle 1_1, 0_2 | 0_1, 1_2\rangle + \text{c.c.}) \ .$$

In this case, the one-atom masers are now acting as "Welcher Weg" detectors, and the interference term vanishes due to the atom–maser quantum correlation.

We note that the more usual Ramsey fringe experiment involves a strong field "$\pi/2$-pulse" interaction in the two regions. This treatment is more involved than necessary for the present purposes.

We conclude this section by emphasizing again that this new and feasible experimental example of wave–particle duality and observation in quantum mechanics displays a feature which makes it distinctly different from the Bohr–Einstein recoiling-slit experiment. In the latter, the coherence, i.e. the interference, is lost due to a phase disturbance of the light beams. In the present case, however, the loss of coherence is due to the correlation established between the system and the one-atom maser. Random-phase arguments never enter the discussion. We emphasize that the argument of the number state not having a well-defined phase is not relevant here; the important dynamics is due to the atomic transition. It is the fact that which-path information is made available which washes out the interference cross terms (for a detailed discussion see [3.29]).

3.7 Summary and Conclusion

Nonclassical light as generated in the one-atom maser is characterized by amplitude fluctuations which are below those of the vacuum state. Such a field prepared with a very narrow photon number distribution is nonclassical because of its non-wave-like character. This property can only be observed when classical fluctuations, e.g. fluctuations in the velocity of atoms or dissipation of photons in the cavity walls, are excluded. The generation of a non-classical field in the one-atom maser is achieved by controlling the rate of photon emissions by the intensity of the field inside the cavity. The intensity is thereby controlled by itself. This scheme uses a back-coupling mechanism which correlates the probability of photon emission with the instantaneous radiation amplitude in such a way that any deviation of the intensity from its mean value is counterbalanced by a correspondingly changed rate of photon emission. Such a quantum non-demolition feedback mechanism is at hand when a single atom with a well-defined velocity interacts with a single mode of the electromagnetic field and the atom–field coupling is described by the Rabi-nutation dynamics. This is—in short—the basic principle that allows one to generate a non-classical maser field.

The possibility of testing the complementarity principle via the one-atom maser is another theme of the present paper: Send an atom prepared in a coherent superposition of two quantum levels through two consecutive one-atom masers. In this way, transfer the population to two neighbouring levels. Is the initial coherence between the two atomic states as manifested in an oscillatory ionization signal—temporal interference—preserved? To elucidate this question, imagine starting from coherent cavity fields with a large mean number of photons. In such a coherent state the fluctuations in photon number are large. Hence the two photons emitted in the transfer process, and caught in the cavities, do not change the field state significantly. One cannot deduce information concerning the transfer by studying the cavity fields. Hence the coherence is preserved giving rise to an interference pattern. However, if the experiment is started from a state of well-defined photon number it is possible to detect the increase, by one, in photon number induced by the population transfer. This destroys the coherence of the state, just as identifying which slit a photon passes through in the Young's double-slit experiment destroys the coherence and hence the interference fringes in that experiment.

References

3.1   For a review see the following articles by S. Haroche, J.M. Raimond: Adv. Atomic Molec. Phys. **20**, (Academic, New York 1985) p. 350; J.A. Gallas, G. Leuchs, H. Walther, H. Figger: Adv. Atomic Molec. Phys. **20** (Academic, New York 1985) p. 413

3.2   W. Ihe, A. Anderson, E.A. Hinds, D. Meschede, L. Moi, S. Haroche: Phys. Rev. Lett. **58**, 666 (1987)

3.3   G. Barton: Proc. Roy. Soc. **A410**, 147 and 175 (1987)

3.4   E.T. Jaynes, F.W. Cummings: Proc. IEEE **51**, 89 (1963)

3.5   D. Meschede, H. Walther, G. Müller: Phys. Rev. Lett. **54**, 551 (1985); for a review see F. Diedrich, J. Krause, G. Rempe, M.O. Scully, H. Walther: IEEE J. Quantum Electron. **24**, 1314 (1988)

3.6   See, e.g. J.H. Eberly, N.B. Narozhny, J.J. Sanchez-Mondragon: Phys. Rev. Lett. **44**, 1323 (1980) and references therein

3.7   H.I. Yoo, J.H. Eberly: Phys. Rep. **118**, 239 (1985); P.L. Knight, P.M. Radmore: Phys. Lett. **90A**, 342 (1982)

3.8   G. Rempe, H. Walther, N. Klein: Phys. Rev. Lett. **58**, 353 (1987)

3.9   For reviews see H.J. Kimble, D. Walls: J. Opt. Soc. Am. **B4**, 1449 (1987); R. Loudon, P.L. Knight: J. Mod. Opt. **34**, 707 (1987)

3.10 First demonstration of photon antibunching: H.J. Kimble, M. Dagenais, L. Mandel: Phys. Rev. Lett. **39**, 691 (1977); Phys. Rev. **A18**, 201 (1978); see also J.D. Cresser, J. Häger, G. Leuchs, M. Rateike, H. Walther: *Dissipative Systems in Quantum Optics* (Springer, Berlin, Heidelberg 1982) p. 21

3.11 First demonstration of sub-Poissonian photon statistics: R. Short, L. Mandel: Phys. Rev. Lett. **51**, 384 (1983)

3.12 First demonstration of squeezing: R.E. Slusher, L.W. Hollberg, B. Yurke, J.C. Mertz, J.F. Valley: Phys. Rev. Lett. **55**, 2409 (1985); for review, see, e.g. H.J. Kimble, D. Walls: J. Opt. Soc. Am. **B4**, 1449 (1987); R. Loudon, P.L. Knight: J. Mod. Opt. **34**, 707 (1987)

3.13 H.J. Carmichael, D.F. Walls: J. Phys. **B9**, 1199 L43 (1976)

3.14 F. Diedrich, H. Walther: Phys. Rev. Lett. **58**, 203 (1987)

3.15 P. Filipowicz, J. Javanainen, P. Meystre: Opt. Comm. **58**, 327 (1986); Phys. Rev. **A34**, 3077 (1986); J. Opt. Soc. Am. **B3**, 906 (1986)

3.16 L. Lugiato, M.O. Scully, H. Walther: Phys. Rev. **A36,** 740 (1987)

3.17 J. Krause, M.O. Scully, H. Walther: Phys. Rev. **A36,** 4547 (1987); J. Krause, M.O. Scully, T. Walther, H. Walther: Phys. Rev. **A39,** 1915 (1989)

3.18 P. Meystre: Opt. Lett. **12,** 699 (1987); P. Meystre: In *Squeezed and Nonclassical Light* ed. by P. Tombesi, E.R. Pike (Plenum, New York 1988), p. 115

3.19 P. Meystre, G. Rempe, H. Walther: Opt. Lett. **13,** 1078 (1988)

3.20 J.J. Slosser, P. Meystre, E.M. Wright: Opt. Lett. **15,** 233 (1990)

3.21 G. Rempe, H. Walther: Phys. Rev. **A42,** 1650 (1990)

3.22 G. Rempe, F. Schmidt-Kaler, H. Walther: Phys. Rev. Lett. **64,** 2783 (1990)

3.23 L. Mandel: Opt. Lett. **4,** 205 (1979)

3.24 See, e.g. D. Bohm: *Quantum Theory* (Prentice Hall, Englewood Cliffs 1951) or M. Jammer: *The Philosophy of Quantum Mechanics* (Wiley, New York 1974)

3.25 A detailed analysis of Einstein's version of the double-slit experiment is given by W. Wootters, W. Zurek: Phys. Rev. **D19,** 473 (1979); see also [3.26]

3.26 For an excellent presentation of the Bohr–Einstein dialogue see Chap. 1 in: J.A. Wheeler, W.H. Zurek: *Quantum Theory and Measurement* (Princeton University Press, Princeton 1983) and in particular, the article by N. Bohr, Discussion with Einstein on Epistemological problems in Atomic Physics

3.27 B.-G. Englert, J. Schwinger, M.O. Scully: Found. Phys. **18,** 1045 (1988); J. Schwinger, M.O. Scully, B.-G. Englert: Z. Phys. **D10,** 135 (1988); M.O. Scully, B.-G. Englert, J. Schwinger: Phys. Rev. **A40,** 1775 (1989)

3.28 M.O. Scully, H. Walther: Phys. Rev. **A39,** 5229 (1989)

3.29 M.O. Scully, B.-G. Englert, H. Walther: Nature **351,** 111 (1991)

Herbert Walther

4. Regioselective Photochemistry in Weakly Bonded Complexes*

S.K. Shin,[a] Y. Chen,[b] E. Böhmer and C. Wittig

With 9 Figures

Photoinitiated reactions in CO_2–HX and N_2O–HI complexes show remarkable entrance channel regiospecificities. In the case of CO_2–HX complexes, H atom approaches from the broadside of CO_2 in CO_2–HBr complexes are more reactive than the relatively end-on approaches in CO_2–HCl complexes. For the case of the N_2O–HI complex, photoinitiated reactions result in a much lower [NH]/[OH] ratio with complexes than under single-collision conditions at the same photolysis wavelength, which illustrates the entrance channel regiospecificity influencing chemically distinct product channels. In addition, deuterium substitution in place of the H atom shows an isotope effect. Future experiments using double resonance techniques are discussed briefly. This will extend our ability to manipulate chemical reactions.

4.1 Introduction

This paper reviews and summarizes our progress over the past six years in the area of *photoinitiated reactions in weakly bonded complexes*. We emphasize particularly (i) the original motivations (most of which are still valid) and fundamental questions and issues that led us into this area, (ii) the highlights and milestones achieved during this period, including a critical assessment of the respective virtues of the different measurements, and (iii) prospects and ideas for further research.

Our initial goal was to study bimolecular chemical reactions with the restricted set of angles and impact parameters for the entrance channel that are available when using weakly bonded binary precursors in the gas phase [4.1]. Since the weak intermolecular forces binding two stable molecules are generally anisotropic, it seemed obvious to examine regiospecific and/or stereospecific propensities. The direct photodissociation of one of the molecules propels a

* This research was supported by the National Science Foundation, Department of Energy and U.S. Army Research Office.

[a] Present address: Department of Chemistry, University of California, Santa Barbara, CA 93106, U.S.A.

[b] Present address: Department of Chemistry, University of California, Berkeley, CA 94720, U.S.A.

fragment towards the other molecule with a distribution of incident angles and impact parameters whose spread is limited at the most fundamental level by the zero-point motions of the intermolecular vibrations. In principle, this affords a high degree of spatial resolution compared with other techniques. This will be discussed in more detail below, but we wish to point out here that although the steric effect in gas-phase reactive scattering has a rich history, it has only been tested *directly* in a few cases.

In those cases where steric effects *were* probed directly in reactive scattering experiments, the results have been quite rewarding. Pioneering experiments with externally applied fields that orient and/or align molecules prior to collision showed that steric effects indeed play an important role in reactive collisions. [4.2, 3]. It was also shown that the results could be counterintuitive [4.4, 5], underscoring the importance of such research. However, in such experiments there are limitations on the extent to which entrance channel geometric properties can be controlled. Because the applied electromagnetic fields fix a laboratory frame as a reference system, molecules that are under the influence of these fields have their axes oriented and/or aligned relative to the laboratory. When such species undergo a collision, this anisotropy is passed on in the center-of-mass (CM) system, for example in a crossed-beam-type experiment. However, impact parameters are not controlled at all, and the best control possible for one collision partner is for the case of a symmetric top having well-defined J, K and M quantum numbers, or conceivably a well-defined wave-packet comprised of such states.

With state-selected symmetric top molecules, steric effects have been observed in measurements best described as elegant and important, but also rather difficult. Here, we refer mainly to the studies of *Bernstein* [4.3, 6–11], *Brooks* [4.2, 4, 5, 12], *Stolte* [4.13–17], *Parker* [4.7, 11, 16, 17], and their coworkers. In these gas-phase experiments, the results have been limited to reactions with large cross sections such as $K + CH_3I \rightarrow CH_3 + KI$, where it was discovered that K-atom attack of the iodine is preferred over K-atom attack of the methyl side [4.4, 5, 12]. However, some very interesting recent results have also been obtained for symmetric top molecules colliding with well-characterized surfaces under UHV conditions [4.18–22]. Namely, the sticking coefficient depends upon the orientation of the molecule initially pointing toward the surface. Counterintuitive results have been reported, in which collisions of NO with Ag(111) exhibit a higher sticking coefficient for the oxygen side approaching the surface than for the nitrogen side, even though an unpaired electron in NO is localized on the nitrogen [4.18, 19]. In contrast, the nitrogen side shows the higher sticking probability than the oxygen side on the Pt(111) [4.19] and Ni(100) [4.20] surfaces. The experimental advantage of the molecule-surface scattering studies is that every incident molecule undergoes a collision. However, this also reveals one of the main difficulties in gas-phase reactive scattering studies using state-selected reagents: the very low fluxes make product state detection almost impossible. This means that reactions having large cross sections are required to insure adequate S/N. Even beyond these practical S/N problems, there are fundamental limitations. For instance, the ability to limit impact parameters is seriously hampered when dealing with large reaction cross sections, where all impact parameters contribute. Also, using state-selected symmetric top molecules, it is hard enough to prepare a single $|J, K, M\rangle$ state, let alone try to construct a useful wavepacket. Thus, while eminently worthwhile, we dont't envision this technique being used by more than a handful of groups.

We believe it is important here to distinguish between (i) reactions involving electronically excited atoms that are produced using polarized laser radiation, in which the excited atomic orbitals are aligned in the laboratory system, and (ii) reactions of essentially ground electronic state reagents (but including the fine structure levels). The latter is the more usual situation in chemistry. In this case, it is the initial positions and movements of the atoms that influence the outcome of the collision. The electron movements control reactivities and are carried on the frames of the reactants rather than perturbed separately by externally applied electromagnetic fields. Some truly excellent experiments have been carried out in which atoms having specifically aligned electron orbitals in the laboratory system react and scatter inelastically, both in the gas phase and in weakly bonded complexes [4.23–25]. We have reviewed some of this work recently [4.26], particularly that of *Soep* and coworkers [4.25, 27–31]. However, in this paper we will discuss only the case of reactions brought about by photodissociation within complexes.

Getting back to our initial goal, we thought that under the right conditions it might be possible to have a weak intermolecular interaction at a sufficient distance to insure essentially complete photodissociation before one of the photofragments encounters the other molecule in the complex. At the same time, we hoped that the anisotropic binding of the ground state complex would serve to align the photofragment along the intermolecular bond. Thus, unlike the corresponding single-collision gas phase experiment, the collision initiated within the complex would be very site-specific. A schematic diagram showing this idealized situation is given in Fig. 4.1.

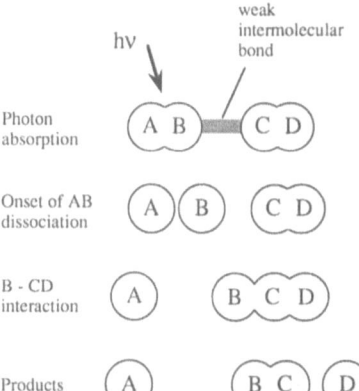

Fig. 4.1. A schematic diagram showing site-specific collisions photoinitiated within the complex

The notion that electronic excitation remains localized on one molecule needs to be examined carefully for each case, since couplings can be very strong if there is a resonance. Even for a case such as direct HX photodissociation, which is known to be very rapid (i.e. 10s of fs), the electronic structure of the excited complex can be affected strongly if the other molecule possesses a nearby electronic state. We have checked this point whenever possible and have encountered no obvious problems so far [4.26]. However, without the use of double resonance (described below) this may seriously limit the number of candidate systems.

Zero-point fluctuations for these floppy systems can be large, especially when the effective mass is ~ 1 amu, and it has been pointed out to us that heavier species will "stay put" more obediently than hydrogen, resulting in better-defined starting conditions. So why are we still using systems in which HX photodissociation propels an H atom at the other molecule in the complex? Firstly, this is how we started. We are not prejudiced against heavier species, but have just not moved on to other attractive systems as quickly as perhaps we should have. Secondly, and we certainly weren't thinking of this when we started these experiments, large-amplitude motion is advantageous in cases where inter-molecular modes can be excited optically, since this offers significantly different spatial distributions than the ground state. This will be discussed further below.

The site-specificity aspect for molecular beam vs. weakly bonded complex conditions is shown schematically in Fig. 4.2. The case of a monoenergetic, prefectly collimated molecular beam interacting with a target molecule is shown in Fig. 4.2a. The molecular beam can, in principle, consist of any atomic or molecular species, but we will only consider atomic hydrogen, since this is what has been used in all our experiments to date. If the target is aligned and/or oriented, there will clearly be entrance channel geometric specificity, but with no control of impact parameters. Figure 4.2b shows the case where a distribution of

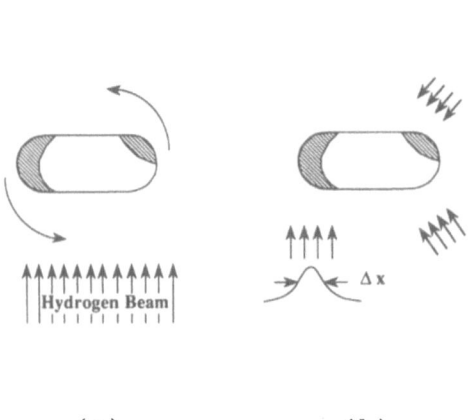

Reactive Parts

(a)   (b)

Fig. 4.2a, b. A schematic diagram showing the site-specificity aspect for **(a)** molecular beam vs. **(b)** weakly bonded complex conditions. The shaded areas indicate reactive parts and Δx denotes a spatial width of an H-atom distribution

H atoms with spatial width Δx is made to attack the molecule from qualitatively different approaches. Some approaches may prove reactive while others are unreactive, as in the case of $H + CO_2 \rightarrow OH + CO$ showing a significant steric effect, and it may also be possible to select chemically distinct products, as in the case of $H + N_2O$, which can yield both $OH + N_2$ and $NO + NH$ products. Although the specificity shown in Fig. 4.2b is greater than that shown in Fig. 4.2a, it requires that nature cooperates by providing the appropriate precursors. But nature is not always cooperative. Thus, the use of weakly bonded precursors is idiosyncratic and probably can't be applied to a large variety of systems. Is this a handicap? Somewhat perhaps, but no experimental approach is truly universal, and with the present method we believe that the best approach is to identify valuable prototypes and study them in great detail.

4.2 End-On CO_2–HCl and Broadside CO_2–HBr

In the case of CO_2–HX complexes, we recorded rotationally resolved IR spectra with a tunable diode laser ($\pm$ 5 MHz accuracy) and discovered that CO_2–HCl and CO_2–HBr have different structures [4.32]. Hereafter, we refer to CO_2–HCl as end on, i.e. OCO–HCl, even though there is very large amplitude zero-point motion in the "hinge" angular cordinate and the vibrationally averaged geometry is not perfectly linear [4.33]. On the other hand CO_2–HBr is T shaped, with the bromine facing the carbon and a θ_{OCBr} angle of $\sim 90°$. Experimental efforts to determine the H-atom position prove a planar equilibrium structure with the CO_2 and HBr axes roughly parallel *on average* [4.34]. This pronounced structural difference between CO_2–HCl and CO_2–HBr can be ascribed to competition between hydrogen bonding and a dispersive interaction: with CO_2–HCl the dominant interaction is hydrogen bonding, while in CO_2–HBr the dominant interaction is dispersive attraction between the CO_2 quadrupole and the polarizable bromine, similar to that observed in CO_2–rare-gas complexes [4.35]. Hereafter, we refer to CO_2–HBr as broadside.

We also carried out electronic structure calculations for the CO_2–HBr ground state geometry (MP2, double zeta plus two d polarization functions, bromine core potential) and found two minima, one for linear OCO–HBr and the other for the T-shaped complex having a planar equilibrium geometry and a C–Br distance and θ_{OCBr} angle that are close to the experimental values [4.36]. The θ_{CBrH} potential is shallow and indicates very large amplitude H-atom bending motion, as shown in Fig 4.3. The classical turning points for the $v = 0$ and 1 levels of the H-atom bending mode are $\sim 28°$ and $\sim 43°$, respectively, relative to the equilibrium position. Thus, even though the average H-atom position shows roughly parallel CO_2 and HBr axes, there is significant probability for finding the H atoms pointing in the direction of the oxygen even for $v = 0$ and particularly for $v = 1$. Figure 4.3 depicts just a single cut on the potential surface, and we note that the probability for finding the H atom

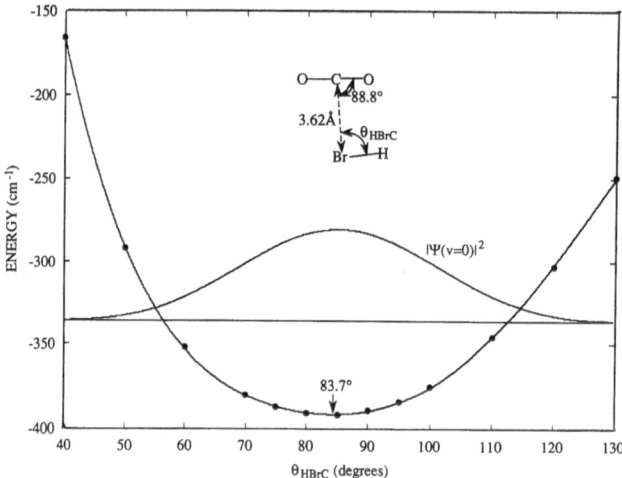

Fig. 4.3 Potential curves as a function of θ_{HBrC} for T-shaped CO_2–HBr (MP2/DZ2P calculation with basis set superposition error corrections) and the zero-point bending amplitude. R_{CBr} and θ_{BrCO} are held at 3.62 Å and 88.8°

pointing in the direction of the oxygen increases when the other intermolecular zero-point motions are taken into account.

For the cases of CO_2–HBr and CO_2–HCl, excitation of the HX chromophore is expected to result in rapid expansion of the HX bond followed by either escape of the H atom or collision with CO_2. The reactive encounters lead to the formation of a HOCO† intermediate, which subsequently decomposes to OH + CO. So one obvious question is: which precursor leads to the more reactive collision of H atoms with CO_2? With CO_2–HCl, nearly all of the H atoms "strike" the oxygen, while with CO_2–HBr, the majority fly out into the vacuum chamber missing or hardly grazing the oxygen. To the extent that the halogen does not participate in the dynamics or chemistry, such a measurement can elucidate the steric effect for the $H + CO_2 \rightarrow OH + CO$ reaction.

In order to pursue this, we carried out experiments both with CO_2–HCl and CO_2–HBr complexes and with their single-collision counterparts in which the relative reaction probabilities were measured [4.37]. This can be made to sound easy. On alternate days, CO_2/HCl/He and CO_2/HBr/He mixtures would be used and the OH yields would be measured. To remove as much as possible the effect of the different HCl and HBr bond strengths on the H-atom kinetic energy, we used 222 nm HBr photolysis and 193 nm HCl photolysis, as shown in Fig. 4.4. For uncomplexed HCl and HBr, the corresponding $H + CO_2$ CM collision energies are 15 390 and 14 220 cm^{-1}, respectively, which are close enough so that the reaction probabilities are similar. Thus, with uncomplexed samples one expects similar OH yields per absorbed photon. This is exactly what was observed.

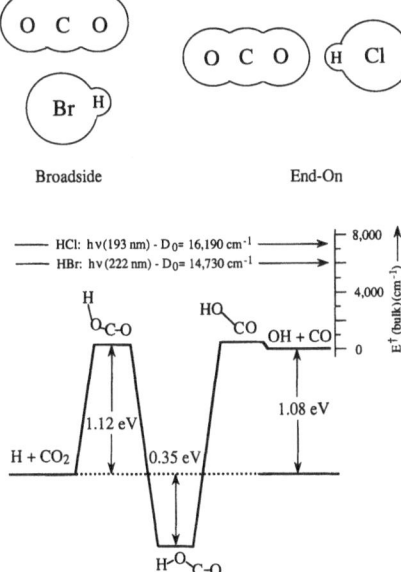

Fig. 4.4 Schematic drawing indicating broadside CO_2–HBr and end-on CO_2–HCl complexes, and an energy diagram showing relevant stationary points for H + CO_2 system. $E^\dagger$ corresponds to the energy available to the OH + CO product channel under bulk conditions (i.e. with the CM transformations taken into account). The quantities $h\nu - D_0$(H–X) are given for HCl and HBr excimer laser photolysis wavelengths that result in similar energies

Next, we carried out the measurements with the complexes, but for meaningful comparisions it was necessary to know the relative CO_2–HCl and CO_2–HBr concentrations. The CO_2HBr^+ and CO_2HCl^+ mass spectrometer signals were comparable to each other, but this can be misleading because CO_2–HX parent species may fragment differently in the mass spectrometer upon ionization. Thus, we measured relative CO_2–HCl and CO_2–HBr concentrations using tunable diode laser spectroscopy. Specifically, (i) relative rotational level populations were determined from the areas of the absorption lines, (ii) rotational temperatures were obtained and were used to calculate rotational partition functions, and (iii) relative CO_2–HCl and CO_2–HBr concentrations were determined by assuming that all such species are in the ground vibrational level and that rotations are described adequately by a temperature. When this was done, we found that comparable amounts of CO_2–HBr and CO_2–HCl were produced in the expansions, which is not surprising. CO_2HX^+ mass spectrometer signals were then calibrated against IR absorption signals and the relationship was seen to be linear over the range of values used in the experiments. Since experimental setups for the spectroscopy and photoinitiated reactions are different (e.g., the former used a slit expansion while the latter used a circular orifice), but nevertheless use similar mass spectrometers (i.e. UTI 100C), we used the calibrated CO_2HX^+ mass spectrometer signals to infer the relative CO_2–HX concentrations in the photoinitiated-reaction experiments.

The net result of the above calibration measurements was to find that comparable amounts of CO_2–HCl and CO_2–HBr were present in the photoinitiated-reaction studies. This is what we had guessed originally, but such guesses are of no obvious value. Next, we did the key experiment and got a big surprise when the measured the relative amounts of OH produced per absorbed photon. CO_2–HBr yielded about 30 times more OH than OCO–HCl. It was a large effect, with so little OH produced from OCO–HCl that we would barely record a LIF signal when focusing the 193 nm excimer laster output (~ 100 mJ) into the sample. Incidentally, this null result is a testimony to our ability to discriminate against unwanted gas-phase $H + CO_2$ collisions that involve expansion-cooled, but uncomplexed molecules. If the difference in reaction probabilities had been small, such as a factor of two, it would have been hard to say that it was outside the uncertainty of the measurement. But such a large order-of-magnitude effect stood out. Note also that with CO_2–HBr, most of the hydrogen atoms do not even strike the oxygen because of the broadside ground state equilibrium geometry.

4.2.1 Product State Distributions

The measurement of nascent product state distributions is tedious. We spent several years doing this for cases such as CO_2–HBr, CO_2–DBr CO_2–HI N_2O–HBr N_2O–HI N_2O–DI CO_2–H_2S and OCS–DBr. It is an essentially useful exercise, because excitations in the various degrees of freedom of different products reflect the ensemble-averaged dynamics. However, it is easy to spend a lot of time on this, and there is a limit to the amount of information likely to accrue from repeated measurements. At this point, we believe that product state distributions should definitely be recorded, but unless an experiment is designed that requires precise, detailed distributions, it is sufficient to view general trends, rather than getting bogged down with details.

A common characteristic of all OH rotational state distributions deriving from complexes is that they appear relatively cold compared with their single-collision counterparts. This varies some from case to case, but the trend is clear. Some years ago we noticed that the majority (i.e. ~ 75%) of the OH could be fit to a distribution very much like that of the single-collision experiments, but with ~ 20% less energy in excess of reaction threshold [4.38]. This seems reasonable in terms of the underdeveloped hydrogen speed characteristic of photoinitiated reactions in complexes, i.e. the *squeezed-atom effect* [4.36, 39]. Unlike the gas phase, where HX photodissociation apportions H and X translational energies in the CM system as per $(M_X/M_{HX})E_{CM}$ and $(M_H/M_{HX})E_{CM}$, respectively, the hydrogen atom expelled from its counterpart in the complex experiences repulsion from CO_2 before the HX bond dissociates completely, and this trapped hydrogen can transmit forces to the heavier fragments. We note that this effect has been observed by *Schatz* and coworkers in trajectory studies [4.40].

The remaining ~ 25% of the OH appear rotationally cold. Since they are distributed over a relatively small number of quantum states, this accounts for the marked shift in the peak of the rotational distribution compared with the corresponding single-collision experiment. We are not certain why some percentage of the OH product has such low rotational excitation, but this could be due to a $BrC(O)OH^\dagger$ short-lived intermediate [4.36] OH + Br collisions in the exit channel, higher-than-binary complexes, etc.

4.2.2 Time-Resolved Product Buildup

Some people say that the frequency domain provides all of the information obtainable using time-resolved, sub-picosecond-resolution methods. This may be true in some abstract mathematical sense, but it is not the case in practice. Take for example the time-resolved study of photoinitiated CO_2–HI complexes carried out by the *Zewail* group [4.41]. In this work, HI was excited using a picosecond pulse and the OH product was monitored using a second picosecond pulse. It is a standard pump/probe experiment, made 'non-standard' by the need to use high-energy, separately tunable, ultraviolet, picosecond pulses. This experiment showed that OH buildup took typically a couple of picoseconds, with longer buildup times occurring at the longer photolysis wavelengths. This is interpreted as proof that hydrogen does not strip away an oxygen atom, but instead that $HOCO^\dagger$ is the species that gives birth to OH. There is no other direct way to obtain this buildup rate, certainly not by frequency domain measurements.

Such experiments can (i) give state-specific reaction rates directly from the reaction time measurements, (ii) provide mechanistic clues regarding direct reactions vs. those proceeding through intermediates, and (iii) offer critical tests for theory. Choices are legion, and in our laboratory we have now developed a sub-picosecond-resolution pump/probe source that we hope will prove suitable for such measurements.

4.2.3 Theoretical Considerations

The above results must be accounted for with any model deemed viable, and one question is whether or not the bromine plays an important role chemically. In the absence of a significant bromine interaction, it may be possible to use the 4-atom HOCO potential surface to explore different H-atom approaches, as has been done by *Schatz* and coworkers [4.42, 43]. Before our discovery that CO_2–HBr is T shaped, we had assumed that the complex is linear [4.1], by analogy with CO_2–HF and CO_2–HCl. For linear OCO–HBr with a 2.34Å O–H distances [4.36] the carbon–bromine distances would be 4.9 Å, and this seemed too long to allow a $BrC(O)OH^\dagger$ intermediate to be formed. Thus, the role of the bromine was assumed to be limited to influencing the development of the H-atom velocity. However, with the broadside precursor, the carbon–bromine distance is only 3.6Å, so we decided to examine the possibility of $BrC(O)OH^\dagger$ formation using ab initio methods. These calculations were carried out at the GVB-RCI level, and the results, while lacking quantitative accuracy, can be used to examine propensities.

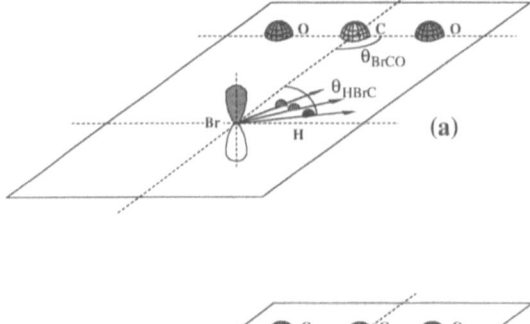

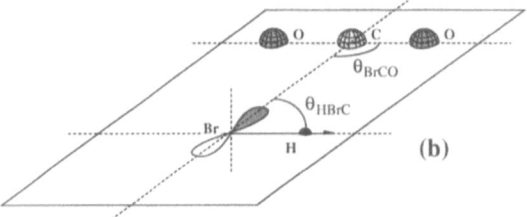

Fig. 4.5 Schematic drawing illustrating interactions of hydrogen and bromine atoms with CO_2 after $n \rightarrow \sigma^*$ HBr excitation, which correlates to atomic hydrogen and a singly occupied bromine p orbital that lies either (a) perpendicular to or (b) in the CO_2–HBr plane

Figure 4.5 indicates schematically the case of $n \rightarrow \sigma^*$ HBr excitation, which correlates to atomic hydrogen and a singly occupied bromine p orbital that lies either in or perpendicular to the CO_2–HBr plane. This simple picture offers guidance, but should not be misconstrued as being accurate. Other orbitals participate and a higher-level electronic structure calculation will be needed to obtain numerically accurate results.

With the bromine p orbital perpendicular to the plane, several H-atom approaches were examined, as shown in Fig. 4.5a. A broad range of approaches are available via the large amplitude zero-point bending, or even hot bands involving the intermolecular modes. If the CO_2 molecule is held at its equilibrium geometry, there is no H-atom capture when the hydrogen attacks the oxygen, i.e. the hydrogen is squeezed between the bromine and the oxygen, experiencing only Br–H and H–O repulsive interactions, and eventually escapes into vacuum. However, the CO_2 does not remain at its equilibrium geometry, since impact from an H-atom collision at the oxygen will induce the bending of the OCO frame and slight expansion of the OC–O bond. Thus, we distorted the CO_2 molecule accordingly and examined different hydrogen approaches. The results are shown in Fig. 4.6. One sees that these distortions do open a reactive window, and although we have not carried out the corresponding dynamical calculations, we believe that the hydrogen atom will be captured in these broadside approaches. In simple chemical terms, they facilitate the formation of the required HOCO intermediate, with its 1.34 Å OC–OH bond length and 127.6° equilibrium θ_{OCO} angle [4.42]. *Schatz* and coworkers have examined broadside collisions for the 4-atom $H + CO_2$ system, and report significant reaction probabilities when the H-atom strikes the oxygen [4.40].

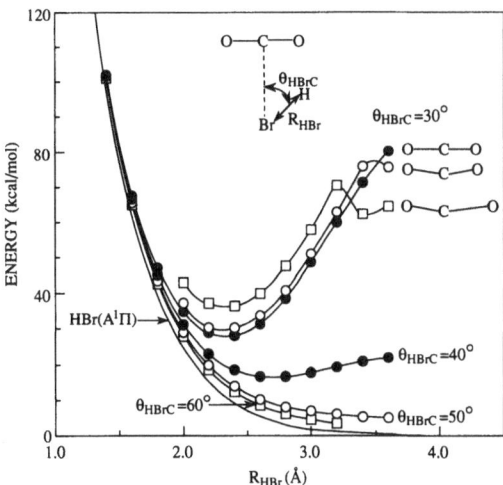

Fig. 4.6 Energies of the excited state of CO_2–HBr complex as a function of R_{CB} and for various values of θ_{HBrC}. For $\theta_{HBrC} = 30°$, the open circles and boxes correspond to distortions which indicate that CO_2 vibrational excitation promotes reactivity

There was no sign of significant bonding between the bromine and the carbon for the case shown in Fig. 4.5a, even *after* the attacking hydrogen atom has formed an OH bond. The HOCO radical has an orbital on the carbon that we thought might bond to a nearby radical, in this case to form a short-lived bromoformic acid intermediate, $BrC(O)OH^{\dagger}$. However, the orientation of the bromine orbital shown in Fig. 4.5a is non-bonding, and there was no attractive interaction. Thus, in this case the role of the bromine is limited to influencing the hydrogen velocity.

For the case shown in Fig. 4.5b, in which there is more of a bonding overlap of the bromine and carbon orbitals, we did find attraction between bromine and HOCO. However, for the large 3.6 A initial Br–C distance, it is not clear that these weak attractive forces can overcome the Br–HOCO recoil velocity caused by the H–Br bond dissociation and the squeezed-atom effect. Also, for the H-atom positions that are most favorable sterically, i.e. large φ, the Br–C bonding overlap is worse because the bromine orbital is twisted in the direction of a non-bonding interaction. Thus, it seems that there is also a large steric effect in this reaction, as for the case shown in Fig. 4.5a, while the formation of a short-lived $BrC(O)OH^{\dagger}$ intermediate remains an unproven possibility.

The linear OCO–HBr configuration was also examined, but not in detail. From the end-on direction, an H-atom collision with CO_2 compresses the nearby CO bond and bending is not especially encouraged. Thus, from a simple chemical point of view, the collision works against forming the HOCO intermediate. This was confirmed by examining the 4-atom HOCO potential surface [4.40], where it was seen that near-linear approaches in which the nearby CO bond is compressed never opened a reactive window. Thus, it is easy to see why end-on approaches are unreactive.

In general, one should consider not only collision-induced distortions of the CO_2 equilibrium geometry, but also the movement of nuclei due to the HOCO potential field, which are generally toward the HOCO equilibrium geometry. However, in the present case the light hydrogen can escape before the heavier nuclei move very much, making this less viable than for cases of more comparable mass combinations.

From the above, we conclude that broadside approaches, in which hydrogen attacks oxygen at small θ_{HBrC} values, are favored sterically, and that end-on hydrogen approaches are relatively inefficient. Referring to Fig. 4.3, as θ_{HBrC} decreases, the probability density $|\Psi(\theta_{HBrC})|^2$ decreases while the reaction probability increases, resulting in some range of θ_{HBrC} values that actually react. It would be nice to determine what these θ_{HBrC} values are. Since our measurements are sensitive to *differences* in reactivities, the CO_2–HBr precursor is not necessarily efficient in producing OH. In fact, our belief is that most of the H atoms from CO_2–HBr probably do not react; it is just the very low CO_2–HCl reactivity that accounts for the large observed difference.

4.3 N_2O–HI: OH + N_2 vs. NH + NO

The H + N_2O system has available to it two chemically distinct product channels, and photoinitiating this reaction in the complex may result in *selectivity* toward one of the channels. Although our intuition tells us that this probably is the case, i.e. attack at the oxygen side yielding OH + N_2 and attack at the nitrogen side yielding NH + NO, it would still be dramatic if the weak binding energy of the complex ($\sim 200 \text{ cm}^{-1}$) could result in a marked bias in the chemistry of the system.

The geometric structures of N_2O–HBr and N_2O–HI complexes are similar to CO_2–HBr(I) in that the halogen faces the central nitrogen and bonding is believed due to a dispersive interaction. We have determined the N_2O–HBr structure and surmise that N_2O–HI is qualitively similar [4.44]. However, it has not yet been possible to locate the hydrogen experimentally. Electronic structure calculations indicate that the equilibrium structure is planar, with the hydrogen favoring the oxygen over the terminal nitrogen [4.45]. Again, we introduce the caveat that higher-level calculations are needed. Initial accounts of this work have been published, and a large manuscript is in preparation [4.46], so here we present a summary of our results for this system.

In the N_2O–HI complex, the H atom can attack either at the oxygen side or at the nitrogen side depending on the initial N_2O–HI geometry. However, geometrically specific entrance channels may or may not carry selectivity through to products, depending on the nature of the potential surface and its accessibility from the starting point of the weakly bonded complex, as made clear in the CO_2–HX cases discussed above. When there is no long-lived intermediate involved in the reaction, it is easy to see why site-specific attack might result in products that reflect initial bias. However, if a bond part of the potential successfully facilitates the formation of a long-lived vibrationally excited intermediate, this may not be the case. Such systems have the uncanny ability to erase memory of the entrance channel. Of even greater subtlety, the above consideration might also depend on energy.

The $H + N_2O$ system exemplifies the above issue. Large-scale ab initio calculations have located all relevant stationary points on the $H + N_2O$ potential surface [4.47, 48]. H atom attack at the oxygen site accesses directly the $OH + N_2$ exit channel, which is downhill $\sim 28\,400\,\mathrm{cm}^{-1}$ from the top of the barrier, with no chance for a long-lived intermediate once past the barrier. On the other hand, $NH + NO$ production is via a $HNNO^\dagger$ intermediate, which may fragment by a unimolecular decomposition mechanism. In addition, this $HNNO^\dagger$ intermediate can yield $OH + NO$ via the H-atom migration from the nitrogen to the oxygen. It has been previously suggested from experimental rate constant vs. temperature measurements that the low-energy reaction mechanism involves initial addition of the H atom to the terminal nitrogen, followed by a shift to the oxygen end and concomitant expansion of the N–O bond to yield $OH + NO$ [4.47, 48].

In order to test product selectivity, we carried out experiments with N_2O–HI and N_2O–DI complexes and their single-collision counterparts, in which the relative product branching ratios were measured. $H \rightarrow D$ substitution in the complex does not change its equilibrium structure, but reduces the bending amplitude. This can be used to more sharply define the position of hydrogen in the complex and to enhance product selectivity, if the reaction carries its entrance channel specificity through to products. However, if reaction proceeds via an $HNNO^\dagger$ intermediate, then deuterium substitution suppresses hydrogen migration from the nitrogen to the oxygen by the kinetic isotope effect, thus favoring $ND + NO$ over $OD + N_2$. Since the above considerations may depend on the reaction energy, we have chosen 245 nm HI photolysis and 240 nm DI photolysis to provide identical reaction energies for both systems (i.e. the reaction energy under the bulk condition is $(44/45)(127/128)[h\nu(245\,\mathrm{nm}) - D_0(\mathrm{H–I}) + RT] + E_{\mathrm{int}}(N_2O)$ for photolysis and $(44/46)(127/129)[h\nu(240\,\mathrm{nm}) - D_0(\mathrm{D–I}) + RT] + E_{\mathrm{int}}(N_2O)$ for DI photolysis).

4.3.1 Product State Distributions

NH $(v = 0)$ rotational state distributions from single-collision experiments can be ascribed temperatures for all photolysis wavelengths examined: 245–266 nm. These temperatures vary linearly with $h\nu - D_0(\mathrm{HI})$ as shown in Fig. 4.7, extrapolating to zero at $11\,600\,\mathrm{cm}^{-1}$, close to $\Delta H_0(12\,100\,\mathrm{cm}^{-1})$ [4.49, 50]. A rotational temperature derived from ND $(v = 0)$ rotational state distribution at the 240 nm DI photolysis fits well to the linear relationship shown in Fig. 4.7. Thus, the $NH + NO$ channel probably proceeds via an $HNNO^\dagger$ intermediate [4.47, 48], resulting in statistical or near-statistical rotational state distributions. With complexes, the NH(D) rotational temperature is slightly colder than its bulk counterpart at the same photolysis wavelength, but of the same qualitative character. When placed on the plot of rotational temperature vs. available

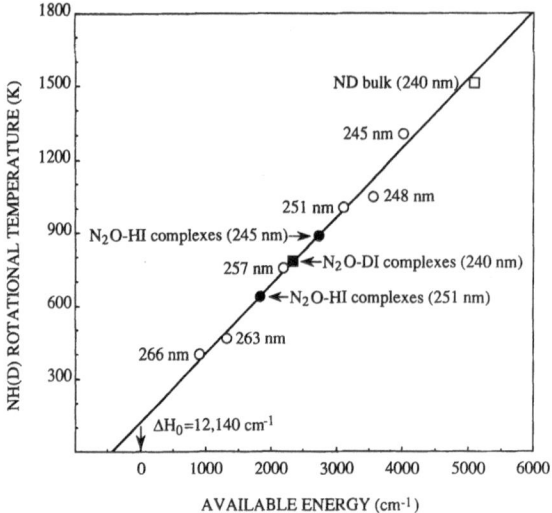

Fig. 4.7 Open circles indicate observed NH(D) rotational temperatures vs. energy available for ND(D) + NO bulk reaction products. Note that 300 K samples contain ~ 470 cm^{-1} of V, R energy, while complexes have essentially no internal energy except zero point, and are relative to free N$_2$O + H(D)I. The darkened symbols indicate the observed rotational temperatures for complexes (the horizontal axis is meaningless). A straight line extrapolation comes close to the 0 K enthalpy change for the H + N$_2$O system

energy for bulk reaction products, it corresponds to an available energy somewhat smaller than the bulk. Note that 300 K bulk samples have 410 cm^{-1} of rotational energy in HI + N$_2$O, which is not available in complexes. In addition, the N$_2$O–HI binding energy must be overcome. The remaining small difference can be easily accounted for by the squeezed-atom effect [4.39] The deuterium substitution for the bulk reaction increases the ND rotational temperature because the D-atom reaction products have 1100 cm^{-1} more available energy than the H-atom reaction products due to the zero-point energy difference. With deuterium substituted complexes, the ND rotational temperature corresponds to an available energy much smaller than both the bulk and the unsubstituted complex. This significant decrease in available energy with deuterium substitution is in accord with the squeezed-atom effect since the energy transmitted to the heavier fragments may be proportional (in some way) to the mass of the hydrogen. The NH $[v = 1]/[v = 0]$ ratios are ~ 0.16, 0.10 and 0.05 for 240, 245 and 246 nm N$_2$O–HI photolyses, respectively. The ND $[v = 1]/[v = 0]$ ratio is 0.16 for 240 nm N$_2$O–DI photolysis.

OH(D) deriving from bulk conditions has abundant internal excitation–not surprising given the large available energy. Once past the reaction barrier(s), there is no long-lived intermediate, so product state distributions can be biased by the large reverse barrier. OH(D) internal excitation is essentially independent of photolysis wavelength in the range 240–270 nm [4.51, 52]. For the case

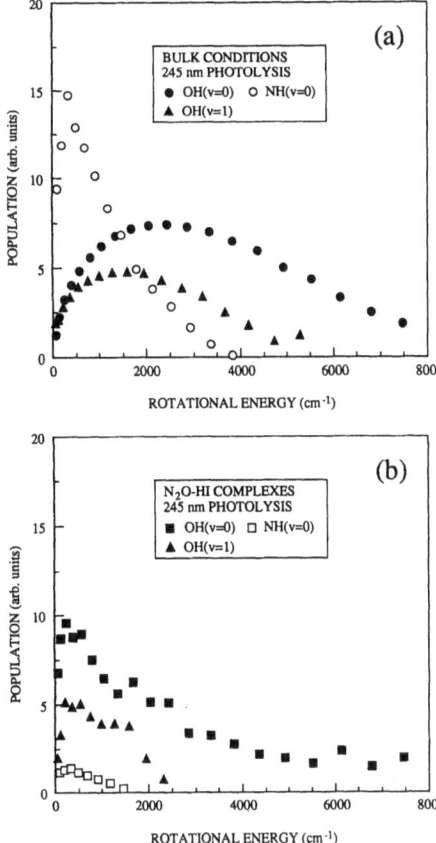

Fig. 4.8 Branching ratio measurements showing relative NH and OH ($v = 0, 1$) rotational populations for (a) bulk and (b) complexed conditions, using 245 nm HI photolysis (the (a) and (b) entries are normalized to one another by setting the OH($v = 0$) populations equal). From these data, we conclude that with N_2O–HI complexes, the [NH]/[OH] branching ratio is lower than its bulk counterpart (i.e. factor of approximately 12), at least for the observed OH levels

shown in Fig. 4.8a, the $v = 0$ rotational distribution extends to the LIF detection limit imposed by $A^2\Sigma$ predissociation, although it is falling for energies above 5000 cm^{-1}. The $[v = 1]/[v = 0]$ ratio is ~ 0.5 for OH at 245 nm HI photolysis and ~ 0.6 for OD at 240 nm DI photolysis, neglecting $v = 0$ rotational levels above 8000 cm^{-1}. Vibrational levels above $v = 1$ must be detected by using sequence bands, because of $A^2\Sigma$ predissociation. It is possible that the amount of OH(D) presently undetected (i.e. $v \geqslant 2$) is comparable to the combined $v = 0$ and 1 populations. In contrast to the NH(D) result, OH(D) deriving from complexes is much colder rotational than its bulk counterpart, both $v = 0$ and 1. This may be due to interaction with the nearby I atom, although details are unclear. The $[v = 1]/[v = 0]$ ratio is ~ 0.5 for OH with 245 nm HI photolysis and ~ 0.5 for OD with 240 nm DI photolysis.

4.3.2 The Product Branching Ratio: OH + N$_2$ vs. NH + NO

Perhaps the most intriguing observation is the change in the relative NH(D) and OH(D) populations between bulk and complexed conditions for the same photolysis wavelength. Note the rather small NH population in Fig. 4.8b compared to Fig. 4.8a. The (a) and (b) entries are normalized to one another by setting the OH ($v = 0$) populations equal. The [NH]/[OH] ratio (for the observed OH levels) drops by a factor of 11 in going from bulk to complexed conditions. For the deuterium case, the [ND]/[OD] ratio drops by a factor of 18 in going from bulk to complexed conditions. This observed isotope effect suggests that the majority of the OH + N_2 production may proceed via a direct abstraction mechanism in complexes without involving an HNNO† intermediate and H-atom migration. In addition, it strongly suggests that entrance channel selectively is carried over to products, i.e. *the reaction is highly selective.*

4.4 Future Experiments

We are always in a state of designing experiments, and would like to discuss here a few obvious possibilities. Firstly, the time-resolved measurements of *Zewail* and coworkers are very important [4.41], and it is inconceivable that such measurements will not continue to expand in scope, sophistication and elegance. Particularly in combination with double-resonance techniques (see below), it is possible that real-time studies could be carried out for trimers, different geometric isomers, etc.

Secondly, we believe the time is right for double-resonance techniques to be applied to the study of photoinitiated reactions in complexes. There are many possibilities, but let us consider just those appropriate to the CO_2–HX and N_2O–HX systems. By pumping the HX-stretch vibration in the complex, it is possible to spectroscopically 'select' the complex of interest, and to simultaneously shift the HX ultraviolet absorption to longer wavelengths, as shown in Fig 4.9. The doule-resonance method has been around for over half a century and there is nothing fundamentally new about the proposed measurements. However, recent developments in tunable coherent sources now make it practical to obtain up to 10 mJ of tunable infrared radiation using reliable technology. Specifically, optical parametric oscillators (OPOs) now work better than ever, thanks mainly to improved crystal growth, and can be pumped with 1.06 µm Nd:YAG lasers and their 532 nm and 355 nm harmonics [4.53]. OPO energies of several mJ are more than adequate for saturating HBr transitions, and with an OPO bandwitdth that overlaps a Q branch, it may be possible to give a large fraction of the complexes one quantum of HBr-stretch excitation. Figure 4.9 shows that this will shift the HBr UV absorption to longer wavelengths and this is a basis for initiating reaction from only the selected molecules. If an overtone is pumped, the shift in UV absorption to longer wavelengths is so dramatic that there may be no background from unexcited complexes.

Such experiments can be used to select (i) binary complexes, discriminating effectively against higher-than-binary complexes, (ii) trimers in cases where absorption features can be identified and excited selectively, (iii) geometrical

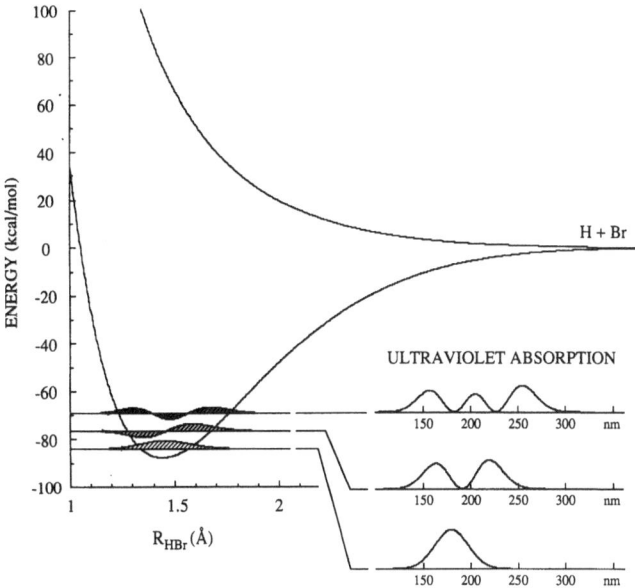

Fig. 4.9 Potential curves for the ground ($^1\Sigma$) and excited ($^1\Pi$ states of free HBr (GVB-DCCI/DZP calculation) and vibrational wavefunctions for the ground state ($v = 0, 1, 2$) and their ultraviolet absorption cross sections (in arbitrary units), assuming no change in electronic moment with R_{HBr}

isomers, and (iv) rotational states of the complexes. With the high OPO energies now available, even the HBr first overtone can be almost saturated, and with large absorption coefficients at wavelengths as long as 248 nm, as shown in Fig. 4.9, it may be possible to prepare and dissociate a significant number of complexes. In addition, tunable HBr photolysis is possible in the practical wavelength region (280–230 nm). A forseeable practical limitation is predissociation of the excited complexes. Although many excited complexes live long enough for ns-duration lasers to be used, each case must be scrutinized separately to be sure that predissociation does not remove excited complexes before HBr photodissociation can be effected. Laser pulses of several picosecond duration would overcome just about any predissociation rate, while still providing enough spectral selectivity to isolate species. However, these experiments will not be so easy. Nevertheless, the double resonance experiments *can* be done if one is willing/able to invest the time and resources.

The use of overtones is particularly attractive if it can be applied to complexes with HF. By shifting the HF absorption spectrum out of the VUV into a more workable region, it will be possible to separate the absorption features of the two molecules in the complex. Furthermore, HF is attractive because it can hydrogen bond to different parts of one molecule. For example, N_2O–HF has two geometric isomers, one with the hydrogen bonded to the oxygen ($v_{HF} = 3878\ cm^{-1}$) and the other with the hydrogen bonded to the terminal nitrogen ($v_{HF} = 3900$ $^{-1}$) [4.54]. By spectrally selecting one or the other isomer and then photodissociating HF, it may possible to study chemical selectivity with hitherto unavailable selectivity. Because HF absorbs deep in the VUV, it will probably be necessary to excite the second HF overtone, which will have a small absorption coefficient.

An elegant experiment that we hope to carry out someday is to excite a combination band that contains one or more quanta of HX stretch and one quantum of an intermolecular vibration, e.g., hydrogen in-plane bend for the case of CO_2–HBr. In this case, only those molecules having one quantum of the intermolecular mode will be detected, so the effect of intermolecular excitation can be elucidated. This would constitute changing the initial conditions of the photoinitiated reaction in a sophisticated way.

4.5 Conclusion

The measurements proposed above are made difficult mainly by the need for the right coherent sources. The experimental challenges of generating high-energy, tunable, picosecond or sub-picosecond, temporally synchronized, etc. pulses are formidable and may understandably frighten away many researchers. But there is no reason to believe that the development of the needed sources would not occur. On the contrary, there is every reason to be optimistic about future laser developments. After all, how many people would have predicted the present array of laser-based tools 25 years ago when the dye laser was first invented. So we not only thank Fritz Schaefer and Peter Sorokin for inventing the dye laser, but also for spawning the entire field that has evolved with, and consequent to, this seminal invention, and which continues to promote and inspire our scientific inquiries.

Acknowledgements. The authors thank S.W. Sharpe, R. A. Beaudet, W. A. Goddard, III, G. Hoffmann and C. Jaques.

References

4.1   C. Wittig, S. Sharpe, R.A. Beaudet: Acc. Chem. Res. **21**, 341 (1988)
4.2   P.R. Brooks, E.M. Jones: J. Chem. Phys. **45**, 3449 (1966)
4.3   B.J. Beuhler, R.B. Bernstein, K.H. Kramer: J. Am. Chem. Soc. **88**, 5332 (1966)
4.4   P.R. Brooks: Science **193**, 11 (1976)
4.5   P.W. Harland, H.S. Carman, Jr., L.F. Phillips, P.R. Brooks: J. Chem. Phys. **93**, 1089 (1990)
4.6   K.H. Kramer, R.B. Bernstein: J. Chem. Phys. **42**, 767 (1965)
4.7   D.H. Parker, K.K. Chakravorty, R.B. Bernstein: J. Phys. Chem. **85**, 466 (1981)
4.8   S.R. Gandhi, T.J. Curtiss, Q.X. Xu, S.F. Chai, R.B. Bernstein: Chem. Phys. Lett. **132**, 6 (1986)
4.9   S.R. Gandhi, Q.X. Xu, T. J. Curtiss, R. B. Bernstein: J. Phys. Chem. **91**, 5437 (1987)

4.10 R.B. Bernstein, S.E. Choi, S. Stolte: J. Chem. Soc. Faraday. Trans. II **85**, 1097 (1989)
4.11 D.H. Parker, R.B. Bernstein: Ann. Rev. Chem. Phys. **40**, 561 (1989)
4.12 G. Marcelin, P.R. Brooks: J. Am. Chem. Soc. **97**, 1710 (1975)
4.13 D. van den Ende, S. Stolte: Chem. Phys. Lett. **76**, 13 (1980)
4.14 S. Stolte: Ber. Bunsenges Phys. Chem. **86**, 413 (1982)
4.15 S. Stolte: In *Atomic and Molecular Beam Methods*, ed. by G. Scoles (Oxford University Press, New York Vol. 1, Chap. 25, 1987)
4.16 D.H. Parker, H. Jalink, S. Stolte: J. Phys. Chem. **91**, 5427 (1987)
4.17 H. Jalink, G. Nicolasen, S. Stolte, D.H. Parker: J. Chem. Soc. Faraday Trans. II **85**, 1115 (1989)
4.18 A.W. Kleyn, E.W. Kuipers, M.G. Tenner, S. Stolte: J. Chem. Soc. Faraday Trans. II **85**, 1337 (1989)
4.19 M.G. Tenner, E.W. Kuipers, W.Y. Langhout, A.W. Kleyn, G. Nicolasen, S. Stolte: Surf. Sci. **236**, 151 (1990)
4.20 G.H. Felcher, M. Volkmer, B. Pawlitzki, N. Bowering, U. Heinzmann: Vacuum **41**, 265 (1990)
4.21 T.J. Curtiss, R.S. Mackay, R.B. Bernstein: J. Chem. Phys. **93**, 7387 (1990)
4.22 S.I. Ionov, M.E. LaVilla, R.S. Mackay, R.B. Bernstein: J. Chem. Phys. **93**, 7406 (1990)
4.23 C.T. Rettner, R.N. Zare: J. Chem. Phys. **75**, 3636 (1981)
4.24 C.T. Rettner, R.N. Zare: J. Chem. Phys. **77**, 2416 (1982)
4.25 C. Jouvet, M.C. Duval, B. Soep, W.H. Breckenridge, C. Whitham: J. Chem. Soc. Faraday Trans. II. **85**, 1133 (1989)
4.26 S.K. Shin, Y. Chen, S. Nickolaisen, S.W. Sharpe, R.A. Beaudet, C. Wittig: In *Advances in Photochemistry*, ed. by D.H. Volman, G.S. Hammond, D.C. Neckers (Wiley, New York 1991) Vol. 16, p. 249.
4.27 C. Jouvet, B. Soep: Chem. Phys. Lett. **96**, 426 (1983)
4.28 W.H. Breckenridge, C. Jouvet, B. Soep: J. Chem. Phys. **84**, 1443 (1986)
4.29 A. Zehnacker, M.C. Duval, C. Jouvet, C. Lardeux-Dedonder, D. Solgadi, B. Soep, O. Benoist-D'Azy: J. Chem. Phys. **86**, 6565 (1987)
4.30 C. Jouvet, M.C. Duval, B. Soep: J. Phys. Chem. **91**, 5416 (1987)
4.31 M.C. Duval, O. Benoist-D'Azy, W.H. Breckenridge, C. Jouvet, B. Soep: J. Chem. Phys. **85**, 6324 (1986)
4.32 S.W. Sharpe, Y.P. Zeng, C. Wittig, R.A. Beaudet: J. Chem. Phys. **92**, 943 (1990)
4.33 J.A. Shea, W.G. Read, E.J. Campbell: J. Chem. Phys. **79**, 614 (1983)
4.34 Y.P. Zeng, S.W. Sharpe, S.K. Shin, C. Wittig, R.A. Beaudet: J. Chem. Phys. (1991) to be published
4.35 R.W. Randall, M.A. Walsh, B.J. Howard: Faraday Discuss. Chem. Soc. **85**, 1 (1988)
4.36 S.K. Shin, C. Wittig, W.A. Goddard: III, J. Phys. Chem. in press (1991)
4.37 S. KI. Shin, Y. Chen, D. Oh, C. Wittig: Philos. Trans. R. Soc. Lond. A **332**, 362 (1990)
4.38 G. Hoffman, D. Oh, Y. Chen, Y.M. Engel, C. Wittig: Israel J. Chem. **30**, 115 (1990)
4.39 C. Wittig, Y.M. Engel, R.D. Levine: Chem. Phys. Lett. **153**, 411 (1988)
4.40 K. Kulda, G.C. Schatz: J. Phys. Chem. in press (1991)
4.41 N.F. Scherer, C. Sipes, R.B. Bernstein, A.H. Zewail: J. Chem. Phys. **92**, 5239 (1990)
4.42 G.C. Schatz, M.S. Fitzcharles, L.B. Harding: Faraday Discuss. Chem. Soc. **84**, 359 (1987)
4.43 G.C. Schatz, M.S. Fitzcharles: In *Selectivity in Chemical Reactions*, ed by J. C. Whitehead (Kluwer, Dordrecht 1988)
4.44 Y.P. Zeng, S.W. Sharpe, D. Reifschneider, C. Wittig, R.A. Beaudet: J. Chem. Phys. **93**, 183 (1990)
4.45 S.K. Shin, C. Wittig, W.A. Goddard III: unpublished
4.46 S.K. Shin, E. Böhmer, Y. Chen, C. Wittig: In preparation
4.47 P. Marshall, A. Fontijn, C.F. Melius: J. Chem. Phys. **86**, 5540 (1987)
4.48 P. Marshall, T. Ko, A. Fontijn: J. Phys. Chem. **93**, 1922 (1989)
4.49 K.M. Ervin, P.B. Armentrout: J. Chem. Phys. **86**, 2659 (1987)
4.50 W.R. Anderson: J. Phys. Chem. **93**, 530 (1989)

4.51 G. Hoffmann, D. Oh, H. Iams, C. Wittig: Chem. Phys. Lett. **155,** 356 (1989)

4.52 G. Hoffmann, D. Oh, C. Wittig: J. Chem. Soc. Faraday Trans. 2, **85,** 1141 (1989)

4.53 C.L. Tang, W.R. Bosenberg, T. Ukachi, R.J. Lane, L.K. Cheng: Laser Focus World, Oct. 1990, p. 107

4.54 D.J. Nesbitt: Chem. Rev. **88,** 843 (1988)

Curt Wittig

5. Novel Resonator Design for Femtosecond Lasers

V. Magni, S. De Silvestri, A. Cybo-Ottone, M. Nisoli and O. Svelto

With 6 Figures

We present a general analysis of resonators with multiple foci. The optical stability is characterized by means of a simple analytical formulation. An analysis of the sensitivity of the resonator to external perturbations in terms of misalignment sensitivity, defines new criteria in the resonator design, which provide new configurations different from the traditional nearly confocal ones. A few experiments on a resonator configuration with low misalignment sensitivity for hybridly mode-locked dye lasers are reported. The problem of pulse stabilization for these lasers is also considered.

5.1 Introduction

Resonators with internal foci are used in lasers for the generation of femtosecond light pulses. Additive pulse mode-locked (APM) solid state lasers and Kerr lens mode-locked (KLM) Ti:sapphire lasers operate with resonators with one internal focus, while hybridly mode-locked and cw colliding pulse mode-locked (CPM) dye lasers use resonators with two internal foci. The knowledge of the mode profile inside the resonator is extremely important for the optimization of the pulse shortening mechanism in KLM lasers considering the spatial mode changes, due to self-focusing processes, and in dye lasers considering nonlinear effects (saturation and self-phase modulation) in the gain medium and in the saturable absorber. All the above processes are strictly related to the behavior of the mode transverse profile inside the resonator. Since the early work of *Kogelnik* et al. [5.1] on a folded three mirror cavity, many papers [5.2–7] have been published on complex resonators made by several mirrors that evidenced the existence of different optical stability regions. These analyses also addressed the optimization of the mode spot size in the crucial positions inside the resonator (namely, for dye lasers the gain and absorber position). However these works were dealing with numerical simulations rather than analytical solutions, which always refer to specific resonator configurations and do not allow the definition of general and simple rules for the resonator design. A second important aspect for laser operation in the femtosecond time domain is the sensitivity of the resonator to external perturbations (namely, mechanical vibrations) that can cause displacement and tilting of the mode. These effects can be described in terms of the resonator misalignment sensitivity which describes

Topics in Applied Physics, Vol. 70
Dye Lasers: 25 Years Ed.: Dr. Michael Stuke
© Springer-Verlag Berlin Heidelberg 1992

the stability of the resonator mode axis against external perturbations [5.8, 9]. The influence of the misalignment sensitivity on the behavior of femtosecond lasers is related to (i) random displacement of the mode axis from the pumped area in the gain medium causing fluctuations in the output power and (ii) random displacement of the mode axis in the region of the dispersive prisms causing instability in the pulse duration.

In this work we present a general analysis of resonators with multiple foci. The structure of the stability zones is clearly evidenced and understood by means of an analytical formulation. Furthermore, an analysis of the sensitivity of the resonator to external perturbations in terms of misalignment sensitivity defines new criteria in the resonator design, which provide new configurations different from the traditional near-confocal one. An analysis of one- and two-focus resonators is presented. A few experiments on a new resonator configuration for hybridly mode-locked dye lasers are reported. Finally, the issue of cavity length stabilization for these lasers is considered.

5.2 One-Folding Resonators

Let us consider the case of a generic linear resonator with one focal region determined by two focusing mirrors (Fig. 5.1a). The optical stability and the mode characteristics of such a resonator are mainly determined by the distance between the two focusing mirrors, which must usually be controlled to a

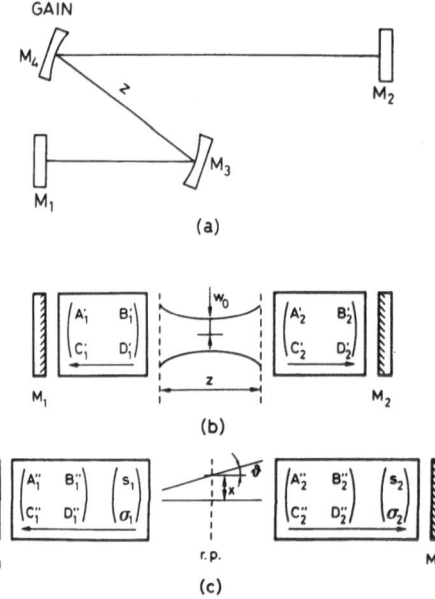

(a)

(b)

(c)

Fig. 5.1. (a) Single folding linear resonator: mirrors M_3 and M_4 are of 100 mm radius of curvature and separated by a distance z, the distance from mirror M_1 to mirror M_3 is 335 mm and the distance from mirror M_4 to mirror M_2 is 1065 mm; (b) ray matrix picture of the resonator in (a), the arrows indicate the direction at which the matrices correspond; (c) displacement x and tilting angle ϑ of the mode axis at the reference plane (r.p.), the misalignment is represented by the vectors (s_1, σ_1) and (s_2, σ_2)

precision of a tenth of a micrometre. To study this type of resonator we consider the ray matrix equivalent of the resonator, shown in Fig. 5.1b. The distance z represents the separation between the two focusing mirrors of the folding configuration. The matrices describe the propagations towards the two end mirrors, which are assumed to be flat, and include the folding mirrors and all the other optical elements contained in the resonator. If the end mirrors were curved, they should be intended to be resolved in a plane mirror and a lens (included in the matrix) of focal length equal to the mirror radius of curvature. To study the optical stability of the resonator on the basis of the scheme of Fig. 5.1b we use the technique of the one-way ray matrix of the resonator [5.9, 10]. If we define as A, B, C, D the elements of the ray matrix relevant to the propagation from mirror M_1 to mirror M_2, the condition for the optical stability is simply provided by the following relationship:

$$ABCD < 0 .\tag{5.1}$$

The A, B, C, D matrix is called the "resonator one-way matrix". The stability condition given by (5.1), which differs from that obtained by the usual round trip ray matrix, can be easily handled to derive a simple relationship for the stability limits. From (5.1) we find that the stability limits are obtained by equating to zero each element of the $ABCD$ matrix. The generic matrix element $J(A, B, C, D)$ is a function of z and has the following expression:

$$J = a_j (z - z_j) ;\tag{5.2}$$

the coefficients a_j and z_j are reported in Table 5.1. From (5.2) and Table 5.1 it is evident that the resonator presents four stability limits as a function of z. These limits define two distinct stability regions of equal width Δz given by

$$\Delta z = \min(|z_A - z_B|, |z_A - z_C|) .\tag{5.3}$$

At the stability limits the gaussian mode of a stable resonator degenerates in spherical waves originating from the two end mirrors or having a wavefront

Table 5.1 Coefficients for the matrix elements given by (5.2)

J	a_J	z_J
A	$C'_1 A'_2$	$-\dfrac{D'_1}{C'_1} - \dfrac{B'_2}{A'_2}$
D	$A'_1 C'_2$	$-\dfrac{B'_1}{A'_1} - \dfrac{D'_2}{C_2}$
B	$A'_1 A'_2$	$-\dfrac{B'_1}{A'_1} - \dfrac{B'_2}{A'_2}$
C	$C'_1 C'_2$	$-\dfrac{D'_1}{C'_1} - \dfrac{D'_2}{C'_2}$

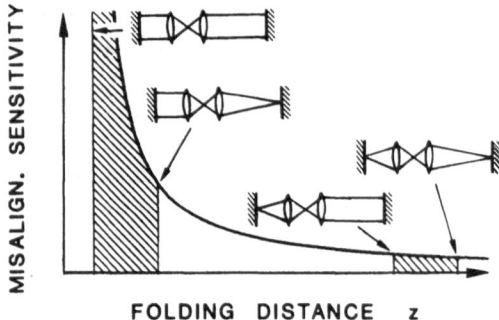

Fig. 5.2. Qualitative behavior of the misalignment sensitivity of the resonator as function of z. The resonators show the mode configuration at the edges of the stability zones

MISALIGN. SENSITIVITY

FOLDING DISTANCE z

matching the curvature of the end mirrors [5.9], therefore the stability limits and the mode behavior at the stability limits can be obtained, in practical cases, by simple geometrical considerations. Figure 5.2 shows the mode configuration at the stability limits as a function of z for a one-folding resonator. The existence of two stability regions is clearly evident; each of the stability limits can be easily identified with the vanishing of an element of the one-way matrix of the resonator. To obtain (5.3), one of the two optical systems in Fig. 5.1b was shifted by varying z; this implies that the resonator length does not remain fixed. This is a strict requirement, however, in the case of hybridly mode-locked lasers. Since the width of the stability regions, ranging usually from 0.5 to 2 mm, is very small compared with the resonator length of about 2 m, the mode profile is in practice completely unaffected by such changes in the resonator length.

By ray matrix calculation, it can be shown that the spot size at the beam waist of the gaussian TEM_{00} mode inside the folding mirrors vanishes at the stability limits and, as a function of z, presents a symmetric behavior in the two stability zones reaching a maximum value w_0. The width of the stability zone is simply related to w_0 by

$$w_0^2 = (\lambda/2\pi)\Delta z \ . \tag{5.4}$$

The previous relationship is particularly useful to calculate the mode spot size at the beam waist. Since the spot size falls rapidly to zero near the stability limits, we can assume that w_0 is representative of the spot size for values of z within the stability regions.

As it can be seen from (5.4), w_0 is the same in both the stability regions; this ensures that in both stability regions the laser operates with the same output power and all nonlinear effects are present at the same extent. Therefore, as far as the mode profile is concerned, the two stability regions are completely equivalent. A criterion to choose between the stability regions can be found in the resonator sensitivity to misalignment, a parameter which accounts for the sensitivity to external perturbations. If some of the resonator elements are misaligned, in any transversal plane in the resonator the mode axis shifts and/or rotates with respect to the axis of a perfectly aligned resonator. The misalignment sensitivity is then defined as the ratio between the displacement or the steering angle of the mode axis and the value of the parameter that generates the misalignment. The misalignment of an optical component can be included in the matrices by adding a 2×1 vector to that obtained by multiplying the input ray vector by the corresponding matrix [5.11]. The elements of the misalignment vector give the position and slope of the output ray when the input ray coincides with the reference axis of the system. The position x and the angle ϑ of the mode axis in a given plane (see Fig. 5.1c), with respect to the mode axis of a perfectly aligned resonator, are determined by a self-consistency equation expressing the fact that the mode axis is the unique ray that retraces itself after one round trip in the resonator:

$$\begin{bmatrix} x \\ \vartheta \end{bmatrix} = -\frac{1}{C} \begin{bmatrix} D_2'' & D_1'' \\ -C_2'' & C_1'' \end{bmatrix} \begin{bmatrix} \sigma_1 \\ \sigma_2 \end{bmatrix} . \tag{5.5}$$

The expressions for both x and ϑ turn out to be inversely proportional to the C element of the one-way matrix previously defined. Since the C element vanishes at one of the stability limits, the mode axis shifts and rotates by a quantity that diverges if the resonator operates close to this limit. The qualitative behavior of misalignment sensitivity is shown in Fig. 5.2. It must be noted that the foldings of all linear resonators so far reported in the literature (including the popular three mirror folded cavity) are set in the nearly confocal configuration. This corresponds to a configuration which falls very close to the stability limit $C = 0$: therefore these lasers operate in a region where the misalignment sensitivity diverges.

The misalignment sensitivity has been experimentally investigated by using a synchronously pumped Rhodamine 6G dye laser. The pumping source was a cw mode-lacked Nd:YAG laser operating at a repetition rate of 100 MHz. The mode axis displacement in the long resonator arm of Fig. 5.1a for a tilting of mirror M_1 was measured for various folding mirror distances z in the two

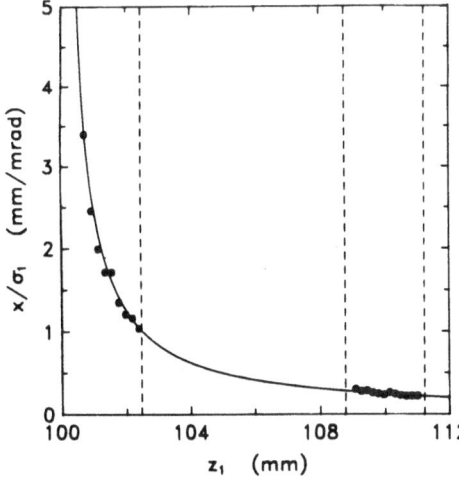

Fig. 5.3. Ratio of the mode axis displacement in the long arm of the resonator to the tilting angle of mirror M_1 as a function of the folding mirror distance z

stability zones. The experiment was performed by inserting an iris in front of mirror M_2 and measuring the tilting angle σ_1 of mirror M_1 that stops the laser action. Figure 5.3 shows the plot of the iris radius divided by σ_1 compared to the calculated values of the displacement of the mode axis to the tilting angle of the mirror, given by (5.5). The divergence of misalignment sensitivity in the stability region corresponding to the nearly confocal configuration is clearly demonstrated.

5.3 Two-Folding Resonators

The resonator commonly used in femtosecond dye lasers (Fig. 5.4a) contains two pairs of focusing mirrors, one for the gain medium and the other for the saturable absorber. To analyze this structure we have considered the resonator shown in Fig. 5.4b made by three optical systems enclosed between the two plane mirrors M_1 and M_2 and separated by two regions of lengths z_1 and z_2 representing the free propagation of the beam within the focusing mirrors. Again, the condition for the optical stability of the resonator can be easily derived from the one-way ray matrix from mirror M_1 to M_2. The expression of each matrix element is given by:

$$ J = b_j[(z_1 - z_{1j})(z_2 - z_{2j}) - 1/C_3^2] \,, \tag{5.6} $$

where $J = A, B, C, D$ and the coefficients are reported in Table 5.2. It is apparent from (5.6) that in a plane (z_1, z_2) the stability limits are defined by four hyperbolas generated by translation of the hyperbola $z_1 z_2 = 1/C_3^2$; this greatly simplifies the construction of the stability diagram. In Fig. 5.5 we show the stability diagram of the resonator reported in Fig. 5.4a as a function of z_1 and z_2; the dashed area represents the regions of optical stability for the resonator.

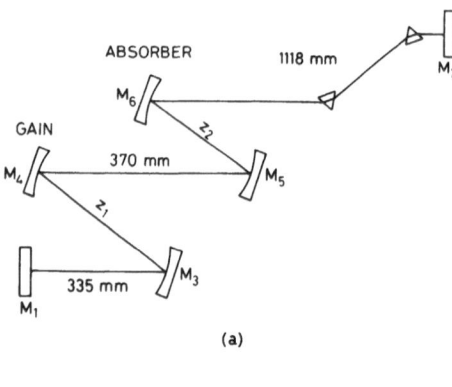

(a)

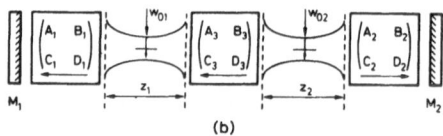

(b)

Fig. 5.4. (a) Two folding linear resonator: mirror M_3 and M_4 are of 100 mm radius of curvature and separated by a distance z_1, mirrors M_5 and M_6 are of 50 mm radius of curvature and separated by a distance z_2, the distance from mirror M_1 to M_3 is 335 mm, the distance from mirror M_4 to M_5 is 370 mm and the distance from mirror M_6 to M_2 is 1118 mm; (b) Ray matrix picture of the resonator in (a)

Table 5.2 Coefficients for the matrix elements given by (5.6)

J	b_J	z_{1J}	z_{2J}
A	$C_1 A_2 C_3$	$-\dfrac{D_1}{C_1} - \dfrac{A_3}{C_3}$	$-\dfrac{B_2}{A_2} - \dfrac{D_3}{C_3}$
D	$A_1 C_2 C_3$	$-\dfrac{B_1}{A_1} - \dfrac{A_3}{C_3}$	$-\dfrac{D_2}{C_2} - \dfrac{D_3}{C_3}$
B	$A_1 A_2 C_3$	$-\dfrac{B_1}{A_1} - \dfrac{A_3}{C_3}$	$-\dfrac{B_2}{A_2} - \dfrac{D_3}{C_3}$
C	$C_1 C_2 C_3$	$-\dfrac{D_1}{C_1} - \dfrac{A_3}{C_3}$	$-\dfrac{D_2}{C_2} - \dfrac{D_3}{C_3}$

From Fig. 5.5 the existence of a variety of resonator configurations is apparent, different from the usual confocal one that can be used for laser operation. Again the resonator sensitivity to misalignment can provide a criterion for choosing the most appropriate operating region within the stability diagram. The dashed line shown in the stability diagram of Fig. 5.3 represents the hyperbola defined by $C = 0$. Along this line the resonator misalignment sensitivity goes to infinity. It must be noted that all linear resonators with two foldings so far reported in the literature for hybridly mode-locked lasers are set in the nearly confocal configuration. Thus the stability diagram of Fig. 5.5 can be used as a guide for the optimization of the misalignment sensitivity by considering new resonator configurations far from the line corresponding to $C = 0$.

A second important feature of the stability diagram of Fig. 5.5 is the possibility of evaluating the spot sizes at the beam waist within the two foldings. If we draw horizontal (or vertical) straight lines on the stability diagram, the intercepts of these lines with the stability limits always define two segments of

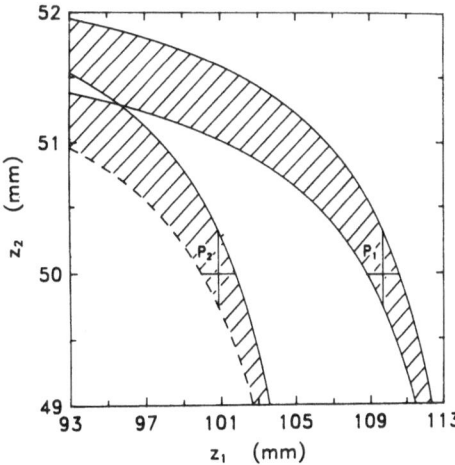

Fig. 5.5. Stability diagram for the resonator of Fig. 5.1a vs. z_1 and z_2. The shaded area represents regions of optical stability for the resonator. The stability limit corresponding to $C = 0$ is indicated by a dashed line. The point P_1 and P_2 indicate respectively the low- and high-MS resonator configurations used in the experiments

equal amplitude Δz, which are directly related to the mode spot size in the corresponding folding. If we consider z_2 constant and vary z_1, the mode spot size at the beam waist in this second folding drops to zero at the stability limits and reaches a maximum value w_{01} within the stability zone given by

$$w_{01}^2 = (\lambda/2\pi)\Delta z \ . \tag{5.7}$$

From (5.7) it can be seen that the beam waist in each folding is proportional to the square root of the width of the stability zone relevant to that folding. For a certain working point in the stability diagram the ratio between the cross-sectional area of the mode in the two foldings (saturation ratio) can easily be calculated by

$$w_{01}^2/w_{02}^2 = \Delta z_1/\Delta z_2 \ , \tag{5.8}$$

Δz_1 and Δz_2 being the widths of the stability zones along two straight lines drawn parallel to the axes and crossing at the working point. The stability diagram of Fig. 5.5 can be used for the laser optimization by separately adjusting the resonator misalignment sensitivity (MS) and the saturation ratio to reach optimum conditions for laser operation.

On the basis of these considerations, it is possible to design a new resonator configuration that operates in a stability region free of divergence of MS. One of these low-MS resonators is represented by point P_1 in Fig. 5.5. The traditional resonator configuration (high-MS configuration), is represented in Fig. 5.5 by point P_2. Since the mode spot sizes in the gain and the saturation ratio are the same in both low- and high-MS configurations, the average output power and pulse duration will be the same for the two configurations. The advantage of the low-MS configuration is a much lower sensitivity to external perturbations.

The effects of the misalignment sensitivity in the low- and high-MS configurations have been experimentally investigated in a hybridly mode-locked Rhodamine 6G dye laser with DODCI as saturable absorber. The dye laser was synchronously pumped by the second harmonic of a cw mode-locked Nd:YLF laser operating at 76 MHz. The resonator is shown in Fig. 5.4a. The average output power (140 mW) and the pulse duration (60 fs) measured in the low- and high-MS configurations were the same, as expected from the above considerations. To study the influence of the misalignment sensitivity on the laser behavior, the second harmonic of the average output power was measured in both low- and high-MS configurations. The average second harmonic power is very sensitive to fluctuations in energy and duration of the output pulses. Figure 5.6 shows the behavior of the second harmonic power as measured in the (a) low- and (b) high-MS configurations after optimum alignment. It is evident that the fluctuations on the second harmonic are larger by a factor of almost five in the high-MS configuration. Since the mechanical structure of the resonator is the same for the two configurations, the increase of the amplitude of the fluctuations in the high-MS configuration are definitely due to a larger misalignment sensitivity. This demonstrates that the effects of external perturbations will be amplified in the high-MS configuration compared with the low-MS one.

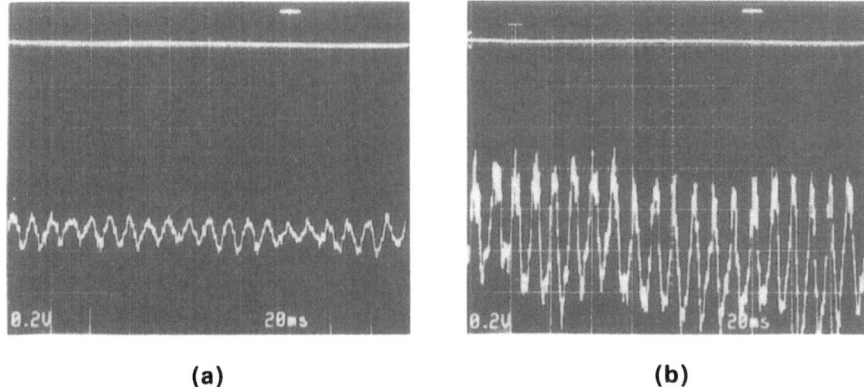

(a)   **(b)**

Fig. 5.6a, b. Average second harmonic power measured with the laser operating in: (a) low-MS and (b) high-MS configurations

The previous considerations are extremely important for an active stabiliz-ation of the output pulse fluctuations. Previous techniques of laser stabilization used as a reference the shape of the pulse spectrum [5.12] or the average second harmonic power [5.13]. The results obtained with these techniques show residual peak-to-peak fluctuations respectively of 5% and 10% in the average output power. The technique based on the second harmonic works by the hypothesis that fluctuations in the second harmonic signal are only due to variations of the cavity length caused by external perturbations. Since in the low-MS configuration the mode axis displacement and tilting are small, we expect that fluctuations in the second harmonic power are primarily due to cavity length variations. On the other hand, in the high-MS configuration beside the same amount of cavity length fluctuations, mode displacement and tilting are present to a larger extent, acting as a source of pulse fluctuations. To experimentally confirm the previous considerations, we have stabilized the laser cavity length in the low- and high-MS configurations. This was achieved by supporting mirror M_2 on a piezoelectric stack. In the low-MS configuration, the residual fluctuations of the output average power were of the order of one precent and the long-term stability was excellent. In the case of the high-MS configuration, larger residual fluctuations (about 10%) are found and the electronic feedback loop tends to unlock spontaneously.

5.4 Conclusions

In conclusion, we believe that these results show that nearly confocal resonator configurations should always be avoided and that more stable configurations with the same output power and pulse duration can be found. The stability diagram provides information on the resonator misalignment sensitivity and, at the same time, on the mode behavior in the crucial points of the resonator. These results can be applied to one or two internal folding resonators, such as those used respectively for femtosecond solid state and dye lasers.

Acknowledgements. This work has been partially supported by the National Research Council of Italy under the "Progetto Finalizzato" on Telecommunications.

References

5.1   W. Kogelnik, E.P. Ippen, A. Dienes, C.V. Shank: IEEE J. Quantum Electron. **QE-8,** 373 (1972)
5.2   C.E. Wagstaff, M.H. Dunn, A.I. Ferguson, S.J. Bastow: Opt. Commun. **25,** 379 (1978)
5.3   K.K. Li, A. Dienes, J.R. Whinnery: Appl. Opt. **20,** 407 (1981)
5.4   K.K. Li: Appl. Opt. **21,** 967 (1982)
5.5   K.S. Budil, I.A. McIntyre, C.K. Rhodes: Opt. Commun. **64,** 279 (1987)
5.6   E. Cojocaru, T. Julea, N. Herisanu: Appl. Opt. **28,** 2577 (1989)
5.7   W. Castner Jr., J.J. Korpershoek, D.A. Wiersma: Opt. Commun. **78,** 90 (1990)
5.8   S. De Silvestri, L. YuPu, V. Magni, O. Svelto, In: *Ultrafast Phenomena VI*, ed. by T. Yajina, K. Yoshihara, C.B. Harris, S. Shionoya, Springer Ser. Chem. Phys. Vol. 48 (Springer, Berlin, Heidelberg 1988) p. 33
5.9   V. Magni, S. De Silvestri, A. Cybo-Ottone: Opt. Commun. **82,** 137 (1991)
5.10   V. Magni: J. Opt. Soc. Am. **A 4,** 1962 (1987)
5.11   A. Gerrard, J.M. Burch: *Introduction to Matrix Methods in Optics* (Wiley, London 1975) pp. 106–108, 286–291
5.12   J. Chesnoy, L. Fini: Opt. Lett. **11,** 635 (1986)
5.13   M.D. Dawson, D. Maxson, T.F. Boggess, A.L. Smirl: Opt. Lett. **13,** 126 (1988)

Orazio Svelto

6. Distortion of Femtosecond Pulse Fronts in Lenses

Z. Bor and Z.L. Horváth

With 4 Figures

The discovery of dye lasers [6.1] had a decisive impact on the development of various fields of modern spectroscopy. One such field is ultrafast laser spectroscopy, which started in 1968 with the first mode-locked operation of dye lasers [6.2], and which resulted in 6-fs-long pulses [6.3] from CPM lasers. Ultrashort pulses also opened new perspectives in the development of high-power nonlinear optics, plasma physics, and atomic and nuclear physics.

A common problem with ultrashort light pulses is that the spatial profile of a femtosecond pulse front can suffer considerable distortion upon propagation through a lens [6.4–9]. As shown by geometrical optical ray tracing [6.4, 5], the delay between the pulse and phase fronts for an arbitrary ray entering the lens at a distance of a from the optical axis can be calculated as [6.5]

$$\Delta T(a) = \frac{a_0^2 - a^2}{2cf^2} \, \lambda \, \frac{df}{d\lambda} \, , \tag{6.1}$$

where $2a_0$ is the diameter of the lens, $df/d\lambda$ is the longitudinal chromatic aberration of the lens and c is the speed of light.

The validity of (6.1) has already been experimentally verified by measuring the pulse front distortion using the time of flight interferometer [6.10]. The behavior of the pulse front is especially interesting at the near vicinity of the focus. Here, as geometrical optical ray tracing has revealed, the spatial profile of the pulse front can be very complicated and may even form a loop [Ref 6.4, Fig. 3].

In this paper an exact wave optical analysis of propagation of femtosecond pulse fronts through lenses is described.

6.1 Diffraction of a Femtosecond Pulse by a Dispersive Lens

Let us consider a lens having a focal length f_0 at wavelength λ_0 (Fig. 6.1). The material of the lens has a dispersion of $dn/d\lambda_0$ resulting in a longitudinal chromatic aberration of

$$f(\lambda) = f_0 + \frac{df}{d\lambda_0} (\lambda - \lambda_0) = f_0 - \frac{f_0}{n_0 - 1} \frac{dn}{d\lambda_0} (\lambda - \lambda_0) \, . \tag{6.2}$$

Topics in Applied Physics, Vol. 70
Dye Lasers: 25 Years Ed.: Dr. Michael Stuke
© Springer-Verlag Berlin Heidelberg 1992

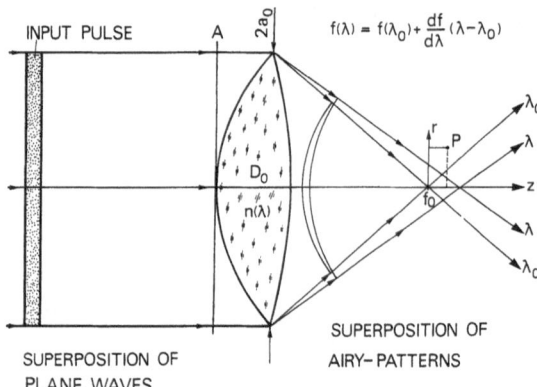

INPUT PULSE

A

$f(\lambda) = f(\lambda_0) + \dfrac{df}{d\lambda}(\lambda - \lambda_0)$

SUPERPOSITION OF
PLANE WAVES

SUPERPOSITION OF
AIRY–PATTERNS

Fig. 6.1. A lens converts the monochromatic plane waves into Airy-type diffraction patterns. Due to the longitudinal chromatic aberration of the lens, each Fourier component of the input pulse has its own focal point on the axis. The pulse front after the lens can be calculated as the superposition of Airy patterns of the Fourier components

The electric field of an undisturbed femtosecond pulse front before the lens can be represented as the superposition of monochromatic plane waves

$$E_i(t) = \frac{1}{2\pi} \int_{-\infty}^{\infty} F(\omega) e^{i\omega t} \, d\omega \; , \tag{6.3}$$

where

$$F(\omega) = \int_{-\infty}^{\infty} E_i(t) e^{-i\omega t} \, dt \tag{6.4}$$

is the Fourier transform of the input pulse having an electric field of $E_i(t)$. The electric field of the input pulse is seen to have Gaussian temporal pulse shape

$$E_i(t) = E_0 e^{-2 \ln 2 [t/\tau]^2} \cos \frac{2\pi c}{\lambda_0} t \; , \tag{6.5}$$

where τ is the duration (FWHM) of the intensity of the pulse. The lens converts the monochromatic plane waves (6.3) into spherical-like waves having radius of curvature $f(\lambda)$ converging to the focal point. The aperture of the spherical-like waves just behind the lens is $2a_0$.

The intensity distribution of one monochromatic component can be calculated by solving the proper diffraction integrals [6.11]. Following the notation of [6.11], the electric field in point P (Fig. 6.1) can be calculated as

$$E(P) = -\frac{i\omega \, a_0^2 \, E_0}{2cf^2} e^{i\omega(\frac{z}{c} - t)} [C(u, v) - i \, S(u, v)] \; , \tag{6.6}$$

where u and v are the dimensionless coordinates of point P defined as

$$u = \frac{\omega}{c} \left[\frac{a_0^2}{f^2} \right] z \; , \tag{6.7}$$

$$v = \frac{\omega}{c} \frac{a_0}{f} r \ , \tag{6.8}$$

where z and r are the coordinates of point P. Throughout this article we assume perfect cylindrical symmetry; consequently the use of two spatial coordinates is sufficient. The functions $C(u, v)$ and $S(u, v)$ can be calculated using the Lommel functions as described in [6.11]. Obviously for a lens, allowing also for chromatic aberration, the electric field of the pulse around the focus of the lens can be calculated as the superposition of the fields of individual spectral components:

$$E(z, t) = \frac{a_0}{4\pi c} \int_{-\infty}^{\infty} \frac{\omega F(\omega)}{f(\lambda)} e^{i\omega[z/c - t]} [C(u, v) - i \, S(u, v)] \, d\omega \ , \tag{6.9}$$

where

$$u = \frac{\omega}{c} \frac{a_0^2}{f^2(\lambda)} \left[z - \frac{df}{d\lambda_0} (\lambda - \lambda_0) \right] \ , \tag{6.10}$$

$$v = \frac{\omega}{c} \frac{a_0}{f(\lambda)} r \ . \tag{6.11}$$

In (6.9) a new origin of time t was selected so that the peak of the input pulse was seen to reach the plane A at a time of

$$t_0 = -\frac{D_0 \left(n_0 - \lambda_0 \dfrac{dn}{d\lambda_0} \right) + f_0}{c} \ , \tag{6.12}$$

where D_0 is the axial thickness of the lens and $c/(n_0 - \lambda_0 \, dn/d\lambda_0)$ is the group velocity [6.12, 13] of the light in the material of the lens. (Thus $- t_0$ is the propagation time of a pulse from plane A (Fig. 6.1) to the focal point along the optical axis.)

6.2 Results of the Calculations

Superposition of the diffraction patterns [i.e., (6.9)] has been numerically integrated using the parameters corresponding to the lens studied in [6.4] (i.e., $\lambda_0 = 0.249$ nm, $f_0 = 150$ mm, $a_0 = 40$ mm, $n_0 = 1.50799$, $\lambda_0 \, dn/d\lambda_0 = -0.1375$ (fused silica), $D_0 = 10.4988$ mm, $c = 0.299792$ mm/ps, $df/d\lambda_0 = 163.0895$ mm/μm). The input pulse was supposed to have a duration of 100 fs (FWHM in intensity). The intensity was calculated as $I = |E|^2$.

6.2.1 Distortion of the Pulse Front

Figure 6.2 shows the results of the calculations. In the upper part the topographic plots of the pulse fronts are shown for the moments $- 9$, $- 3$ and 3 ps,

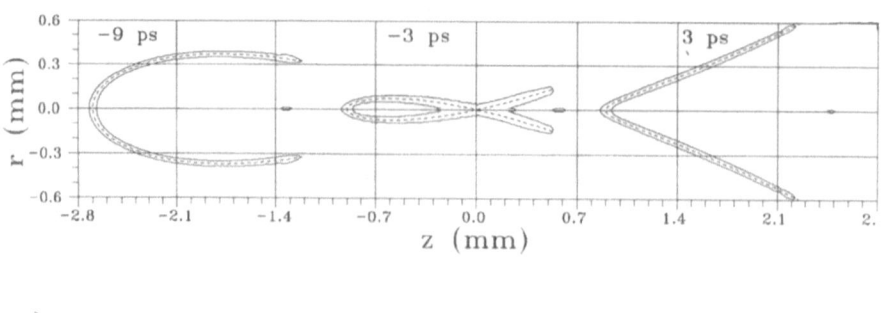

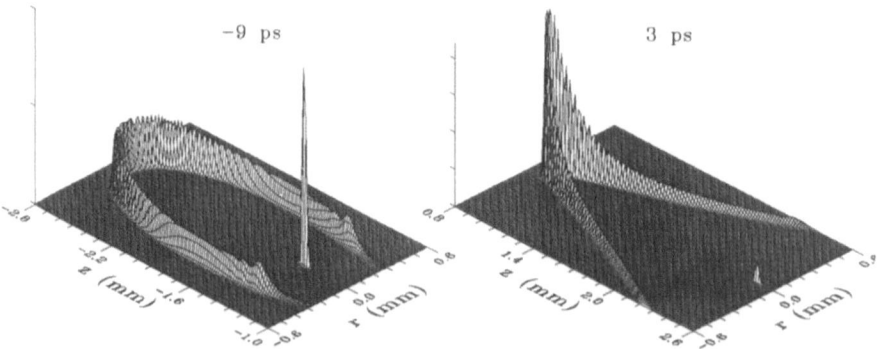

Fig. 6.2. Topographic plots of the position of the pulse front; the moments for − 9, − 3 and 3 ps are calculated from the superposition of the diffraction integrals. The broken line shows the position of the peak of the pulse front as calculated from the geometrical optical theory. The lower part shows the two-dimensional plots of the pulse fronts for the moments − 9 and 3 ps. Note the boundary wave pulse propagating along the optical axis

respectively. Within the topographic lines the broken line shows the position of the peak of the pulse front calculated from the geometrical optical theory (see [Ref. 6.4, Fig. 6.3]). The figure proves the validity of the geometrical optical theory concerning the position of the pulse front.

6.2.2 Boundary Wave Pulse on the Optical Axis

The most unexpected result of the wave optical calculation is that, together with the horseshoe-shaped main pulse front, an additional narrow pulse is propagating along the optical axis (lower part of Fig. 6.2). This pulse was not predicted by geometrical optics and will be referred to as the *boundary wave pulse.*

Our preliminary interpretation of the boundary wave pulse is the following. According to [6.14], in many diffraction problems the Fresnel–Kirchoff diffraction integral after proper transformation can have such a form that the electric field is represented as the sum of a direct field and the field of the boundary diffraction wave. The latter may be thought of as arising from scattering of the incident radiation by the boundary of the aperture [6.14]. The boundary waves in our case have circular symmetry; thus, due to interference, significant intensity is resulted only on the optical axis in a spot having a diameter of the order of $d_0 \approx \lambda_0 f_0 / 2 a_0$. Note that the boundary wave pulse has a completely different physical origin than the Sommerfeld and Brillouin-type of precursor (forerunner) pulse [6.15] and therefore they are not to be confused.

6.2.3 Pulse Shape in the Focus

The temporal shape of the pulse in the focal point $[I(z = 0, r = 0, t)]$ resembles a square pulse with a duration $\Delta T_0 = 4.75$ ps (Fig. 6.3). This is in agreement with $\Delta T_0 = 4.81$ ps [6.4], which can also be calculated from

$$\Delta T_0 = \frac{a_0^2}{2 c f_0^2} \lambda_0 \frac{df}{d\lambda_0} = -\frac{D_0}{c} \lambda_0 \frac{dn}{d\lambda_0},$$ (6.13)

as the total delay between the axial and marginal ray [see (6.1)]. A simplified but plausible interpretation of the constant intensity is that an observer located in the focal point sees that light is arriving from the lens only in the time interval of $-\Delta T_0 < t < 0$. For $t = -\Delta t_0$ a thin bright ring seems to appear on the periphery of the lens. With increasing time, the diameter of the bright ring seems to shrink. At $t = 0$, the bright ring contracts to a point at the center of the lens. For time $t > 0$ the bright ring disappears from the surface of the lens. The thickness of the ring is determined by the input pulse duration and the shape of the delay between the phase and pulse front. However, it can also be shown [6.16] that the area of the rings is constant in time. This is why the intensity observed in the focal point is constant over the time of approximately ΔT_0.

6.2.4 Intensity Distribution in the Focal Plane

Figure 6.4a shows the radial intensity distribution in the focal plane ($z = 0$) calculated from (6.9). The interference rings in Fig. 6.4 somewhat resemble the Airy pattern; however, there are two essential differences, namely that with

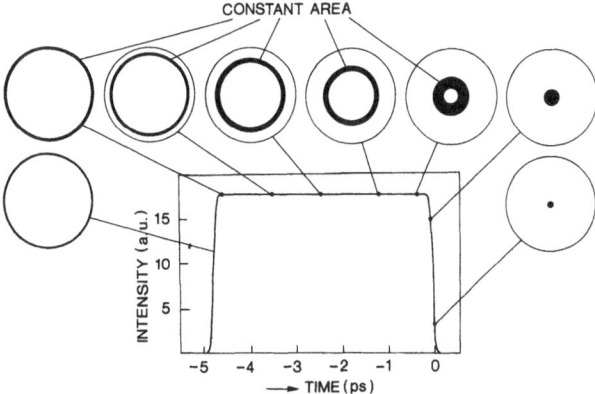

Fig. 6.3. Pulse shape in the focus. The rise and fall time of the pulse is approximately equal to the duration of the input pulse. The duration of the square-like pulse is determined by the total delay ΔT_0 [(6.13)]. The rings indicate that part of the lens aperture which seems to be illuminated by the short pulse. (The observer is thought to be located in the focal point)

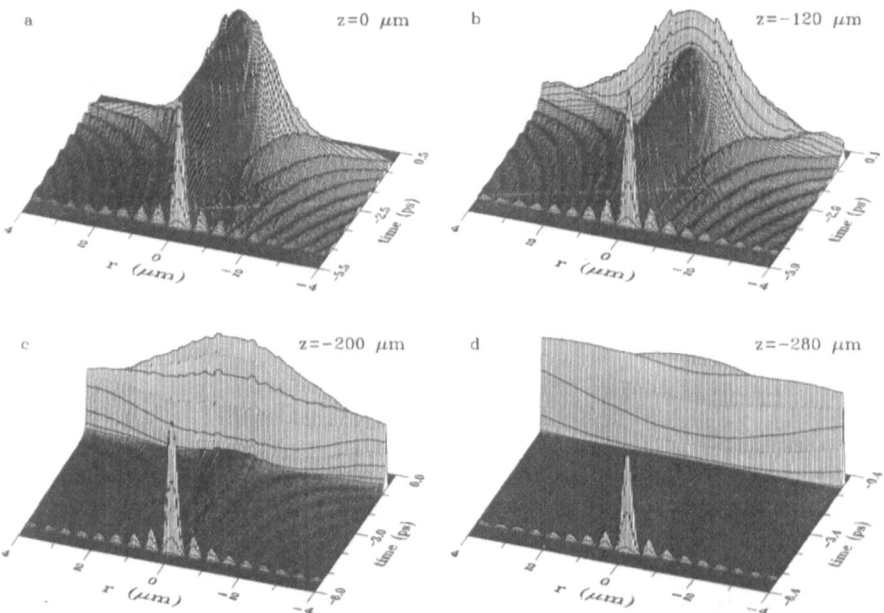

Fig. 6.4. The radial intensity distribution in planes $z = 0, -120, -200$ and $-280\,\mu m$ before the focal plane. The pattern observed in the focal plane (a) results from diffraction from the bright rings shown in the inset of Fig. 6.3

increasing order number of the maxima of the rings, the intensity decays much slower than is expected from an Airy pattern. Besides, the radius of the rings is increasing with time.

A simplified interpretation of the observed pattern is the following. As already discussed in Sect. 6.3.3, an observer located in the focal point sees that for the interval of time $-\Delta T_0 < t < 0$ a thin bright ring with shrinking diameter appears on the surface of the lens (see inset in Fig. 6.3.). The pattern observed in the focal plane (Fig. 6.4a) results from diffraction from the bright rings.

6.2.5 Intensity Distribution in a Plane Neighboring the Focal Plane

Figure 6.4b, c, d show the intensity distribution in planes $z = -120, -200$ and $-280\,\mu m$ before the focal plane. (The spatial resolution of this figure in the radial direction is much higher than the one used in Fig. 6.2, allowing one to see finer details of the boundary wave pulse.) The temporal separation between the boundary wave pulse and the main pulse is ΔT_0 i.e. 4.81 ps [see (6.13)]. The boundary wave pulse has the same duration (100 fs) as the input pulse.

As discussed in Sect. 6.3.2, the boundary wave pulse may be thought of as arising from scattering of the incident radiation by the boundary of the aperture

[6.14]. Consequently, the radial distribution should correspond to the diffraction pattern from a narrow ring. The solution of such a diffraction problem might be found for example in [6.17]. Applying those results, it can be shown [6.16] that the radial distribution should obey

$$\frac{I_b(r)}{I_b(r=0)} = \left[\frac{J_0\left(\frac{2\pi a_0}{\lambda_0 f_0} r\right)}{\frac{2\pi a_0}{\lambda_0 f_0}} \right]^2 .$$

(6.14)

Detailed analysis showed that (6.14) is in full agreement with the radial intensity distribution shown in Fig. 6.4.

6.3 Conclusions

The propagation of a 100-fs-long pulse through a lens has been described using a wave-optical description. The pulse front after the lens can be calculated as the superposition of the electric field of the Airy-type diffraction patterns of the Fourier components of the input pulse.

The wave-optical description supports the validity of the geometrical optical theory, describing the shape of the pulse front (Fig. 6.2.).

The intensity distribution in the focal plane (Fig. 6.4) is different from the Airy pattern. The diffraction pattern consists of rings with continuously increasing radius. The most unexpected result is that in front of the main pulse, a boundary wave pulse also appears. The radial intensity distribution of the boundary wave pulse supports the assumption that it results from the interference of the boundary waves originating from the aperture of the lens. Our preliminary calculations show that the boundary wave pulse appears also for lenses having no chromatic aberration and for mirrors. We suppose that apodization of the input beam leads to considerable reduction of the intensity of the precursor pulse.

A detailed theoretical and experimental analysis of the boundary wave pulse will be given in [6.16].

Acknowledgements. The authors are greatly indebted to Prof. F.P. Schäfer for his continuous support of this research. This work has been supported by the OTKA Foundation (No 3055 and 3056) and the Hungarian – U.S. Science and Technology Joint Fund. (No 061/90).

References

6.1   F.P. Schäfer (ed): *Dye Lasers*, Topics Appl. Phys. Vol. 1, 3rd edn. (Springer, Berlin, New York, Heidelberg 1990)
6.2   W. Schmidt, F.P. Schafer: Phys. Lett **26A**, 558 (1968)
6.3   R.L. Fork, C.H. Brito Cruz, P.C. Becker, C.V. Shank: Opt. Lett. **7**, 483 (1987)
6.4   Z. Bor: J. Mod. Opt. **35**, 1907 (1988)
6.5   Z. Bor: Opt. Lett. **14**, 119 (1989)
6.6   S. Szatmari, G. Kühnle: Opt. Commun. **69**, 60 (1988)
6.7   H. Staerk, J. Ihlemann, A. Helmbold: Laser und Optoelektronik **20**, 6 (1988)
6.8   Z. Bor: Digest of papers of the XVII International Conference on Quantum Electronics, 21–25 May 1990 Anaheim, California, p. 256
6.9   A. Schuster: Phil. Mag. **7**, 509 (1894)
6.10 Z. Bor, Z. Gogolak, G. Szabo: Opt. Lett. **14**, 862 (1989)
6.11 M. Born, E. Wolf: *Principles of optics* (Pergamon, Oxford 1975) p. 435
6.12 Z. Bor, B. Racz: Opt. Commun. **54**, 165 (1985)
6.13 L. Cohen, C. Lin: Appl. Opt. **16**, 3136 (1977)
6.14 M. Born, E. Wolf: *Principles of optics* (Pergamon, Oxford 1975) p. 449
6.15 L. Brillouin: *Wave Propagation and Group Velocity* (Academic, New York 1960)
6.16 Z.L. Horváth, Z. Bor: To be published
6.17 E.N. Leith, G. Collins, I. Khoo, T. Wynn: J. Opt. Soc. Am. **70**, 141 (1980)

Zsolt Bor   Z.L. Horváth

7. Infrared Dye Lasers for the Wavelength Range 1–2 μm

T. Elsaesser and W. Kaiser

With 8 Figures

Dye lasers represent a standard source of coherent light in the visible spectral range, where the spectral bandwidth, the temporal characteristics of the laser emission and the power can be varied over many orders of magnitude. Laser dyes emitting in the visible frequently exhibit fluorescence quantum yields of 0.5 to 1 with lifetimes of the fluorescing S_1 state in the nanosecond regime [7.1]. The excited fluorescent state is predominantly deactivated by *radiative* transitions to the ground state making dyes a favorable active material for a variety of laser systems.

Laser action of dye molecules in the infrared wavelength region between 1 and 2 μm is more difficult to achieve. Compounds with a S_0–S_1 transition in this range show very small fluorescence quantum yields of 10^{-4} to 10^{-5}, corresponding to lifetimes of the fluorescent state of several picoseconds [7.2–5]. Fast *radiationless* internal conversion to the S_0 state represents the dominating mechanism of depopulation of the S_1 state. Due to the rapid ground state recovery, infrared dyes have found application as saturable absorbers in mode-locked and Q-switched solid-state lasers. In infrared dye lasers, highly judicious and efficient pumping schemes are necessary to obtain strong stimulated emission, in spite of the high radiationless transition rates.

In recent years, different dye laser systems emitting intense pulses beyond 1 μm have been realized. In particular, pico- and subpicosecond pulses tunable over a broad range have been generated with dye lasers pumped by short pulses from mode-locked Nd lasers. In this case, the duration of the pump pulses is close to the lifetime of the emitting S_1 state of the molecules, resulting in a stable laser output of high power.

In this paper, we present a review on infrared dye lasers working in the wavelength range from 1 to 2 μm. We first discuss briefly the photophysical properties of infrared dyes. The main part of the article is devoted to laser systems generating short infrared pulses of a duration from several hundreds of femtoseconds up to nanoseconds. Different laser configurations will be presented with emphasis on the production of ultrashort light pulses. Finally, nonlinear frequency conversion of the dye laser output to longer wavelengths in the midinfrared is reported.

7.1 Photophysics of Infrared Dyes

The spectral position of the S_0–S_1 absorption and emission bands of organic dye molecules is related to the size of their π-electron chromophore. According to

Topics in Applied Physics, Vol. 70
Dye Lasers: 25 Years Ed.: Dr. Michael Stuke
© Springer-Verlag Berlin Heidelberg 1992

the free electron model, the S_0–S_1 transition of a chain-like planar chromophore
– as, e.g., realized in cyanine dyes – shifts to the red with increasing length, i.e.,
with the number of conjugated double bonds [7.6, 7]. For this reason, the
molecular structure of most dyes absorbing and emitting beyond a wavelength
of 1 μm is larger than of dyes for the visible spectral range, typically having a
molecular weight of 500 to 1500.

In Fig.7.1, the structure and the electronic spectra of a widely used infrared
dye, the polymethine pyrilium compound no. 5, are presented [7.4, 8]. The
S_0–S_1 absorption (A) of a solution of the dye in dichloroethane peaks at 1.09 μm
with a maximum extinction coefficient of $\varepsilon \approx 1.4 \times 10^5$ M^{-1} cm^{-1}. Fluorescence (F) occurs with a small Stokes shift of approximately 600 cm^{-1} and a low
quantum yield of roughly 10^{-4}. The approximate mirror symmetry of absorption and emission bands points to a relatively rigid geometry of the dye
molecules. The dashed area represents the effective gain profile of a highly
concentrated solution (concentration $c = 10^{-3}$ M) serving as the active
medium of traveling wave dye lasers which will be discussed in Sect. 7.2.3. It
should be noted that the oscillator strength of the S_0–S_1 transition of laser dyes
is of the same order of magnitude for the infrared and the visible spectral range,
i.e. similar gain coefficients are expected.

A large number of dyes have S_0–S_1 absorption bands between 1.0 and 1.1 μm
[7.4, 5, 8]. In addition, several groups of polymethine cyanine dyes absorbing
and emitting at even longer wavelengths have been synthesized [7.9–11]. In
Fig.7.2, the structure and the absorption spectrum of the compound S501 is
presented. The maximum absorption occurs at $\lambda = 1.42$ μm with an extinction

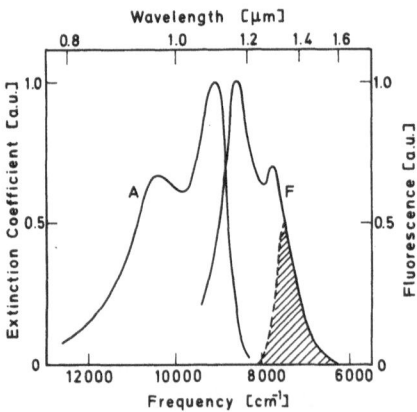

Fig. 7.1 Upper part: Molecular structure of the
infrared dye no. 5 with a ClO_4^- counterion.
Lower part: $S_0 \rightarrow S_1$ absorption (A) and
$S_1 \rightarrow S_0$ fluorescence (F) spectra of dye no. 5
dissolved in dichloroethane (solid lines). The
dashed area represents the effective gain profile
of a highly concentrated dye solution (concentration $c = 10^{-3}$ M). The emission at short
wavelengths is suppressed by reabsorption

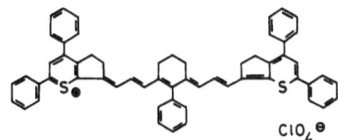

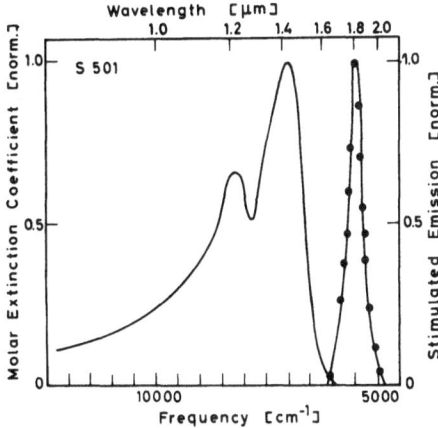

Fig. 7.2. Normalized absorption (left) and picosecond stimulated emission (right) of dye S501 pumped in a traveling-wave geometry. The stimulated emission peaks at 1.8 μm with a spectral bandwidth of $320 \, \text{cm}^{-1}$

coefficient of $\varepsilon \approx 7 \times 10^4 \, \text{M}^{-1} \, \text{cm}^{-1}$. Stimulated emission of this dye was found at very long wavelengths from 1.8 to 2 μm [7.11] (see Sect.7.2.3).

Excited state lifetimes of infrared dyes were determined in picosecond pump–probe experiments, where either the repopulation of the electronic ground state or the time-dependent gain by stimulated emission of the S_1 levels was monitored. The two techniques give identical kinetics of the excited state [7.12]. The decay times of the S_1 state frequently range between 2 and 50 ps [7.2–5, 11, 12]. As an example, results for dye S501 dissolved in o-dichlorobenzene are depicted in Fig.7.3. The sample was excited by a 4-ps pulse from a mode-locked Nd:glass laser ($\lambda = 1.054 \, \mu\text{m}$) and the resulting transmission increase was probed by a weak pulse at the same wavelength. In Fig. 7.3, the change of transmission is plotted versus delay time between the two pulses (points). The monoexponential decay of the bleaching reveals a S_1 lifetime of 12.5 ps (solid line). The triangles (dashed line) represent the cross-correlation function of pump and probe pulses, i.e., they indicate the time resolution of the experimental system.

The decay times of the S_1 level are very short compared to the radiative lifetime of the $S_1 \rightarrow S_0$ transition. The latter may be estimated from the electronic spectra [7.13] and lies in the nanosecond regime. The picosecond lifetimes point to the rapid radiationless depopulation of S_1, resulting in the low fluorescence quantum yields mentioned earlier. Internal conversion to the electronic ground state represents the main mechanism of radiationless deactivation, as is obvious from the identical S_1 *depopulation* (gain) and S_0 *repopul-*

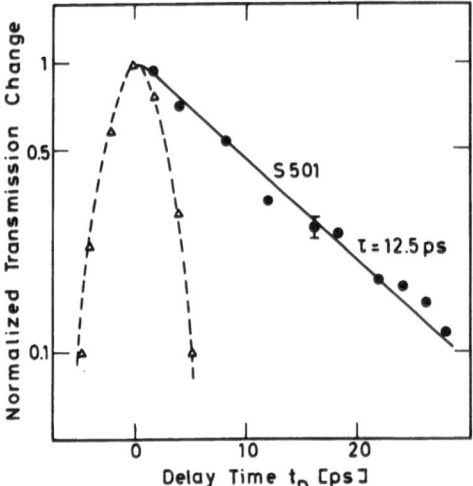

Fig. 7.3. Picosecond bleaching of the electronic ground state of dye S 501 dissolved in o-dichlorobenzene. The normalized change of transmission $\Delta T/T_0 = (T - T_0)/T_0$ (T, T_0 : transmission of the sample with and without excitation) is plotted as a function of delay time between pump and probe pulses at $\lambda = 1.054\,\mu m$ (points). The solid line gives a monoexponential decay with a time constant of 12.5 ps. The triangles (dashed line) represent the cross-correlation function of pump and probe pulses

ation (ground state recovery) times [7.12]. Strong internal conversion is facilitated by the large number of vibrational degrees of freedom, some of which couple very effectively to the electronic S_1 state of the dye. In non-rigid compounds, low frequency torsional motions of subgroups in the molecular structure enhance the radiationless decay rates. Intersystem crossing to the triplet manifold which is characterized by nanosecond time constants plays a minor role for the depopulation of S_1.

The photostability of infrared dyes is an important prerequisite for application in lasers and requires special attention. Large chain-like polymethine chromophores tend to be less stable than their smaller counterparts emitting in the visible. However, incorporating part of the conjugated chains into 5 and 6-membered ring systems considerably improves the photochemical stability (see Figs. 7.1 and 7.2, top). The rapid deactivation of the excited singlet levels within tens of picoseconds reduces the population of reactive states (e.g. triplets) and thus diminishes potential decomposition. It should be emphasized that high purity of the solvents is very important for the stability of the dye solutions. Contamination with water or metal ions has to be avoided. Solutions of infrared dyes prepared with spectrograde solvents of high purity show a photostability similar to dye samples for the visible spectral range.

7.2 Infrared Dye Lasers

In the past, various types of dye lasers for the wavelength range from 1 to 2 µm have been realized. We first report on nanosecond dye lasers in Sect. 7.2.1. Lasers for pico- and subpicosecond pulses will be discussed in detail in Sects.

7.2.2 and 7.2.3. Finally, nonlinear frequency conversion of the dye laser output to the mid-infrared will be discussed.

7.2.1 Nanosecond Infrared Dye Lasers

Q-switched Nd : glass or Nd : YAG lasers working at a wavelength around 1 µm were used to excite highly concentrated samples of different polymethine dyes [7.14–18]. Here the duration of the nanosecond pump pulse exceeds the excited state lifetime of the laser dyes by several orders of magnitude. Consequently, the maximum population density of the excited state is much smaller than the density of absorbed photons. The population inversion between excited and ground state follows in time the envelope of the pump pulse. At high intensities, the dye molecules are repeatedly excited during the pump pulse.

In the experiments of [7.14], several percent of the nanosecond pump light with a peak intensity of 1 to 30 MW cm^{-2} are converted to amplified spontaneous emission (ASE) at a fixed wavelength in the range from 1.1 to 1.2 µm. The spectral bandwidth of the emission is relatively wide between 10 to 20 nm. A serious limitation of this system consists in the rapid decomposition of the dye molecules which was reported to occur after several hundred pump pulses of megawatt power.

Tunable nanosecond pulses with megawatt peak powers were generated in dye lasers pumped by a Q-switched Nd : YAG laser ($\lambda = 1.064$ µm) [7.15–18]. The repetition rate was 10 Hz. A longitudinally pumped dye cell of a length of 0.5 to 1 cm was placed in a resonator containing prisms [7.15] or a grating [7.16–18] as tuning element. The different systems provide pulses tunable between 1.1 and 1.4 µm with a bandwidth of less than 0.1 nm. The output of the dye laser amounts to several percent of the pump power. The intensity fluctuations of the pulses are roughly 10%. The photochemical stability of the dye solution was substantially improved in comparison to earlier experiments resulting in a stable operation of the dye laser for some tens of hours.

Pulses from nanosecond dye lasers working in the near infrared were down-converted to longer wavelengths in the medium infrared by parametric frequency mixing in nonlinear crystals. In Sect. 7.2.4, this technique will be discussed in detail.

7.2.2 Synchronously Pumped Picosecond and Subpicosecond Systems

Two types of synchronously pumped infrared dye lasers have been reported in the literature [7.19–27]:

1) Intense pulse trains generated with low repetition rate in a passively (or active–passive) mode-locked Nd : YAG laser ($\lambda = 1.064$ µm) synchronously pump a dye laser system [7.19–22]. Each pulse train, which has a total energy in the mJ range, comprises 10 to 15 pulses of a duration of 20 to 35 ps, separated by several nanoseconds. The output of the broadly tunable dye laser has a total energy of several tens of µJ.

2) Actively mode-locked Nd:YAG lasers emitting a continuous train of picosecond pulses pump a linear cavity with a dye jet as the active medium [7.24–27]. These systems give cw pulse trains of milliwatt average power and a repetition rate of approximately 100 MHz. The round trip time of the dye laser cavity in both schemes is made identical to the nanosecond time interval between successive pump pulses. Transversal and longitudinal pumping of the active medium was investigated.

In the first case, the output of the dye laser builds up within 10 to 15 round trips in the cavity due to the limited number of pump pulses. A high gain of 1 to 5 cm^{-1} per single pass of the dye solution is required to generate an intense train of output pulses. Laser emission tunable between 1.08 and 1.12 μm was achieved with a linear cavity containing a prism for wavelength selection (active material: DNDTPC: 3,3′-diethyl (9, 11, 15, 17-dineopentylene-6,7,6′,7′-dibenzo) thiapentacarbocyanine perchlorate) [7.19]. The output consists of 4 to 5 pulses of a total energy of 4 mJ and a duration of 10 ps. A somewhat wider tuning range from 1.15 to 1.25 μm was realized in a similar configuration with the dye no. 26 as the amplifying material [7.20]. Here the duration of the output pulses was 5.5 ps, considerably shorter than the pump pulses of 21 ps. The short-lived population inversion in the dye that follows the picosecond envelope of the individual pump pulses and the high gain per round trip result in a temporal narrowing of the generated pulses.

Picosecond pulses which are tunable over a very wide wavelength interval of several hundred nanometers were produced with ring lasers pumped by an active–passive mode-locked Nd:YAG laser [7.21, 22]. Pulses traveling in a ring cavity pass the dye solution only when the active medium is excited by the pump pulses. This behavior is in contrast to linear systems where each round trip is connected with a passage through the unpumped and thus absorbing material. The reduced reabsorption in the ring laser permits an extension of the tuning range to shorter wavelengths. The dyes 9860 [7.2] and no. 5 give pulses of several tens of μJ tunable from 1.08 to 1.4 and from 1.18 to 1.53 μm, respectively. Variation of the wavelength was accomplished by a grating that was part of the ring cavity. A single reflection from the grating leads to a small temporal broadening of the pulses (see Sect. 7.2.3) which is partly compensated by gain narrowing during amplification. The resulting pulse duration of 27 ps is close to that of the pump pulses (35 ps). Pulse shortening was observed with a wavelength selector of subtractive dispersion, i.e. by twofold reflection from the grating [7.22]. Here the duration of the output pulses has a value of 10 ps. Measurements of the cross-correlation function between single pulses from the dye laser and from the pump laser (which were selected by electrooptic shutters) gave an upper limit of 3 ps for the temporal jitter demonstrating the precise synchronization of the two laser systems.

We now discuss systems synchronously pumped by a cw mode-locked Nd:YAG laser with a high repetition rate of 80–100 MHz. The properties of these lasers are strongly influenced by the short excited state lifetime of the dye molecules. For pump pulses long compared to the decay time of the S_1 state, high average pump powers of several watts are necessary to generate an output in the milliwatt regime. The dye laser is operated close to the lasing threshold with a net gain of several percent per round trip. The population inversion in the dye follows in time the picosecond pump pulse, in contrast to dyes for the visible, where – due to the nanosecond lifetime of the S_1 state – a long-lived population inversion is established. It is interesting to compare the pulse amplification for dyes with long and short lifetimes. In the linear cavity of the dye laser, two counter-propagating pulses starting from the dye jet pass the active medium after reflection from one of the end mirrors, and meet again after a complete round trip. First, we discuss the net amplification of the two pulses when the population inversion persists for several nanoseconds. The pulse reflected by the end mirror close to the dye jet is amplified since it arrives after several hundred picoseconds in the active medium. On the other hand, the gain is depleted at the later arrival time of the second pulse. In this way, just one stable pulse train builds up in the cavity.

A different behavior is found in the infrared dye laser. Here the gain exists only during the picosecond pump pulse and the two pulses traveling in the cavity experience a similar amplification. As a result, either two pulse trains or a bistable operation mode are found, where one single pulse train of changing propagation direction occurs [7.25, 27].

Infrared dye lasers have been pumped either by the original pulse train from the Nd:YAG laser [7.24, 25] or by pulses shortened with the help of a fiber-grating compressor [7.26]. The respective pulse durations are approximately 70 ps and 5 ps. The output of the first cw mode-locked infrared dye laser was tunable from 1.2 to 1.32 μm with a maximum power of 10 mW and a pulse duration of approximately 3 ps (dye no. 26 in benzyl alcohol) [7.24]. Nearly transform-limited pulses of 6 ps and lower intensity fluctuations were produced with an improved system [7.25]. Pumping with compressed pulses gives tunable infrared pulses of a duration of 300 fs and an average power of up to 10 mW [7.26].

7.2.3 Picosecond Dye Lasers Pumped in a Traveling-Wave Geometry

The traveling-wave pumping scheme allows the generation of single picosecond pulses with peak intensities of several MW/cm^2 and tunability over a wide spectral range [7.11, 28–30]. The principle of this technique is depicted in Fig. 7.4. A single picosecond pulse from a mode-locked Nd:laser passes the diffraction grating GR1. Reflection from the grating introduces a continuous time delay across the horizontal beam diameter which is due to the difference of optical path lengths traveled by the different parts of the beam [7.31]. The time delay is adjusted to the traveling time of stimulated emission in the dye cell G pumped transversally by this pulse. The condition for exact synchronization of pump and emission is $\tan \gamma = n$, where γ represents the angle between the normal to the propagation direction and the tilted wave front; n is the refractive

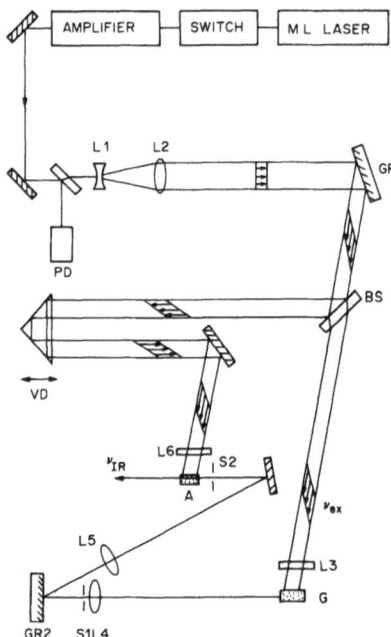

Fig. 7.4. Schematic diagram of a picosecond infrared dye laser with traveling-wave geometry. The pumping beam generated in a mode-locked Nd laser is expanded by L_1, L_2 and subsequently reflected from the grating GR_1 (600 lines/mm), creating a continuous temporal delay across the horizontal beam diameter. In this way, the pump light and the strong stimulated emission traveling in the generator G are synchronized. Tunable pulses are generated by selecting a spectrally narrow emission component from G (by the frequency filter consisting of L_4, GR_2, L_5 and S_2) with subsequent amplification in the dye cell A

index of the dye solution. In this way, the energy of the pump pulse is used very effectively resulting in strong amplified spontaneous emission (ASE) of the dye.

A component of narrow spectral width is selected from the ASE output of the first dye cell by a frequency filter (made up of grating GR2, lenses L4, L5, and slits S1, S2) and amplified in a second cell A which is excited by the remainder of the pump beam. Here the concentration of the dye solution and, consequently, the gain is reduced in comparison with the first dye cell, in order to suppress ASE in the amplifier cell. The optical path length through the frequency filter is compensated for by the delay line VD. The wavelength of the amplified pulses is tuned by rotating the grating GR2.

Numerous infrared dyes were investigated with the traveling-wave setup of Fig. 7.4. Here we discuss data for dye S501 (solvent *o*-dichlorobenzene) pumped by single pulses of a duration of 4 ps from a mode-locked Nd : glass laser. The spectrum of ASE from the first dye cell is plotted on the right-hand side of Fig. 7.2. The stimulated emission peaks at 1.8 μm with a spectral bandwidth of 320 cm^{-1}. The emission is located on the long-wavelength side of the fluorescence spectrum. At shorter wavelengths, the gain is suppressed by the strong reabsorption of the dye solution. Approximately 1% of the pumping energy of 1 mJ is converted to the output pulse. The threshold for laser action lies at pumping energies around 20 μJ.

The tuning curve of the complete system is presented in Fig. 7.5, where the energy of the amplified pulse is plotted versus wavelength. The very broad tuning profile extends from 1.7 to 2 μm, the longest wavelengths ever realized

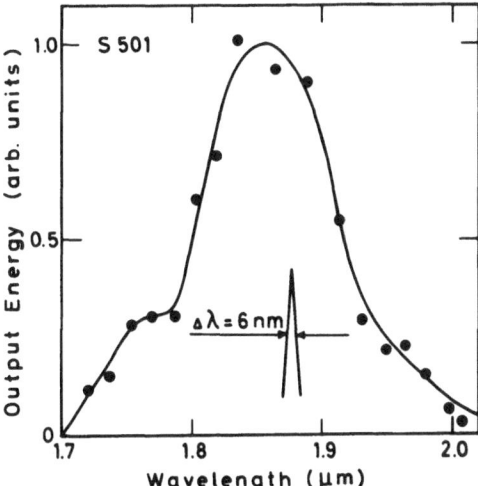

Fig. 7.5 Tuning curve of a picosecond traveling-wave dye laser with dye S501 as the active medium. The energy of the amplified output pulses is plotted against wavelength. The wavelength interval from 1.7 to 2 μm is covered continuously with pulses of a bandwidth of $\Delta\lambda = 6$ nm

with a dye laser. The tuning range is considerably wider than the spectrum of amplified spontaneous emission from the first dye cell. This finding is due to a strong saturation of the gain in the amplifier producing intense output pulses even for a very weak input signal. The bandwidth of the amplified pulses of $\Delta\lambda = 6$ nm is determined by the tuning element (GR2) and is thus identical for all spectral positions.

Approximately 5% of the pumping pulse is converted to the narrow band output at the maximum of the tuning curve, corresponding to an energy of several tens of μJ per pulse. ASE makes a contribution of less than 2% to the total output energy. The highly collimated beam from the dye laser has a divergence of less than 5 mrad. The duration of the tunable pulses is close to that of the pump pulse of 4 ps.

The parameters of the traveling wave dye laser were studied in more detail with dye no. 5 as the active material (solvent dichloroethane) which was pumped by 20-ps pulses from a Nd:YAG laser [7.29, 30]. The absorption and fluorescence spectra of this compound are depicted in Fig. 7.1. In Fig. 7.6b, the energy of the output from the first dye cell G (length 1 cm, concentration $c \approx 10^{-3}$ M) with maximum at 1.34 μm is plotted as a function of the pumping energy E_p (points). Stimulated emission is found for $E_p > 20$ μJ. Starting from threshold, the output energy rises exponentially with the pumping energy and begins to saturate around $E_p = 300$ μJ. At $E_p = 1000$ μJ, the number of pumping photons absorbed per cm³ of the dye solution exceeds the concentration of dye molecules by a factor of 10, resulting in strong saturation of the gain. In this range, several percent of the pumping light is converted to amplified spontaneous emission. Traveling wave dye lasers are operated in this range of pumping energy to generate intense pulses of high stability of emission power. The duration of the pulses was determined for $E_p = 1000$ μJ measuring the cross correlation with the pump pulse. In Fig. 7.7a, the sum frequency signal is shown

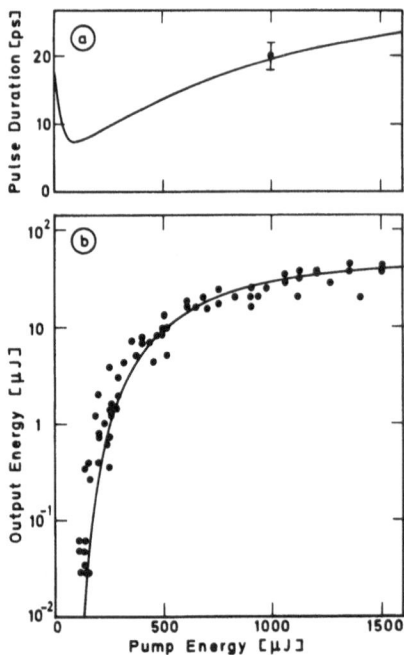

Fig. 7.6. (a) Calculated duration of the output pulse of a single traveling-wave dye cell as a function of the energy E_p of the pump pulse (dye no. 5 dissolved in dichloroethane). The point represents the experimental result for $E_p = 1$ mJ. (b) Spectrally integrated output energy versus pumping energy. Experimental points and calculated curve (solid line). Note the strong saturation of the output pulse

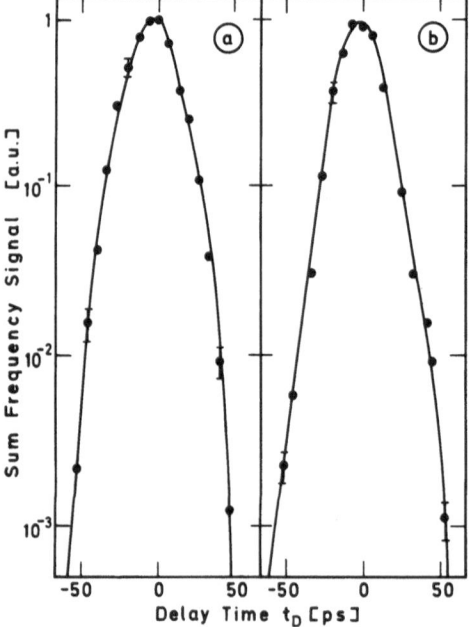

Fig. 7.7. (a) Cross-correlation trace of the pulse from a traveling wave dye cell ($\lambda = 1.34$ μm) and of the pump pulse from a Nd:YAG laser ($\lambda = 1.064$ μm, points). The sum frequency signal is plotted against delay time between the two pulses. A similar duration of the two pulses of 20 ps is derived from the data. (b) Cross correlation of amplified pulses from the generator/amplifier system (depicted in Fig. 4, $\lambda = 1.3$ μm) with the pump pulses from the Nd:YAG laser. The respective pulse durations are 20 ps

as a function of delay time between the two pulses. The temporal half-widths of the pump and the stimulated emission are very similar having a value of 20 ps (point in Fig. 7.6a).

Pulses tunable from 1.18 to 1.4 μm were generated with the complete traveling-wave system (cf. Fig. 7.5) using dye no. 5 [7.28, 30]. In Fig. 7.7b, the cross-correlation trace of the amplified output pulses and the pump pulses from the Nd:YAG laser is presented. The duration of both pulses amounts to 20 ps with a jitter of less than 3 ps between them. Excellent synchronization with negligible jitter is also found for two independent systems of traveling wave dye lasers pumped by the same Nd:YAG laser. The latter system enables picosecond pump–probe studies with pulses independently tunable in the near infrared [7.32, 33].

Infrared pulse generation in a single traveling cell was analyzed by a rate equation model. The kinetics of the gain after picosecond excitation was simulated in a four-level scheme, where the final state of the lasing transition (level 4) was split into 30 sublevels to account for the spectral properties of the ASE emission. Details of the calculation and the relevant parameters of the dye molecules are given in [7.29]. The duration of the pump pulse of 20 ps is much longer than the S_1 lifetime of the dye of 2.7 ps [7.4]. As a consequence, the gain rises rapidly with the leading edge of the pump pulse and decays within 20 ps. The solid lines in Fig. 7.6a, b give the calculated pulse duration and output energy versus pump energy, respectively. In the regime of small signal gain ($E_p < 100$ μJ), the population inversion in the dye is not affected by stimulated emission and the output energy rises exponentially. The initial spontaneous emission of an energy of approximately 10^{-13} J is amplified by several orders of magnitude. In this range, the duration of stimulated emission is shorter than the pump pulse. Due to gain narrowing, the pulse duration has a minimum value of 8 ps around $E_p = 70$ μJ. A further increase of E_p leads to saturation of the gain and a temporal broadening of the output pulse.

The spectrum of stimulated emission is strongly influenced by reabsorption in the highly concentrated dye solution. The dashed area in Fig. 7.1 gives the effective gain profile of the sample. The onset of stimulated emission at higher pumping energies leads to a narrowing of the gain profile at the long wavelength side and a concommittant reduction of the bandwidth. The spectrum at short wavelengths changes only slightly with E_p, since spectral narrowing is compensated by the bleaching of the active medium after excitation. At $E_p > 1$ mJ, reabsorption of vibrationally hot molecules in the electronic ground state reduces the emission intensity at short wavelengths [7.29].

We stress the favorable properties of the pulses from picosecond traveling wave dye lasers, i.e. tunability over a wide range, high energies of several tens of μJ per pulse, and low intensity fluctuations. These features allow efficient nonlinear frequency conversion, as will be discussed in the next section.

Finally, it should be mentioned that traveling-wave excitation has also been reported for dye lasers in the visible spectral range [7.34–38]. A grating or prisms were used to accomplish synchronization of pump light and stimulated emission. Pulses of a duration below 100 fs were generated by femtosecond excitation with amplified pulses from a colliding-pulse mode-locked dye laser [7.37, 38].

7.2.4 Nonlinear Frequency Conversion with Infrared Dye Lasers

In this section, we briefly discuss techniques of nonlinear frequency conversion by which pulses in the medium infrared are produced with the help of near infrared dye lasers. Nanosecond and picosecond pulses, which are tunable in the wavelength range from 3 to 19 μm, were generated by parametric down-conversion in various nonlinear crystals. In the parametric process, a tunable pulse from the dye laser is mixed with a second pulse at a fixed shorter wavelength to generate the difference frequency. Phase matching of the nonlinear interaction is achieved by proper alignment of the crystal and by different polarizations of the three pulses. In the type-I scheme, the pulse at the shortest wavelength is polarized extraordinarily (e), whereas the dye laser and the generated mid-infrared wave have ordinary (o) polarization (in short e + o → o). The type-II scheme gives an output at the difference frequency with extraordinary polarization (e + o → e). In addition to phase matching, exact synchronization of the two input pulses is required to obtain intense mid-infrared light with low intensity fluctuations.

Down-conversion of nanosecond pulses using a Q-switched Nd:YAG laser to generate the frequency-fixed second pulse has been reported in [7.17, 18]. Frequency mixing in GaSe (thickness 0.5 to 0.9 cm) gives pulses tunable between 9 and 19 μm. The mid-infrared pulses at 15 μm have an energy of approximately 25 μJ for input pulses of 8 mJ at 1.06 μm and 150 μJ at 1.145 μm. Nearly identical conversion efficiencies are found with type-I and type-II phase matching. A similar system of down-conversion in $AgGaS_2$ crystal generates mid-infrared pulse between 5 and 11 μm [7.18]. At 6 μm, approximately 1% of the Nd:YAG laser power is converted to the long-wavelength pulse of a duration of 8 ns.

Intense picosecond pulses in the medium infrared were produced by difference frequency mixing with traveling wave dye lasers (c.f. Sect. 7.2.3) [7.30]. Single picosecond pulses from a mode-locked Nd:YAG or Nd:glass laser pump a traveling-wave system producing a tunable input pulse for the parametric process. These near-infrared pulses are mixed with part of the Nd:laser output in a $AgGaS_2$ crystal of a length of 1 to 2 cm (collinear type-I phase matching). Using dye no. 5 as the active medium of the dye laser, mid-infrared wavelengths between 3.5 and 10 μm are generated. The bandwidth of the pulses is determined by the spectral width of the input from the traveling-wave laser and has a value of 7cm^{-1}, independent of the spectral position in the mid-infrared. In Fig. 7.8, the photon conversion efficiency of the mixing process is plotted as a function of wavelength. The absolute efficiency values give the fraction of photons at 1.06 μm which are converted to mid-infrared photons. The data of Fig. 7.8 were obtained with 10^{16} photons at 1.06 μm. High conversion efficiencies of up to several percent are observed over a wide wavelength range. The drop of conversion at short wavelengths is due to the decrease in the dye laser output at

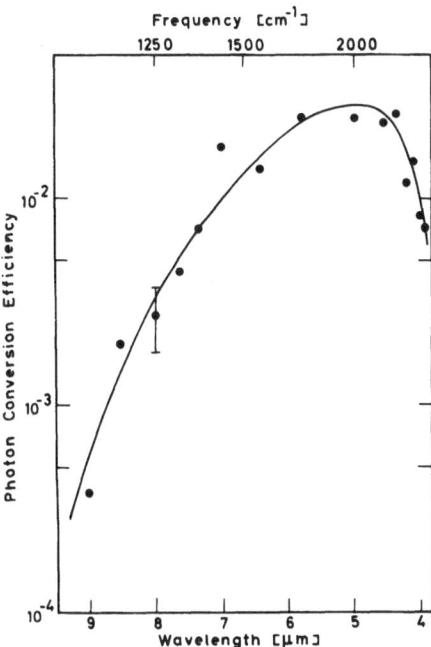

Fig. 7.8. Parametric down-conversion of picosecond pulses from a traveling-wave dye laser in a nonlinear AgGaS$_2$ crystal. The photon conversion efficiency is plotted as a function of the wavelength of the generated mid-infrared pulses

the corresponding wavelengths in the near infrared. At long wavelengths, the effective nonlinearity of the nonlinear crystal and thus the conversion efficiency decreases. The duration of the mid-infrared pulses was determined by measuring the cross correlation with the second harmonic of the Nd:laser. Pulse durations of 2 ps and 8 ps at 5 µm were found for systems pumped by Nd:glass and Nd:YAG lasers, respectively. Recently, two independently tunable mid-infrared pulses were generated by parallel operation of two traveling-wave lasers and two down-converters. The two light sources are synchronized with a jitter of less than 500 fs for pumping by a Nd:glass laser and less than 2 ps for the Nd:YAG laser. The wide tuning range, the high conversion efficiency with low intensity fluctuations, and the accurate synchronization of different sources of coherent radiation make down-conversion of picosecond infrared dye lasers a versatile technique to generate ultrashort pulses in the medium infrared.

In conclusion, the results reviewed in this article demonstrate the great potential of infrared dye lasers for the generation of nanosecond, picosecond, and subpicosecond pulses in the near infrared. Efficient laser dyes of high photochemical stability are available for the entire wavelength range from 1 to 2 µm. Down-conversion of the dye laser pulses by means of nonlinear frequency mixing gives access to the mid-infrared with wavelengths of up to 20 µm. The pico- and subpicosecond laser systems have found extensive application in time-resolved infrared spectroscopy, where ultrafast processes in condensed phases have been studied in great detail. In semiconductors, hot electrons and holes

[7.32. 39–41], exciton bleaching [7.33], nonequilibrium optical phonons [7.42] and intersubband scattering [7.43–45] were investigated. Picosecond infrared experiments also gave information on vibrational redistribution in large molecules [7.46] and on excited state photoreactions [7.47]. These examples illustrate the relevance of infrared dye lasers for spectroscopic studies.

Acknowledgements. We gladly acknowledge the important contributions of H.J. Polland to the development of traveling-wave dye lasers emitting in the infrared.

References

7.1   K.H. Drexhage in: In *Dye Lasers*, ed. by F.P. Schäfer, Topics Appl. Phys. Vol. 1 (Springer, Berlin, Heidelberg 1973) p. 144
7.2   D. von der Linde, K.F. Rodgers: IEEE J. Quantum Electron. **9**, 960 (1973)
7.3   R.J. Scarlet, J.F. Figueira, H. Mahr: Appl. Phys. Lett. **13**, 7 (1968)
7.4   B. Kopainsky, W. Kaiser, K.H Drexhage: Opt. Commun. **32**, 451 (1980)
7.5   B. Kopainsky, P. Qiu, W. Kaiser, B. Sens, K.H. Drexhage: Appl. Phys. **B29**, 15 (1982)
7.6   H. Kuhn: Fortschr. Chem. Org. Naturstoffe **16**, 169 (1958); **17**, 404 (1959)
7.7   H. Kuhn, W. Huber, G. Handschig, H. Martin, F. Schäfer, F. Bär: J. Chem. Phys. **32**, 467 (1960)
7.8   G.A. Reynolds, K.H. Drexhage: J. Org. Chem. **42**, 885 (1977)
7.9   K.H. Drexhage: Unpublished results
7.10   B. Sens: Thesis, Universität-Gesamthochschule Siegen 1984
7.11   H.J. Polland, T. Elsaesser, A. Seilmeier, W. Kaiser, M. Kussler, N.J. Marx, B. Sens, K.H. Drexhage: Appl. Phys. **B32**, 53 (1983)
7.12   A. Seilmeier, B. Kopainsky, W. Kaiser: Appl. Phys. **22**, 355 (1980)
7.13   S.J. Strickler, R.A. Berg: J. Chem. Phys. **37**, 814 (1967)
7.14   G.G. Dyadyusha, I.P. Il'chisin, Yu. L. Slominskii, E.A. Tikhonov, A.I. Tolmachev, M.T. Shpak: Sov. J. Quantum Electron. **6**, 349 (1976) and references therein
7.15   K. Kato: IEEE J. Quantum Electron **14**, 7 (1978)
7.16   K. Kato: Appl. Phys. Lett. **33**, 509 (1978)
7.17   J.L. Oudar, Ph. J. Kupecek, D.S. Chemla: Opt. Commun. **29**, 119 (1979)
7.18   K.Kato: IEEE J. Quantum Electron. **20**, 698 (1984)
7.19   A. Ferrario: Opt. Commun. **30**, 85 (1979)
7.20   W. Kranitzky, B. Kopainsky, W.. Kaiser, K.H. Drexhage, G.A. Reynolds: Opt. Commun. **36**, 149 (1981)
7.21   H. Lobentanzer, H.J. Polland: Opt. Commun **62**, 35 (1987)
7.22   H. Lobentanzer: Opt. Commun. **71**, 175 (1989)
7.23   M. Leduc: Opt. Commun. **31**, 66 (1979)
7.24   A. Seilmeier, W. Kaiser, B. Sens, K.H. Drexhage: Opt. Lett. **8**, 205 (1983)
7.25   H. Roskos, S. Opitz, A. Seilmeier, W. Kaiser: IEEE J. Quantum Electron. **22**, 697 (1986)
7.26   P. Beaud, B. Zysset, A.P. Schwarzenbach, H.P. Weber: Opt. Lett. **11**, 24 (1986); see also Refs. 7.40, 7.41
7.27   J.D. Harvey, A. Seilmeier, H. Roskos: Proc. Phys. **12**, 134 (1986)
7.28   T. Elsaesser, H.J. Polland, A. Seilmeier, W. Kaiser: IEEE J. Quantum Electron. **20**, 191 (1984)
7.29   H. Lobentanzer, T. Elsaesser: Appl. Phys. **B41**, 139 (1986)
7.30   T. Elsaesser, H. Lobentanzer, A. Seilmeier: Opt. Commun. **53**, 355 (1985)
7.31   R. Wyatt, E.E. Marinero: Appl. Phys. **25**, 297 (1981)
7.32   T. Elsaesser, R.J. Bäuerle, W. Kaiser, H. Lobentanzer, W. Stolz, K. Ploog: Appl. Phys. Lett. **54**, 256 (1989)

7.33 H. Lobentanzer, W. Stolz, K. Ploog, R.J. Bäuerle, T. Elsaesser: Solid State Electron. **32**, 1875 (1989)
7.34 Zs. Bor, S. Szatmari, A. Müller: Appl. Phys. **B32**, 101 (1983)
7.35 S. Szatmari, F.P. Schäfer: Opt. Commun. **49**, 281 (1984)
7.36 J. Hebling, J. Klebniczi, P. Heszler, Zs. Bor, B. Racz: Appl. Phys. **B48**, 401 (1989)
7.37 J. Hebling, J. Kuhl: Opt. Lett. **14**, 278 (1989)
7.38 J. Hebling, J. Kuhl: Opt. Commun. **73**, 375 (1989)
7.39 M. Woerner, T. Elsaesser, W. Kaiser: Phys. Rev. **B41**, 5463 (1990)
7.40 H. Roskos, B. Rieck, A. Seilmeier, W. Kaiser: Appl. Phys. Lett **53**, 2406 (1988)
7.41 H. Roskos, B. Rieck, A. Seilmeier, W. Kaiser, G.G. Baumann: Phys. Rev. **B40**, 1396 (1989)
7.42 T. Elsaesser, R.J. Bäuerle, W. Kaiser: Phys. Rev. **B40**, 2976 (1989)
7.43 A. Seilmeier, H.J. Hübner, G. Abstreiter, W. Schlapp, G. Weimann: Phys. Rev. Lett. **57**, 1375 (1987)
7.44 R.J. Bäuerle, T. Elsaesser, W. Kaiser, H. Lobentanzer, W. Stolz, K. Ploog: Phys. Rev. **B38**, 4307 (1988)
7.45 R.J. Bäuerle, T. Elsaesser, H. Lobentanzer, W. Stolz, K. Ploog: Phys. Rev. B40, 10002 (1989)
7.46 H.J. Hübner, M. Woerner, W. Kaiser, A. Seilmeier: Chem. Phys. Lett., 182, 315 (1991)
7.47 T. Elsaesser, W. Kaiser: Chem. Phys. Lett. **128**, 231 (1986)

Wolfgang Kaiser   Thomas Elsaesser

8. Propagation of Femtosecond Light Pulses Through Dye Amplifiers

B. Wilhelmi[1]

With 13 Figures

This chapter discusses the physical processes occurring in the gain media of femtosecond dye amplifiers. In particular, the influence of relaxation and line broadening on energy extraction, pulse duration and chirp production is investigated. For optimum operation stretched pulse amplification is recommended, where the duration of the stretched input pulses should be longer than the cross relaxation time of the gain medium, which for typical dyes is estimated to be on the order of some hundred femtoseconds. With appropriately tailored dye mixtures very broad pulse spectra can be transmitted, even in the regime of strong depletion and high energy extraction, and further progress in reducing the minimum pulse duration of powerful light pulses seems possible.

8.1 Introduction

During the last decade femtosecond light pulses have become a convenient tool to excite, measure and control ultrafast processes in fundamental science as well as in technology [8.1, 2].

The temporal resolution reached is on the order of 10^{-14} s and exceeds the most advanced electronic techniques by some orders of magnitude. Among fundamental processes that have been investigated on a femtosecond time scale are molecular vibrations and phonons and their decay, electronic relaxation processes in molecules and solids, phase transitions, charge and energy transfer in molecules, as well as isomerization and reorientation of molecules. Femtosecond light pulses have been successfully applied for the manipulation of photochemical processes. Techniques have been developed to test and control optoelectronic and electronic devices. Nowadays, the shortest electron and X-ray pulses are produced by applying femtosecond light pulses.

Such attractive applications became possible through the development and improvement of femtosecond pulse sources. Various oscillators are now available yielding pulses with durations below 100 fs and operating from the UV to the IR spectral region. Unfortunately, spectral tunability is restricted to narrow ranges. Pulses from these oscillators usually exhibit energies on the order of

---

[1] The author is now with the Jenoptik Carl Zeiss JENA GmbH, Zentralbereich Forschung u. Entwicklung, Carl-Zeiss-Str. 1, O-6900 Jena, Germany.

Topics in Applied Physics, Vol. 70
Dye Lasers: 25 Years Ed.: Dr. Michael Stuke
© Springer-Verlag Berlin Heidelberg 1992

some nanojoules or even less at repetition rate $\leqslant 100$ MHz. Although these pulse characteristics enable one to apply very sensitive detection and measuring techniques, the pulse energy is too low for many measuring problems. Moreover, frequency conversion and pulse shaping, in particular pulse compression (see, e.g., [8.3]) by which means pulses as short as 6 fs [8.4] could be obtained, require more powerful pulses. Moreover, strong input light pulses are required in nonlinear optical frequency converters which are applied to reach other spectral regions and to tune the pulse wavelength.

Therefore, femtosecond pulse amplifiers have been built where, at the expense of pulse repetition rate, an energy amplification of up to 10^{10}, yielding pulses of tera watt power [8.5], can be obtained. The basic components of such amplification devices are already known from ps and ns pulse amplification. On a fs time-scale, however, new principles had to be applied for the short pulse duration to be maintained. First of all this means compensation of group velocity dispersion (GVD) and avoidance of undesired nonlinear optical processes induced by the high peak intensities. It should be noted that under certain circumstances a combination of both effects can be advantageously utilized for pulse shortening in the amplification process. Simultaneously, new developments and improvements of lasers which serve as pump for fs pulse amplifiers, such as excimer lasers, copper vapor lasers and solid state lasers, enabled and stimulated the design of devices which amplify femtosecond pulses even in the UV and/or at repetition frequencies in the kHz range.

Femtosecond pulse amplifiers differ from each other in the achievable repetition rate and energy amplification; these range from 0.1 Hz to some MHz and from 10 to 10^{10}, respectively. These two parameters cannot be chosen independently of one another. Instead, in present devices the product of repetition rate and pulse energy does not usually exceed 100 mW, i.e. it remains on the order of the mean power of the oscillator. Thus, according to the specific application, one has to compromise between single pulse energy and repetition rate.

As examples, Figs. 8.1 and 8.3 present two devices: First [8.6] a rather complex one suitable for producing high power pulses at various wavelengths and with broad choice of other parameters and second a very small and simple one. Both devices have been used in our laboratories for generating powerful and tunable light pulses. Figure 8.1 shows the complex arrangement in which a CPM dye laser (counterpropagating pulse mode-locked) pumped by an argon ion laser provides the input pulses. The four amplifier stages consist of dye cells pumped by a XeCl laser synchronized to the pulse train from the laser. The amplifier stages are decoupled by the use of saturable absorbers (RG5 filter glasses) and space–frequency filtration.

A total energy amplification of about 10^7 was achieved, with the gain of the individual stages decreasing from 5000 in the first to about 10 in the last. After amplification the pulses pass through a compressor which consists of highly dispersive SF glass prisms. Under optimum operating conditions the amplified and compressed pulses are even shorter than the bandwidth-limited input

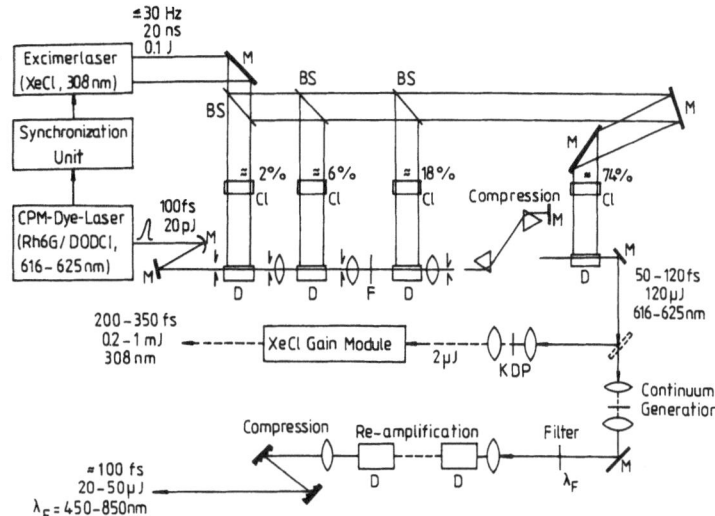

Fig. 8.1. Complex femtosecond device consisting of CPM laser oscillator, four-stage amplifier, pulse compressor, second harmonic generator, eximer amplifier, spectral continuum generator, dye amplifier and compressor

pulses. This additional compression can be explained by nonlinear optical self-phase modulation in the amplifier (particularly in the last stage). We will come back to that point later on (Sect. 8.4).

For the generation of high-power pulses in the UV (308 nm) the red pulses (616 nm) are frequency-doubled in a 1 mm KDP crystal (conversion efficiency 5%) and amplified in the second module of the excimer laser (single pass gain 50–100, double-pass gain 200–500). Due to the finite bandwidth of the XeCl gain medium, the output pulses are longer than the input pulses and have durations of $\tau_L \approx 300$ fs in the case of a weak saturated amplification (single pass) and $\tau_L \approx$ fs for stronger gain depletion (double-pass).

Tunability from the blue to the near IR spectral region is achieved by focusing amplified (red) pulses in a suitable non-linear optical material (e.g., carbon tetrachloride, CCl_4) to generate a fs broadband continuum (see Fig. 8.2) and by filtering out the desired wavelength range. For the spectral selection interferential filters can be used, but also prism or grating arrangements [8.7, 8] are possible. The latter offers the possibility of simultaneous spectral filtering and adjustment of GVD [8.7]. The spectral filtering after continuum generation is connected with an energy loss which can be compensated in additional dye stages pumped by the second excimer gain module of the EMG 150. After amplification to some ten microjoules, the pulses are recompressed to about 100 fs.

Compared with the complex amplifier described so far, Fig. 8.3, shows a very simple and small device [8.9] which is used for the wavelength transformation of

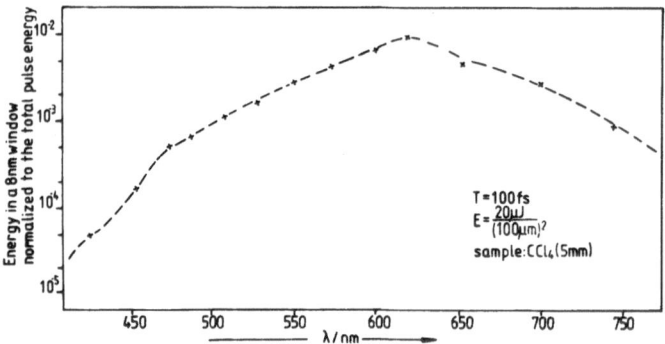

Fig. 8.2. Spectrum of the femtosecond spectral continuum

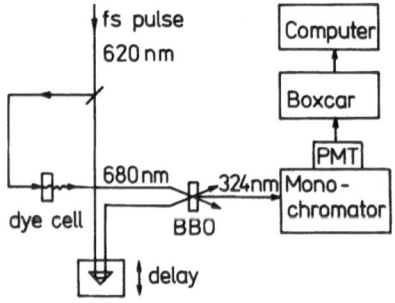

Fig. 8.3. Simple and small femtosecond one-cell device used for the frequency conversion of ultra-short light pulses by self-phase modulation (continuum generation) and amplification

powerful dye laser pulses. The main part consists of only one short (1 mm) dye cell pumped by amplified femtosecond light pulses at a certain wavelength (620 nm). The entrance window of the cell serves as the nonlinear optical wavelength converter where a spectral continuum is generated by self-phase modulation. This continuum is taken as the amplifier input. Here spectral filtering is only achieved by the action of the amplifying dye solution with its specific spectral gain profile. In the example presented here the dye oxazine 1 dissolved in ethanol (3×10^{-4}–1×10^{-3} mol/l) was used, and as a result amplified signal pulses at 680 nm were obtained. First the signal passes together with the pump pulse through the cell in the forward direction, then it is coupled back at the interface between solution and exit window and travels in the opposite direction where further amplification occurs; again it is coupled back at the input interface and so on. Figure 8.4a shows a sequence of amplified pulses generated in this way. The measurement is performed by generating the cross-correlation between signal and pump in the nonlinear optical crystal BBO. The distance between adjacent pulses equals the round trip time in the dye solution. Obviously, the second pulse is the strongest. This means that the additional gain during the described round trip overcompensates the coupling losses; however, the third pulse is weak because the gain is almost depleted by the first and

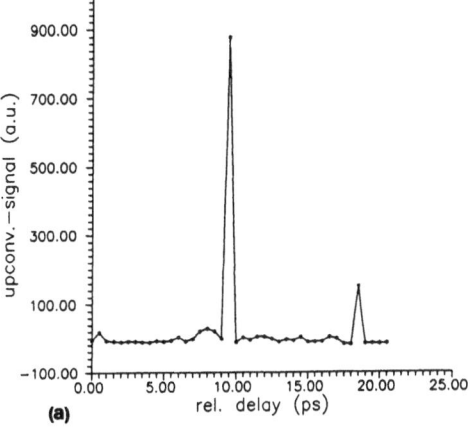

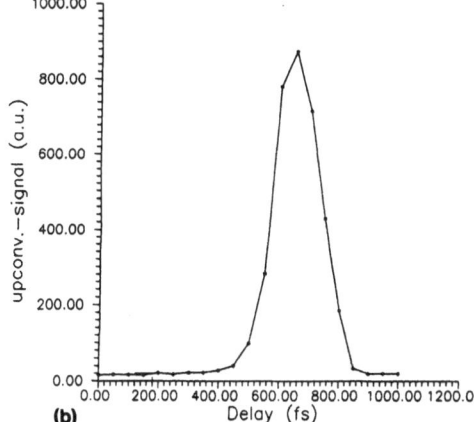

Fig. 8.4a, b. Cross-correlation between the output of the one-cell device and pump radiation measured by nonlinear optical up-conversion: (a) total pulse sequence, (b) main pulse

second pulse and cannot compensate for the losses. Figure 8.4b shows the cross-correlation for the main (second) light pulse of the sequence on an extended scale. Taking into account the duration of the pump, the signal duration is estimated to be somewhat shorter than 100 fs. The advantage of this device with its rather thin dye cell stems from the negligibly small influence of group velocity dispersion between pump and signal on the results. Thus, special measures for the synchronization between pump and signal, which are typical for other traveling-wave amplifiers [8.10, 11], are not required.

From the rough description of these two examples it already becomes obvious that light pulses passing through amplifiers experience the influence of various linear and nonlinear optical effects where the parameters of the output pulses depend on the complex interplay between these effects.

In this chapter we treat the interaction of light pulses with type solutions and discuss the impact of various linear and nonlinear optical effects on the pulse

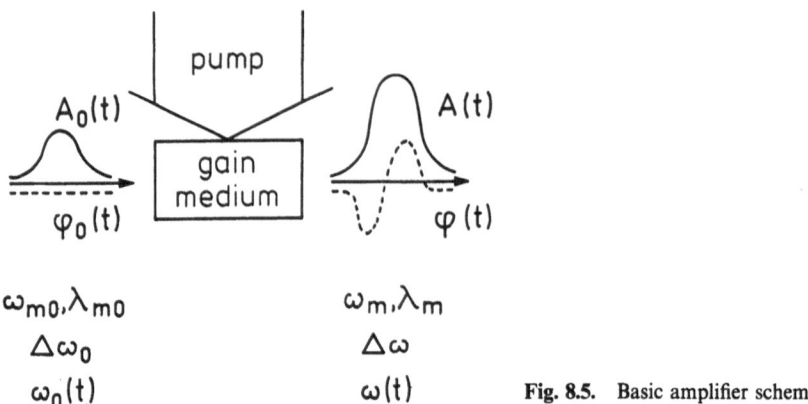

pump

$A_0(t)$ $A(t)$

gain
medium

$\varphi_0(t)$ $\varphi(t)$

$\omega_{m0}, \lambda_{m0}$ ω_m, λ_m

$\Delta\omega_0$ $\Delta\omega$

$\omega_0(t)$ $\omega(t)$ **Fig. 8.5.** Basic amplifier scheme

parameters. In particular we focus at the generation of very short light pulses and their frequency chirp.

The basic scheme of pulse amplification is shown in Fig. 8.5, which might be imagined either as the situation in one amplifier stage or to represent the overall effect of the whole amplifier. The pulses are characterized by their instantaneous intensity $I(t)$ or modulus of field amplitude $A(t)$ (maximum field amplitude $\hat{A}$, duration τ), the instantaneous phase $\phi(t)$ or instantaneous frequency $\omega_L(t) = \omega_L + \Delta\omega_L(t)$, $\Delta\omega_L(t) = (d/dt)\phi(t)$, and the spectral profile $\bar{I}(\omega)$ (frequency and wavelength of maximum ω_m and λ_m, with spectral widths $\Delta\omega_L$ and $\Delta\lambda_L$, respectively). In general all these parameters change during the interaction of the light pulse with the gain medium. We mention immediately that the results obtained for the interaction of the light pulse in a single gain cell are of interest for the understanding not only of pulse amplifiers but also of femtosecond lasers, and in addition they provide a deeper insight into the ultrafast molecular processes occuring in the gain medium.

8.2 Interaction in the Rate-Equation Approximation

When the bandwidth of the input pulses is small compared to that of the gain medium ($\Delta\omega_L \ll \Delta\omega_0$, $\tau_L \gg \tau$, where τ is the phase decay time of the gain transition), the rate equation approximation (REA) can be used and in particular the action of the dye as a spectral filter can be neglected (see, e.g., [8.12]). Then, for light pulses that ar short compared with the lifetime T of the upper level of the amplifying dye, the *Frantz–Nodvick* equations [8.13] completely describe the pulse reshaping.

Qualitatively, the change in the instantaneous intensity caused by gain depletion is shown in Fig. 8.6a. The trailing edge of the pulse is seen to be suppressed. For comparison the impact of saturable absorbers on the instantaneous intensity, which, as described in the introduction, are often used in amplifiers, is demonstrated in Fig. 8.6b. Together with the intensity the in-

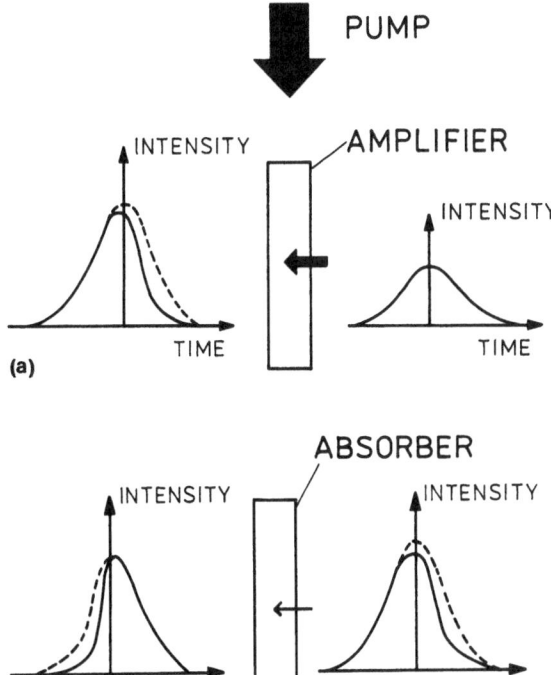

Fig. **8.6a, b.** Schematic presentation of pulse reshaping in nonlinear optical devices: (a) depletable amplifier, (b) saturable absorber

stantaneous phase (frequency) of the light pulse is changed in the interaction process when the laser wavelength λ_L deviates from the line center λ_0. This influence originates from the depletion of dispersion, qualitatively shown in Fig. 8.7a. From Fig. 8.7b we see that the frequency change is quite considerable (some $1/\tau_{L0}$) when the energy of the input pulse per unit area equals the saturation energy per unit area, ($\mathscr{E} = \mathscr{E}_L/\mathscr{E}_{sat} \sim 1$, $\mathscr{E}_{sat} = 1/\sigma$, where σ is the interaction cross section of the transition).

Thus, even in the rate equation approximation and for homogeneously broadened lines, both the instantaneous intensity and the instantaneous frequency may experience considerable changes during amplification, i.e. almost all pulse parameters, and not only the pulse energy are changed. The calculations are in quantitative agreement with the experimental findings for picosecond and subpicosecond pulses ($\tau_L \gtrsim 100$ fs). In these cases the transitions of typical dyes appear to be almost homogeneously broadened and moreover the REA can be applied because of the large bandwidth of the gain curves (~ 50 nm) [8.14].

8.3 Interaction of Very Short Light Pulses with Dye Solutions

With very short light pulses the approximations used in Sect. 8.2 fail. The REA can be dropped without great difficulty as long as the transitions remain

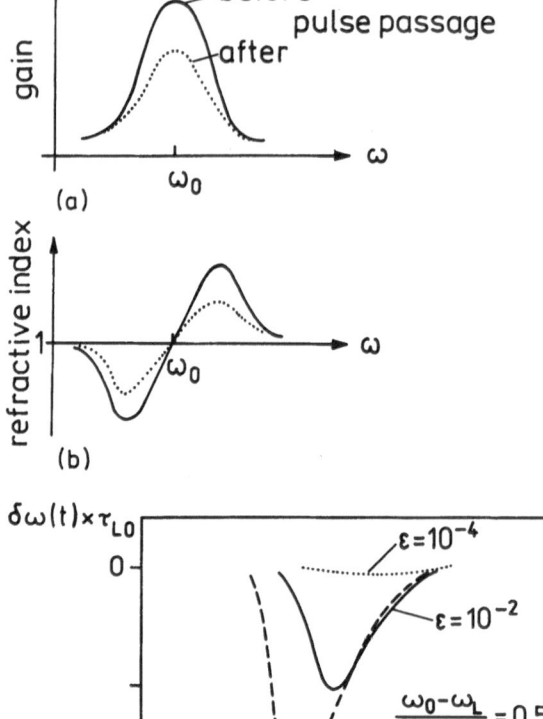

Fig. 8.7a, b. Chirp production by depletable amplification: (a) Gain vs. frequency before and after pulse passage, (b) Chirp produced by depletable amplification of a Gaussian shaped input pulse for various input pulse energies $\mathscr{E} = \mathscr{E}_L/\mathscr{E}_{sat}$

homogeneously broadened (see, e.g., [8.15, 16]). Then the dye is described as a saturable spectral filter, which affects the modulus and the phase of the field amplitude. With femtosecond light pulses the interaction process is not yet completely understood, and in general it cannot be described by the simple terms of homogeneously and inhomogeneously broadened transitions.

Let us first discuss some experimental findings and models established so far to explain the observed phenomena at least qualitatively. From pump excited dichroism *Angel* et al. [8.17] obtained experimental evidence for very fast relaxational channels that connect the radiative (Franck–Condon) states with a variety of nonradiative (dark) states (Fig. 8.8) in the upper and lower electronic level. These relaxational processes are much faster than the vibrational lifetime (k_1, k_1' some $10^{13\,s-1}$, k_{vib} 10^{11}–$10^{12}\,s^{-1}$) (*Angel* et al. [8.17] obtained $k_1 \approx 3 \times 10^{13}\,s^{-1}$ and $k_{vib} \approx 1 \times 10^{12}\,s^{-1}$ for rhodamine 101, and *Heist* et al. [8.6] obtained $k_1 \approx 2 \times 10^{13}\,s^{-1}$ for sulfo-rhodamine).

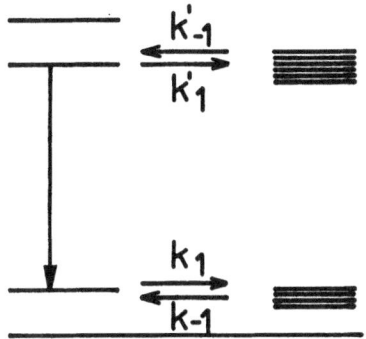

Fig. 8.8. Radiative transition coupled to many other ("dark") transitions

With respect to pulse amplification we may draw the following conclusions from this relaxational model. Because of the very fast relaxational processes thermalization occurs before the signal pulse arrives. When the signal pulse is shorter than $1/k'_{-1}$ it can only interact with the N_r molecules that are initially in radiative states. That means the efficiency is decreased by the factor

$$\frac{N_r}{N_g} = \frac{k'_{-1}}{k'_{-1} + k_1} \, ,$$

where N_g is the total number of molecules in the excited electronic state before the signal pulse arrives. The behavior becomes more complex for $\tau_L \sim 1/k'_{-1}$.

Next we descuss a more specific fluctuational–relaxational model developed to explain the appearance, broadening and decay of spectral holes "burned" into the absorption or gain profile, see Fig. 8.9. *Brito Cruz* et al. [8.18, 19] were the first to obtain evidence for spectral hole burning in dye solutions. They observed in excite-and-probe absorption spectra holes which correspond to individual vibronic transitions $(0 \to 0', 0 \to 1', 1 \to 0', \ldots)$ from the electronic ground level into the first excited singlet level. The holes very rapidly become broadened (100–200 fs), and eventually they disappear and the saturation becomes homogeneous. This behavior can be explained by a simple model describing the interaction of the dye molecule with its surrounding where the electronic

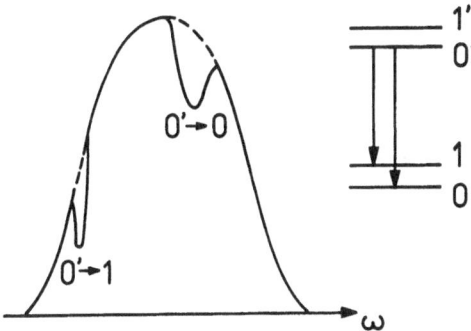

Fig. 8.9. Schematic presentation of holes burned into the spectral gain

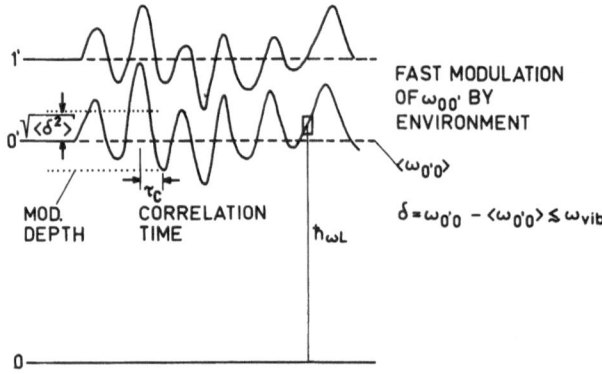

Fig. 8.10. Fast modulation of the electronic transition frequency by environmental influences

transition frequency $\omega_{0'0}$ of the dye molecule is modulated in time under the influence of its neighbors [8.20, 21], see Fig. 8.10. The stochastic modulation processes of the frequency change $\delta(t) = \omega_{0'0}(t) - \omega_{0'0}$ is assumed to obey the relation

$$\langle \delta(t)\delta(0) \rangle = \langle \delta^2 \rangle e^{-t}/\tau_c ,$$

where $\sqrt{\langle \delta^2 \rangle}$ is the modulation depth and τ_C the correlation time of the fluctuation process. This is of course a very rough description of the complex environment around the dye molecule which in reality consists of a huge variety of transitional, rotational and vibrational (phonon) modes of the molecular solution. In spite of this simplification, the model is able to explain all experimental findings mentioned and to provide rough quantitative agreement between experiment and calculations (cf. Fig. 8.11).

We now use this model to discuss femtosecond pulse amplification, i.e. we assume that very short light pulses burn holes into the gain curve (Fig. 8.11). The following conclusions can be drawn:

1) For $\tau_L \gtrsim \tau_C$ the energy extraction decreases with decreasing τ_L as hole burning becomes more and more pronounced, i.e. the molecules outside the spectral range of the light pulse are no longer able to fluctuate into this wavelength region within the pulse duration.
2) For $\tau_L < \tau_C$ (more precisely, $2\pi/\Delta\omega_L < \tau_C$) the energy extraction increases with decreasing τ_L as the holes become broader. The hole width is given by

$$\Delta\omega = \Delta\omega_L / \sqrt{[1 + (\Delta\omega_L/\Delta\omega_{dye})^2]}$$

when the line consists of one vibronic transition only [8.21]. When the gain profile is composed of several vibronic transitions, $\Delta\omega_{dye}$ has to be replaced by the vibration frequency ω_{vib}.

In conclusion, there exists a certain pulse duration τ_L in the range

$$1/\Delta\omega_{dye} < \tau_L < \tau_C,$$

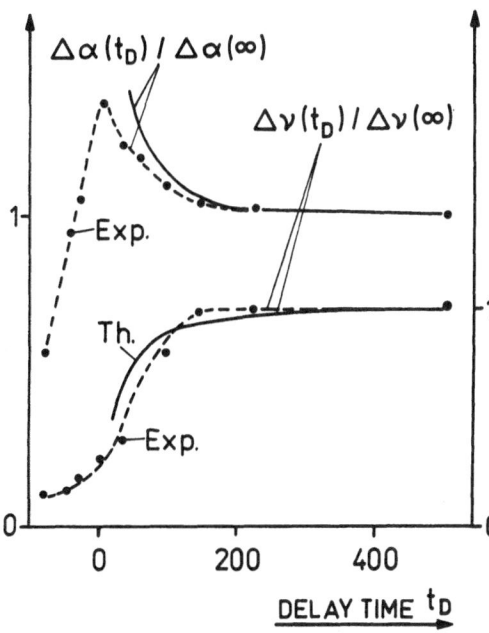

Fig. 8.11. Normalized hole depth and hole width vs. delay between excitation (hole burning) and measurement for cresyl violet. Full lines: calculations [8.21]; dashed lines: experimental results [8.18]. From a comparison between theory and experiment a value τ_C = 150 fs has been obtained

for which the energy extraction in the amplifier takes on its minimum. Such a minimum has been observed by *Noack* [8.22].

It should be mentioned here that the two models discussed so far (radiative-and-dark-state model, fluctuation model) agree in the description of some experimental findings (e.g., decrease of the energy extraction with decreasing pulse duration); however, they differ in the description of other aspects (e.g., existence of a certain pulse duration where minimum energy extraction appears). Until now the experimental findings are not sufficient to give unambiguous support for one model. On the other hand, the relaxational theory has to be refined in order to show the links and the differences between various models.

8.4 Chirped and Stretched Pulse Amplification

Nowadays chirped and stretched pulse amplification is the most advantageous method for generating extremely short and powerful light pulses. First of all chirped and stretched pulse amplification has successfully been applied with solid-state lasers [8.23–27]. In most approaches the aim is to avoid high field strengths in the gain medium which might otherwise result in nonlinear optical interactions disturbing the amplification process. Either the input pulses were stretched only by using linear optical dispersive devices or addtional chirp was produced in conjunction with spectral broadening by nonlinear optical self-phase modulation. In the first case, assuming negligibly small nonlinearities in the amplifier, the pulses can be compressed to the duration of the (bandwidth-limited) initial pulses under ideal conditions; see, e.g., [8.3]. In the second case higher compression factors can be achieved, with the limit being given by the bandwidth of the self-phase-modulated input pulses as long as this bandwidth is small compared with the spectral width of the gain medium. In the linear regime the bandwidth of the output pulses is generally smaller than that of the input pulses because of spectral filtration in the amplifier. Thus gain media with broad spectral gain profiles are required.

Note that nonlinear optical interactions between light pulse and gain medium, in particular self-phase modulation, may lead to an additional spectral broadening. In [8.28] use has been made of this phenomenon for compressing the amplified dye laser pulses (cf. Sect. 8.1) to durations below that of the bandwidth-limited initial pulses. That means chirped pulse amplification has been applied where the chirp is generated mainly in the last amplifier stage not in an additional device in front of the amplifier.

The main reason for stretching ultimately short light pulses in front of dye amplifiers follows from the model considerations and calculations given in Sect. 8.3 and is connected with the complex line broadening processes in solutions. Very short light pulses burn holes into the gain profile and, hence, only a certain group of the excited molecules can take part in the amplification process when the pulse duration is on the order of, or shorter than, the cross relaxation time of the gain transition which describes the refilling of holes and is, in approximation, equal to the correlation time τ_C in the fluctuation model described in Sect. 8.3. On the other hand, when the stretched pulse is long compared with the correlation time τ_C (some hundred femtosecond) all molecules in the excited electronic state take part in the gain process. Moreover, in the same way as in the case of homogeneously broadened transitions, the reduction of electric field strength connected with the stretching process helps to avoid perturbations by other, undesired nonlinear optical effects.

Boyer et al. [8.29] applied the concept of chirped pulse amplification to dye amplifiers and were successful in compressing the amplified pulses to about 16 fs when using the broad spectral gain of dye mixtures. These are the shortest amplified light pulses so far obtained. Self-phase modulation of the 50-fs input pulses was achieved in a short piece of fiber, at the exit of which the pulse duration was about 300 fs. Amplification led to negligibly small additional pulse prolongation.

In spite of these excellent results already obtained, the question arises whether even shorter pulses can be generated by using such concepts. Having in mind the broad gain profile of the dye mixture and the spectral width of the self-phase modulated input pulse, the generation of ultimately short amplified and recompressed pulses with durations of about, or even less than, 5 fs seems feasible.

Difficulties in achieving such extremely short light pulses result mainly from two factors. First, the spectral profile of the light pulses changes during the amplification processes and becomes rugged and narrowed (Fig. 8.12). Second, it

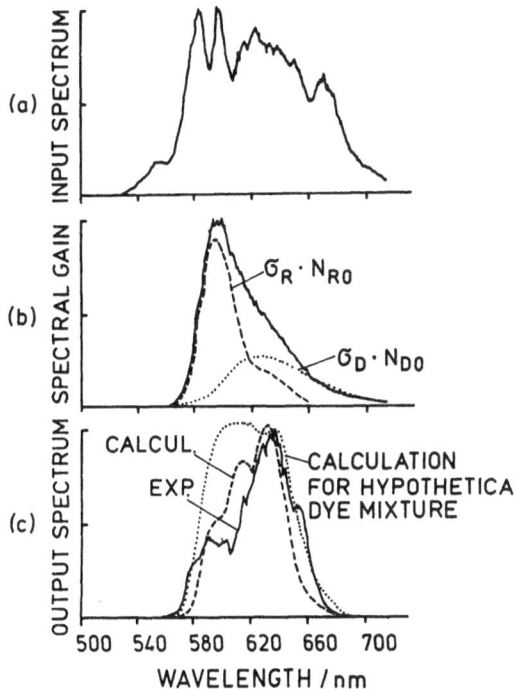

Fig. 8.12a–c. Change of spectral profiles in chirped pulse amplification (experimental results from [8.29]). (a) Spectrum of input pulse [8.29]. (b) Gain profile of amplifying dye mixture (full line) and its components rhodamine 610 (dashed line) and DCM (dotted line). N_{RO}, N_{DO} are the initial occupation densities of the excited levels of rhodamine and DCM, respectively, σ_R, σ_D the emission cross sections of rhodamine and DCM, respectively. (c) Spectrum of output pulse. Full line: experimental results from [8.29]; dashed line: calculations for the dye mixture used in [8.29]; dotted line: calculations for a hypothetical dye mixture with almost no overlap between the spectra of the components (see text)

is almost impossible to construct an ideal compressor that completely compensates for the phase modulation of such structured and spectrally very broad light pulses. Until now, the most successful approaches [8.4, 29] have used combinations of two compression stages and compensate for the change in the instantaneous frequency varying linearly and quadratically with time near the pulse center. By using such an arrangment *Boÿer* et al. [8.29] were able to obtain pulse durations of about 12 fs, but were forced to accept considerable intensity losses.

Here we will deal with the first problem, i.e. the amplification process. We aim at a deeper insight into the interaction between light and matter and the origin of spectral narrowing and spectral substructures.

The input pulses are modeled in the following way. We consider the case of optimum interplay between self-phase modulation and group velocity dispersion in the dispersive nonlinear optical fiber in front of the amplifier [8.30–32] which results in an almost rectangular field-strength pulse, whose frequency increases linearly with time (Fig. 8.13). These prolonged and linearly chirped light pulses act as input pulses of the amplifier. For our calculation these input pulses, which are long compared with τ_C, are temporally decomposed into very short "pulse slices", each of which is characterized by a certain midfrequency, where the first pulse slice has the lowest frequency (longest wavelength) the last one the highest frequency (shortest wavelength). We start with passing the first slice through the amplifier which consists of a mixture of two dyes with

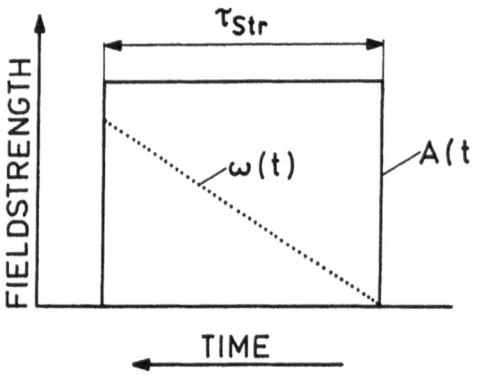

Fig. 8.13. Instantaneous field amplitude and frequency for an ideally chirped and stretched light pulse

different gain profiles. This slice experiences almost the small-signal gain, and it starts to deplete the gain by deexcitation of molecules in the upper level. For the next pulse slice the gain, particularly the part originating from the "long wavelength fluorescent" dye, is already slightly depleted. This depletion is the origin of the red shift of the pulse maximum. The calculated spectrum is seen to agree, at least qualitatively, with the measured one (Fig. 8.12c). Note that the parameters have not been fitted to obtain better agreement between calculations and measurements. It should be mentioned that the sharp decrease in spectral intensity below 580 nm is due only to a minor extent to the strong absorption of ground state rhodamine molecules as assumed in [8.29] and mostly to gain depletion. The strong spectral overlap between the two fluorescence profiles (Fig. 8.12b) leads to early depletion of short wavelength dye and, hence, limits the spectral extension of the amplified pulses on the short wavelength side. For comparison we repeated the calculations using "artificial" fluorescence spectra of the two dyes with almost no spectral overlap where the mixture composed of these dyes exhibits the same gain curve as in the experiment. This procedure results in rather broad and structuresless ("flat") spectra with a half-width of about 70 nm (Fig. 8.12c) and in an instantaneous frequency that changes almost linearly with time.

Thus, in principle, it should be possible to obtain very short compressed light pulses. Of course, the difficult task of finding suitable compressors for such broad spectra remains.

Finally, we mention another reason for using stretched pulse amplification with dye gain media. This second reason originates from the steady-state situation in the occupation of electronic molecular levels under the simultaneous influence of pump radiation and amplified spontaneous emission (ASE) before the short signal pulse arrives. The transition probability for the deactivation of an excited molecule by ASE is given by

$$1/T_{\mathrm{ASE}} = \sigma I_{\mathrm{ASE}} \ ,$$

where I_{ASE} is the photon flux density of ASE and σ is the interaction cross section. From this equation we obtain the following approximation:

$$1/T_{ASE} \approx (1/\tau_F) G \eta_F \frac{\Delta\Omega}{4\pi} \sim 100 \text{ ps} \frac{1}{\eta} \text{ 1 ns}$$

for typical experimental suitations, where τ_F and η_F are the fluorescence lifetime and fluorescence quantum yield respectively; G is the gain in the ASE maximum and $\Delta\Omega$ is the solid angle in which the spontaneous emission experiences the gain G. For $T_{ASE} < \tau_F$ the energy storage time of the amplifier is given by T_{ASE3} and, hence, with long pump pulses ($\tau_{pump} > T_{ASE}$) only the part T_{ASE}/τ_{pump} is available for signal amplification.

In order to increase the energy extraction, the duration of stretched pulses has to be larger than T_{ASE}, and values of τ_{str} on the order of τ_{pump} are desired. Since the pump pulse duration may be as long as some nanoseconds, this second aim of stretched pulse amplification requires stretching factors of about 10 000 and higher, which are difficult to achieve.

Martinez [8.33] proposed the use of stretching devices that consist of grating pairs with a telescope placed between the two gratings and calculated stretching factors of the desired order of magnitude. After amplification a standard grating compressor compensates for the chirp produced in the stretcher.

8.5 Conclusions

The change of pulse parameters during amplification strongly depends on the relaxational and line broadening processes in the gain medium. Hence, the gain medium may be varied in order to obtain pulse parameters that are suitable for certain applications. Nowadays there is strong competition between dye devices and various other types of femtosecond amplifiers: Table 8.1 shows the parameters of the various types of amplifiers available at present and some trends in the construction of high power devices. Dye amplifiers were the first devices that could be used to produce powerful femtosecond light pulses. At present the

Table 8.1. Parameters of various types of amplifiers

Type	Pulse duration [fs]	Repetition rate [Hz]	Pulse energy
Dye	10	10	some mJ
		10^4	some µJ
Excimer	100	10	100 mJ → 1 J → 100 J
Solids:			
Nd:glass	100	1	1 mJ → 1 J → 1000 J
Ti:sapphire	50	10	1 mJ → 0.1 J → 1 J

highest pulse energy is obtained with KrF excimer lasers; in the near future 100 fs devices with 1 J energy will be available, and there are projects in progress to construct 100 fs/100 J amplifiers [8.34].

Recently very rapid progress has been achieved in the field of solid-state femtosecond amplifiers. In particular, Ti : sapphire devices offer the capability of a broad gain curve and high energy storage.

The advantages of dye amplifiers consist in the wide choice of available gain media by using solutions of various dyes and dye mixtures. Thus, gain profiles may be tailored and decisive parameters can be varied over large ranges.

As outlined above, when using the methods of chirped and stretched pulse amplification to avoid hole burning in the gain profile, dye devices provide the potential for the amplification of extremely short light pulses, and in this field they could well play a dominant part in the future.

References

8.1   C.B. Harris, E.P. Ippen, D.A. Mourou, A.H. Zewail (Eds.): *Ultrafast Phenomena VII*, Springer Ser. Chem. Phys. Vol. 53 (Springer, Berlin, Heidelberg 1990)

8.2   E. Klose, B. Wilhelmi (Eds.): *Ultrafast Phenomena in Spectroscopy*, Springer Proc. Phys. Vol. 49 (Springer, Berlin, Heidelberg 1990)

8.3   W. Rudolph, B. Wilhelmi: *Light Pulse Compression* (Harwood, London 1989)

8.4   R.L. Fork, C.H. Brito Cruz, P.C. Becker, C.V. Shank: Opt. Lett. **12**, 483 (1987)

8.5   S. Szatmari, F.P. Schäfer, E. Müller-Horsche, W. Mückenheim: Opt. Commun. **63**, 5 (1987)

8.6   P. Heist, W. Rudolph, B. Wilhelmi: Exp. Tech. Phys. **38**, 163 (1990)

8.7   R.L. Fork: Opt. Lett. **11**, 629 (1986)

8.8   R.N. Thurston, J.P. Heritage, A.M. Weiner, W.J. Tomlinson: IEEE J. Quant. Electron. **QE-22**, 682 (1986)

8.9   D. Grosenick, F. Noack, F. Seifert, B. Wilhelmi: paper presented at IQEC '91, Edinburgh 1991

8.10 J. Hebling, J. Kuhl: In Ref. [8.2] p. 67

8.11 J. Hebling, J. Kuhl: Opt. Lett. **14**, 278 (1989)

8.12 J. Herrmann, B. Wilhelmi: *Lasers for Ultrashort Light Pulses*, 2nd Ed. (Akademie, Berlin 1987; Elsevier, Amsterdam 1987)

8.13 L. M. Frantz, J. S. Nodvik: J. Appl. Phys. **34**, 234, 614 (1963)

8.14 J. C. Diels, W. Dietel, J. J. Fontaine, W. Rudolph, B. Wilhelmi: J. Opt. Soc. Am. B2, 680 (1985)

8.15 V. Petrov, W. Rudolph, B. Wilhelmi: Opt. Quantum Electron. **19**, 337 (1987)

8.16 V. Petrov, W. Rudolph, B. Wilhelmi: Opt. Commun. **64**, 398 (1987)

8.17 G. Angel, R. Gagel, A. Laubereau: Chem. Phys. **131**, 129 (1989)

8.18 C. H. Brito Cruz, R. L. Fork, W. H. Knox, C. V. Shank: Chem. Phys. Lett. **132**, 341 (1986)

8.19 C. H. Brito Cruz, J. P. Gordon, P. C. Becker, R. L. Fork, C. V. Shank: IEEE J. Quantum Electron. **QE-24**, 261 (1988)

8.20 W. Vogel, D. G. Welsch, B. Wilhelmi: Phys. Rev. **A37**, 3825 (1988)

8.21 W. Vogel, D. G. Welsch, B. Wilhelmi: Chem. Phys. Lett. **153**, 476 (1988)

8.22 F. Noack: Private communication

8.23 T. Damm, M. Kaschke, F. Noack, B. Wilhelmi: Opt. Lett **10**, 176 (1985)

8.24 D. Strickland, G. Mourou: Opt. Commun. **56**, 219 (1985)

8.25 D. Strickland, P. Maine, M. Bouvier, W. Williamson, G. Mourou: In *Ultrafast Phenomena V* ed. by G. R. Fleming, A.E. Siegmann, (Springer, Berlin, Heidelberg 1986) p. 38

8.26 P. Maine, D. Strickland, P. Bado, M. Pessot, G. Mourou: IEEE J. Quantum Electron. **QE-24**, 398 (1988)

8.27   W. H. Knox: IEEE J. Quantum Electron. **QE-24**, 388 (1988)
8.28   W. Dietel, E. Döpel, C. Rempel, W. Rudolph, B. Wilhelmi, G. Marowsky, F. Schäfer: Appl.
       Phys. **B46**, 183 (1988)
8.29   G. Boyer, M. Franco, J. P. Chambaret, A. Migus, A. Antonetti, P. Georges, F. Salin, A. Brun:
       Appl. Phys. Lett. **53**, 823 (1988)
8.30   W. J. Tomlinson, R. H. Stolen, C. V. Shank: J. Opt. Soc. Am. **B1**, 139 (1984)
8.31   R. Meinel: Opt. Commun. **47**, 343 (1983)
8.32   B. Wilhelmi: Ann. Phys. **43**, 355 (1986); Rev. Roum Phys. **33**, 995 (1988); and in *Trends in
       Quantum Electronics*, Ed. by I. Ursu, SPIE Proc. 1033, p. 147
8.33   O. E. Martinez: IEEE J. Quantum Electron. **QE-23**, 1385 (1987)
8.34   F. P. Schäfer: Project "SIMBA", Göttingen 1990

Bernd Wilhelmi

9. Terawatt-Class Hybrid Dye/Excimer Lasers

S. Szatmári

With 5 Figures

Dye lasers were not merely an addition to the already long list of lasers. They were the first lasers whose wavelength is easily tunable and – due to the large bandwidth of dyes – dye lasers made it possible to generate ultrashort pulses [9.1, 2].

It is a general tendency that the development of lasers is striving towards ever shorter wavelength. The short-wavelength limit of dye lasers is in the near ultraviolet, which is even more restricted for short pulses. By frequency conversion of dye laser pulses the wavelength range can be extended, but at the expense of significant decrease of the energy and stability. Direct generation of pulses in the UV is possible by excimers with relatively high efficiency, but only at certain wavelengths [9.3]. The bandwidth of the gain curve of excimers is significantly narrower than that of dyes, but still allows the amplification of subpicosecond pulses [9.4]. Among the various excimers XeF, XeCl, KrF and ArF are the most often used and are the main candidates for short pulse amplifiers, having their lasing wavelengths at 351, 308, 248 and 193 nm, respectively. The main drawback of excimers is that they have only been developed for 15 years, and short-pulse experiments were only performed in the last seven years. In excimers the pulse-shortening methods used for solid-state and dye lasers are hardly applicable, therefore excimers are mainly used for amplification of frequency converted short pulses, which are generated by means of dye lasers. The schematics of such a scheme is indicated in Fig. 9.1a, incorporating a short-pulse dye oscillator amplifier, frequency converter, and the UV excimer amplifier. Since the saturation energy of excimers is very similar to that of dyes (several mJ/cm^2) the principles of construction for the dye and excimer amplifier chain are similar. With such a hybrid dye/excimer laser one has a chance to get short ultraviolet pulses with comparable parameters to those generated by dye lasers in the visible. Since excimer amplifiers can easily be scaled up to high energies, this arrangement is not only a simple extension of the dye laser tuning range to a shorter wavelength, but also an effective way for the generation of high-power pulses in the TW range. Until recently such peak powers could only be reached at longer wavelengths by solid-state lasers, using the so-called chirped pulse amplification technique [9.5]. Since the limiting focusable intensity for the same number of photons scales with the third power of frequency, short pulse and short wavelength dye/excimer laser systems are the first candidates to reach intensities in the range $> 10^{20}$ W/cm^2, which were never previously obtained. This capability would certainly open a new strong-field regime of study for matter/field interaction.

Topics in Applied Physics, Vol. 70
Dye Lasers: 25 Years Ed.: Dr. Michael Stuke
© Springer-Verlag Berlin Heidelberg 1992

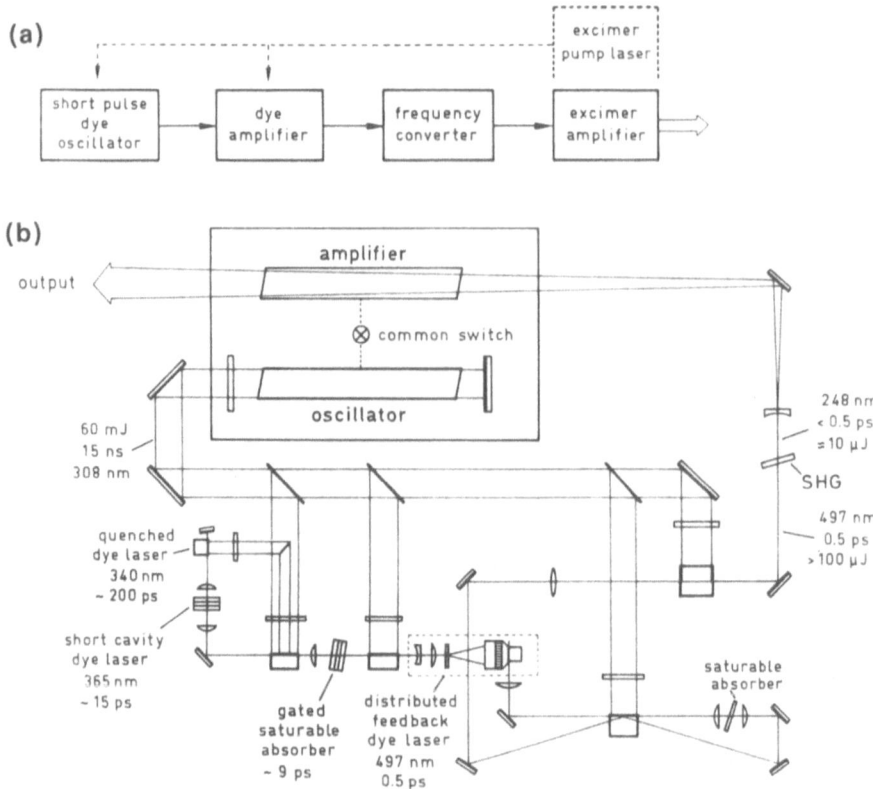

Fig. 9.1. (a) Schematic of a hybrid dye/excimer laser. **(b)** Experimental arrangement, using pulsed dye lasers for short pulse generation

In the following, some physical and technical aspects of a TW-class hybrid dye/excimer amplifier chain are studied, with special emphasis on the minimum achievable pulse duration, ASE contrast, phase and pulse front distortion, and on the highest energy extraction efficiency in excimers.

9.1 Input Pulse Generation by Pulsed Lasers

It is seen from Fig. 9.1a, that if the short pulses are generated in a classical way – by the use of cw mode-locked or CPM dye lasers – the whole arrangement is quite complex, where synchronous operation of numerous lasers is needed [9.6–10]. Due to the limited wavelength range of cw lasers, complicated frequency conversion schemes are necessary to reach certain excimer wavelengths [9.7–10]. These problems do not arise and the arrangement can be significantly be simplified when an excimer laser-pumped pulsed dye laser can be used for short pulse generation. The dashed lines in Fig. 9.1a represent such an arrangement where the same excimer laser is used for pumping as for amplification, providing automatic synchronization between the amplifier and the pulse to be amplified. In this case the only question is how the necessary sub-picosecond pulse duration can be produced starting with a 10-ns long pump pulse. In the case of cw mode-locked dye lasers, significant pulse shortening is achieved as a result of a stepwise evolution of the pulse shape during many round trips. Here stable pulse shortening in the range of 10^4–10^5 must be achieved within a single shot. In order to fulfil this stringent requirement, a special dye laser system has been developed. This laser system has had numerous forms in the last six years. The latest, most advanced version is shown in Fig. 9.1b [9.11, 12]. In this setup, a twin tube EMG 150 Lambda Physik excimer laser is used as a pump laser for pumping a special subpicosecond dye laser amplifier arrangement, and as an amplifier for the amplification of the frequency-doubled output pulses of the dye laser setup. The oscillator channel of the EMG 150 laser is filled with the standard XeCl fill, delivering 80-mJ, 15-ns pump pulses at 308 nm. The pump energy is distributed among the various dye cells as indicated in Fig. 9.1b. The pump beam coupled out by the first two quartz plates is used for pumping a newly developed, simple dye laser setup. This makes use of two cascade dye lasers, two amplifier stages, and a gated saturable absorber (GSA) in between the amplifiers. The combined pulse-forming effect of the oscillators, saturated amplifiers and the GSA results in an output pulse duration of ∼ 8 ps at 365 nm [9.13]. These pulses are then used for pumping the distributed feedback dye laser (DFDL) [9.14, 15] master oscillator. The DFDL used here is a tunable, achromatic arrangement, utilizing a transmission grating, a microscope objective and a special dye cell with the active medium [9.16]. In this DFDL the interference fringes, which are necessary for DFDL operation, are created by imaging a coarse transmission grating onto the inner surface of the dye cell by the use of a microscope objective. This arrangement ensures the creation of a small size, high-visibility interference pattern even with a pump beam of low spatial and temporal coherence. The other advantage of the arrangement is its easy tunability by translation of the transmission grating along the optical axis and by the proper choice of the refractive index of the active medium. In this way the whole visible spectrum can be covered [9.16]. The output pulse is then amplified up to several 100 µJ in a two-stage amplifier of standard design. When the wavelength of the DFDL is set to be twice the wavelength of the excimer amplifier, the amplified and frequency-doubled DFDL pulses can be used as seed pulses for amplification in the excimer amplifier. Frequency doubling is done just before the excimer amplifier. The slightly divergent ultraviolet pulse has an energy of order of ∼ 30 µJ. This pulse is then amplified in the second channel of the EMG 150, which is filled with the appropriate gas mix.

9.2 Short-Pulse Amplification Properties of Excimers

By double-pass amplification, the pulse energy is boosted up typically to 1, 5, and 10 mJ using XeF [9.17], XeCl [9.18] and KrF [9.11] gas fills, respectively. Both the amplitude and pulsewidth stability of the pulses is better than ± 5%. The output pulses exhibit an inherent chirp in all cases. This was the least pronounced for XeF, having practically the same 560-fs pulse duration for the direct output and for the compressed pulse [9.17]. In XeCl, pulse compression resulted in a more significant decrease of the pulse duration from 220 fs [9.18] to 170 fs [9.19]. The pulse shortening is especially pronounced at 248 nm, where

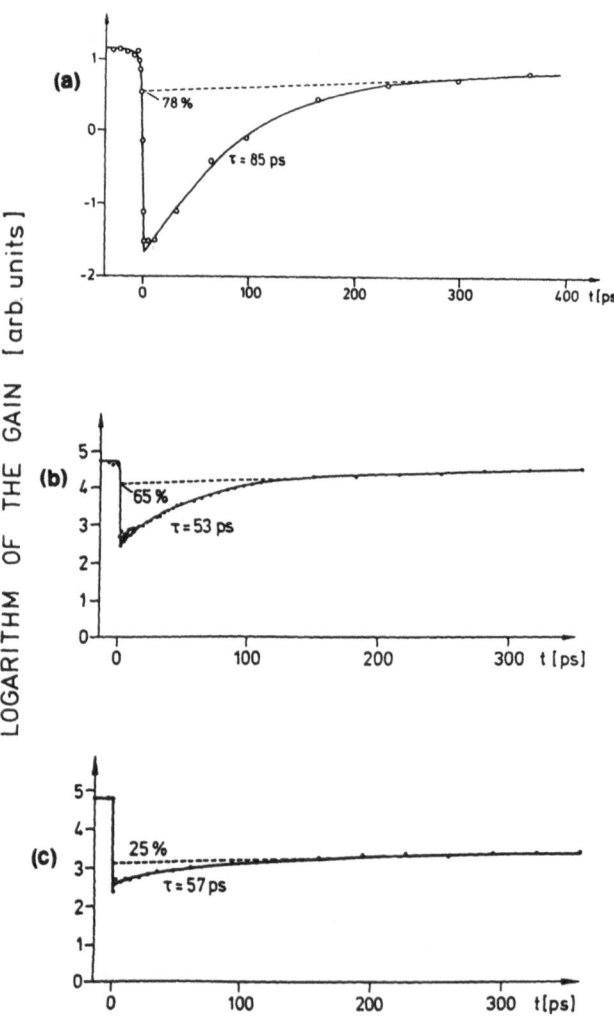

Fig. 9.2a–c. Recovery of the gain in (a) XeF, (b) XeCl and (c) KrF

the typical compressed pulse duration is below 100 fs [9.11, 20]. The shortest pulse duration obtained at 248 nm is 45 fs [9.12].

Because of the lack of a suitable frequency-doubling crystal, the 193 nm wavelength of ArF could only be reached by a more complicated frequency conversion scheme [9.21]. In a double pass amplification scheme – similar to the former experiments – 0.5-mJ pulses are typically obtained, with a compressed pulse duration of $\sim$ 300 fs [9.21].

It is seen from these results that – with the exception of ArF – the shortest possible pulse duration is reached for all the excimers, limited by the gain bandwidth. The different performance of the different excimers is characterized by the different values of the output energies from a given cross section and by the different pulse durations. A quantitative comparison of the excimers as short pulse amplifiers can only be obtained by the measurement of their gain dynamics. The gain-dynamics measurements were performed for XeF, XeCl and KrF using the well-known pump-and-probe technique. These measurements provided a wealth of information on the various relaxation processes contributing to the gain and made it possible to observe quantum beat in stimulated emission for the first time. Figure 9.2a–c, show the result of the gain-dynamics measurement for XeF, XeCl and KrF indicating a $\tau \approx$ 50–90 ps gain recovery with 78, 65, and 25% relative amplitude. It is found that in XeF only 22%, in XeCl 38%, while in KrF 75% of the stored energy is extractable for pulses shorter than $\sim$ 5 ps. This corresponds to the different measured values of the saturation energy density of 0.2, 0.85 and 2 mJ/cm^2, respectively.

From these results we concluded that for physical reasons KrF is superior to XeF and XeCl, and mainly for practical reasons is still more advantageous than ArF as a short pulse amplifier.

9.3 Critical Issues for a TW Excimer Amplifier

The development of the TW, KrF-amplifier chain showed that there are numerous specific problems to be considered when building and operating such a short-wavelength, large-aperture laser system. The most important ones are:

— shape of the phase front (focusibility),
— shape of the pulse front (spatially dependent temporal broadening across the beam),
— nonlinear effects (absorption, self-focusing) at the windows,
— ASE content, nonsaturable absorption (limitations on the cross section),
— limited energy storage time.

The importance of the first two points can be easily understood if one imagines that the output of a 100-fs/TW excimer laser system is a 30-μm-thick slice where the wavelength and the pulse length are only a small fraction of the beam cross section. In this case even the smallest distortions can easily be comparable to the wavelength or to the pulse length. The distortion of the phase front can be solved in principle by the use of well-corrected optics: a condition which is generally difficult to fulfil in practice for large aperture UV beams.

As shown in [9.22, 23], in the case of short-wavelength, femtosecond pulses and large aperture beams, the shape of the phase front and the pulse front can be significantly different, where the difference can easily be comparable to or bigger than the pulse duration. This deviation comes from the difference of phase and group velocities in optical materials, which introduces a delay (ΔT) between the phase and pulse fronts. At the same time, when a short pulse passes through a material, its duration is broadened. This implies that for refractive optical components having different thickness across the beam, the shape of the pulse front changes and the pulse duration becomes spatially modulated [9.22]. This can only be avoided by the use of achromats or preferably by reflective optics.

9.4 Spatially Evolving Chirped-Pulse Amplification

The power density of the unfocused output beam of the high-brightness KrF laser systems is in the range of 10 GW/cm^2. At these high-power densities, nonlinear optical properties of window materials become important [9.24]. Our investigations showed that practically all the UV transparent materials show increased absorption (2- and 3-photon absorption) for pulses at these intensities [9.25]. On the other hand, saturation of the amplifiers – and therefore a certain energy density at the output window – is needed in order to keep extraction efficiency high. This means that in the case of excimer lasers it is also desirable to manipulate the pulse duration – similar to the well-known chirped-pulse amplification (CPA) scheme applied for solid-state lasers [9.5] – but with a moderate (up to 10) stretching/compression factor. For this purpose, a special amplification scheme is developed where the chirp and the pulse duration develops in space [9.26], known as the spatially evolving chirped-pulse amplification scheme (SCPA) [9.27]. The basic idea behind the SCPA is the spatial evolution of the pulse duration after a dispersive element. This can be seen in Fig. 9.3a, where a grating is put into a positively chirped beam. Due to the angular dispersion of the grating, the different spectral components (shown by solid and dotted lines) are separated in a direction perpendicular to the propagation. Since the pulse front is tilted, the above displacement of the spectral components leads to spatial separation of the pulse fronts of the different spectral components also in a direction parallel to the propagation. It means a spatially evolving negative chirp. For the dispersion constant of the single dispersive element, one gets the same expression as is obtained for pulse compressors. This means that by the use of a single dispersive element in a properly chosen position, one can tilt the pulse front and can simultaneously compress the pulse duration of a positively chirped input pulse. It can be shown by simple geometrical considerations that in the plane of optimum pulse

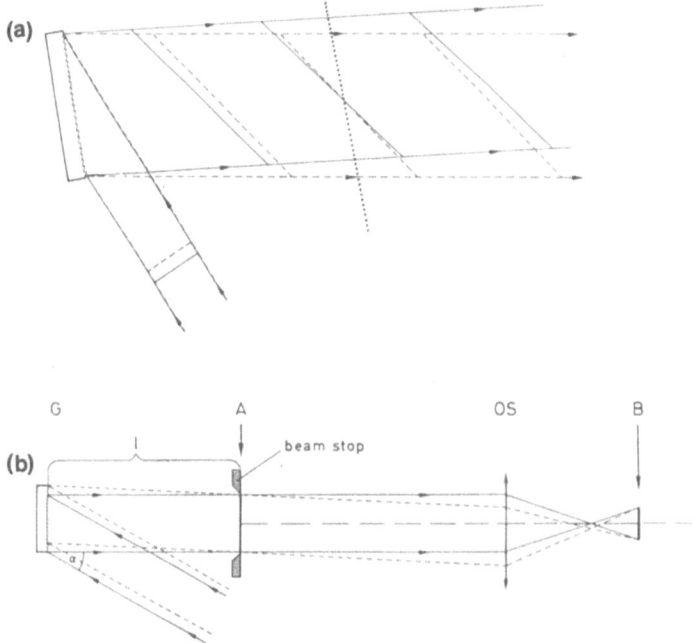

Fig. 9.3. (a) Schematic of the pulse fronts of the different spectral components after diffraction on a grating. **(b)** Schematic arrangement to realize spatially evolving chirped pulse amplification

compression – which is always parallel to the grating – the maximum pulse front tilt introduced by a single grating is limited: not enough for travelling wave excitation (TWE) [9.26]. We have developed several methods how to increase the pulse front tilt. The most straight-forward method is the use of optical imaging, as shown in Fig. 9.3b. The positively chirped incoming beam – after it has been diffracted by grating G – is optimally compressed at plane A, which is then imaged by an optical image systems OS to plane B. Since optical imaging leaves the absolute value of the delay between the two marginal rays unchanged, the increase of the tangent of the pulse front tilt is proportional to the demagnification. Assuming a perfect image system in the position of OS – which has to have the same pathlength between any pair of object and image points for any rays and any sectral components – the same compressed pulse duration is seen by plane B as by plane A, while the pulse duration in between these two planes (seen also by OS) is longer. If an amplifier system is inserted between plane A and OS (see Fig. 9.3b), it will then see long pulse duration and therefore moderate intensities. Moreover, a single lens following the amplifier can generate the ultimate shortest pulse duration at the target plane. The spatial evolution of the chirp and the pulse duration makes it possible to simulate the conventional CPA schemes, however in a much simpler manner: without the use of pulse stretchers and compressors, and only using the previously described spatial evolution of the pulse duration. By the use of the SCPA scheme, it is achieved that the amplifier is working at moderate intensities, while the minimum compressed pulse duration with a variable pulse front tilt can be created at the target plane, using only a single focusing element following the amplifier. This can have great importance in large-scale TWE experiments such as X-ray laser experiments.

9.5 Off-Axis Amplification and Phase-Locked Multiplexing

It has already been shown by gain dynamics measurements why KrF is the first candidate as an efficient amplifying medium. In that case, extraction efficiency was defined as the ratio of the extractable energy for short ($\tau < 5$ ps) and long ($\tau > 100$ ps) pulses, with regard to the experimentally observed ~ 50 ps long recovery of the gain. In practical cases, however, one has to consider more effects, which significantly influence the efficiency, such as ASE, nonsaturable absorption, and limited storage time of the amplifying medium.

In case of a short-wavelength laser system – owing to a short energy storage time – ASE is often a limiting factor. Since at the excimer wavelengths no effective saturable absorbers are available, spatial filtering is the only way to avoid gain depletion, while leaving the ASE content in the direction of the signal unchanged. This content can only be reduced by proper choice of the operational conditions of the amplifier, generally at the expense of efficiency.

Unfortunately, discharge-pumped KrF amplifiers are only efficient when the electric excitation is fast, necessitating a small discharge loop. This limits the cross section of the discharge and prefers elongated, pencil-like pumped geometries. This geometry is not fitted to short-pulse amplifiers, having significant nonsaturable absorption. This problem will be examined in a specific example: in the case of KrF, having $g_0/\alpha = 10$ (where g_0 is a small-signal gain, α is the absorption coefficient). Figure 9.4a shows the local extraction efficiency and the contrast coefficient [defined as $g_{eff}/g_0 - \alpha$, where $g_{eff} = \lim_{\Delta L \to 0} (1/\Delta L) \ln (E_{out}/E_{in})$] as a function of the energy density for KrF [9.28]. Since in many applications the contrast is of comparable importance to the extraction efficiency, the operation of KrF amplifiers generally must be optimized both for efficiency and contrast. However, in this case the optimum operational conditions belong to energy densities, which are around $\varepsilon_{opt} \approx \varepsilon_{sat}$ (see Fig. 9.4a). If optimized operation is required, the momentary stored energy of the amplifier – given by the $E_{stored} = \varepsilon_{sat} A g L$ equation – should not be significantly more than the optimum value of the extractable energy ($E_{out} = \varepsilon_{sat} A$). If we set this ratio to $E_{opt}/E_{stored} = 0.5$ (suggested by the efficiency curves) an upper limit of $gL = 2$ is obtained, for the maximum gain-length product of an "efficiently used" discharge pumped KrF amplifier. This corresponds to $L = 10$ cm, for a typical case, when $g = 0.2$ cm^{-1} [9.21].

It is seen that from the optical point of view a "short" amplifier is ideal; on the other hand, the cross section is limited by the fast pump circuit. These

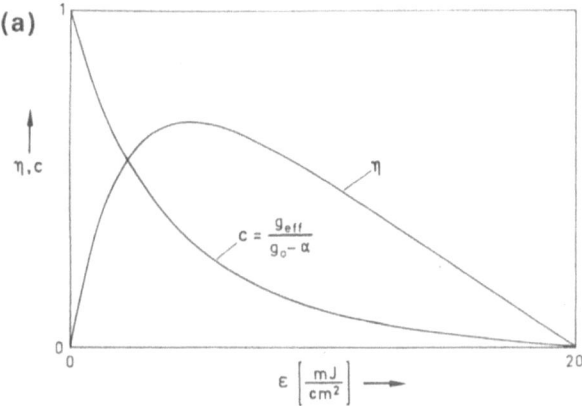

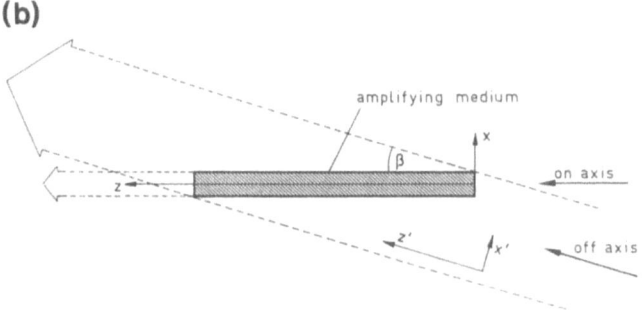

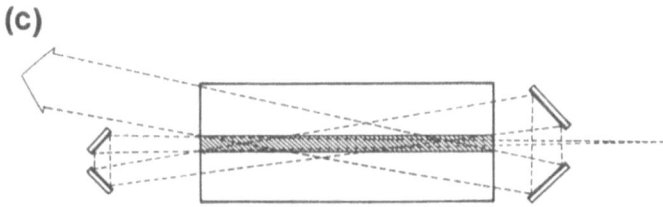

Fig. 9.4. (a) Calculated values of the local extraction efficiency η and the contrast coefficient c as function of energy density in a KrF amplifier. **(b)** The off-axis amplification geometry. **(c)** Experimental arrangement for a three-pass off-axis amplification

requirements would certainly limit the active volume and therefore the maximum stored energy, unless a transformation is found, which converts properly the "geometrical" cross section and length of the amplifier into a new cross section and length, which are seen by the pulse to be amplified. Figure 9.4b shows a simple solution where the effective cross section of the pencil-like discharge-pumped excimer amplifier is increased by tilting the input beam with respect to the geometrical axis of the amplifier [9.28, 29]. This off-axis arrangement is also capable of solving the inherent inhomogeneity problem of these amplifiers, caused by the inhomogeneous transversal distribution of the deposited energy in the discharge. In the off-axis mode, it is the more homogeneous longitudinal distribution, which mainly determines the homogeneity of the output beam. Therefore a more homogeneous, nearly flat-topped distribution can be obtained after amplification.

Another interesting feature of this off-axis amplification scheme is that it makes possible optimization of the operational conditions not only for a single amplification pass, but also for more passes using the same amplifier [9.28, 29]. A practical example for a three-pass arrangement is shown in Fig. 9.4c, where the increasing beam diameter is always optimally matched to the increased effective cross section of the amplifier by proper choice of the divergence angle of the beam, and the off-axis angles of the different passes. It can be achieved that each of the amplification passes are in the earlier defined optimized condition, which makes it possible to extract optimally the stored energy of the same amplifier in subsequent steps.

In excimer amplifiers – due to the short storage time of the active medium – this successive replenishment of the momentarily stored energy is the only way to have access to the whole stored energy, which is given roughly by the $E_{tot} = E_{stored} \, T/\tau$ equation, where T is the gain window and τ is the recovery time of the gain. Since for KrF $T/\tau \approx 10$, significant increase of the extractable energy is expected by successive depletion of the gain. Normally this is done by the optical multiplexing schemes. In optical multiplexing, the beam to be amplified is split into partial beams, which are arranged in the multiplexer to form a pulse train with a separation comparable to the recovery time of the amplifier. After amplification the partial beams are recombined to form again a single pulse (demultiplexing). The key point of any kind of multiplexing is how accurately the recombination of the partial beams can be done. In the conventional multiplexing schemes, recombination is far from that of interferometric accuracy, which imposes severe limitations on the focusibility of the final beam. Synchronism can generally be obtained only for longer than picosecond pulses.

It has been shown that the multiple-pass off-axis amplification technique can substitute optical multiplexing for a limited number of amplification passes (typically 3). However, this is still far from the optimum number of amplification steps, which is around 10 for KrF. For this reason a new multiplexing method is being developed, which does not suffer from the earlier listed shortcomings of the conventional ones. This novel method is based on a common optical arrangement used both for multiplexing and for demultiplexing (Fig. 9.5a). In this way automatic (phase-locked) synchronization of the partial beams is achieved for any alignment of the multiplexer (demultiplexer), and any kind of distortion and/or misalignment is automatically compensated. With this scheme, optimum focusibility is expected for the output, where the minimum focal spot size is already determined by the diffraction limit of the whole, recombined beam. This significantly increases the brightness and therefore the maximum focusable intensity. Figure 9.5b and c show a possible realization for

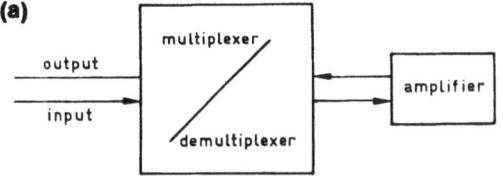

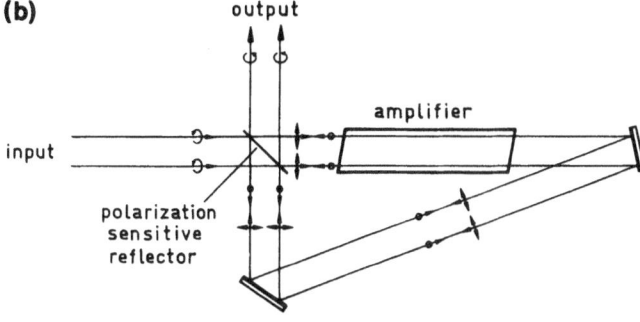

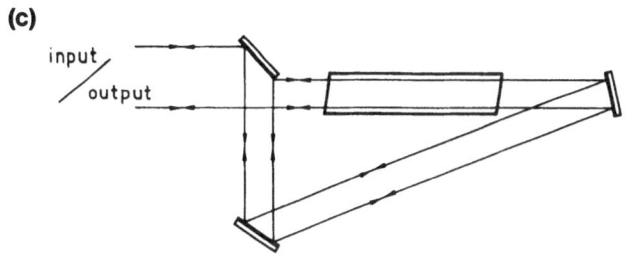

Fig. 9.5. (a) Experimental arrangement, where the same optical device is used both for multiplexing and for demultiplexing. (b, c) Possible realizations, using (b) polarization or (c) geometrical splitting

the case of two multiplexed beams, using a Sagnac-interferometer with polarization (b) or geometrical splitting (c). Each arrangement can be scaled up to multiple beams and for multiple-pass (off-axis) amplification geometries. Details of these questions will be published elsewhere [9.30].

Work is in progress to further develop and finally combine these ideas in order to extract all the stored energy of excimers by a femtosecond pulse in a diffraction limited beam, which could lead to a table-top, TW dye/excimer laser system, giving the highest focused intensity ever achieved by lasers.

Acknowledgements. The author wishes to thank F. P. Schäfer, G. Kühnle, P. Simon, G Almási, H. Gerhardt and Q. Zhao for their contributions to the results presented here.

140   *S. Szatmári*

References

9.1   F.P. Schäfer (ed.): *Dye Lasers*, Topics Appl. Phys., Vol. 1 (Springer, Berlin, Heidelberg 1990) pp. 1–89
9.2   P.P. Sorokin, J.R. Lankand, E.C. Hammond, V.L. Moruzzi: IBM J. Res. Dev. **11** (1967)
9.3   C.K. Rhodes: *Excimer Lasers* (Springer, Berlin, Heidelberg 1979)
9.4   I.A. McIntyre, C.K. Rhodes: J. Appl. Phys. **69**, R1 (1991)
9.5   P. Maine, D. Strickland, P. Bado, M. Pessot, G. Mourou: IEEE J. Quant. Electron. **QE-24**, 398 (1988)
9.6   J.M. Glownia, J. Misewich, P.P. Sorokin: J. Opt. Soc. Am. B **3**, 1573 (1986)
9.7   A.P. Schwarzenbach, T.S. Luk, I.A. McIntyre, U. Johann, A. McPherson, K. Boyer, C.K. Rhodes: Opt. Lett. **11**, 499 (1986)
9.8   J.P. Roberts, A.J. Taylor, P.H.Y. Lee, R.B. Gibson: Opt. Lett. **13**, 734 (1988)
9.9   M. Watanabe, A. Endoh, N. Sarukura, S. Watanabe: In Ultrafast Phenomena VI, Springer Ser. Chem. Phys. Vol. 48 ed. by T. Yajima, K. Yoshihara, C.B. Harris, S. Shinoya (Springer, Berlin, Heidelberg 1988) p. 87
9.10   J.R.M. Barr, N.J. Everall, C.J. Hooker, I.N. Ross, M.J. Shaw, W.T. Toner: Opt. Commun. **66**, 127 (1988)
9.11   S. Szatmári, F.P. Schäfer: Opt. Commun. **68**, 196 (1988)
9.12   S. Szatmári, F.P. Schäfer: *Excimer Lasers and Applications*, SPIE Vol. 1023 (1989) p. 71
9.13   S. Szatmári: Opt. Quantum Electron. **21**, 55 (1989)
9.14   H. Kogelnik, C.V. Shank: Appl. Phys. Lett. **18**, 152 (1971)
9.15   Zs. Bor: IEEE J. Quantum Electron. **QE-16**, 517 (1980)
9.16   S. Szatmári, F.P. Schäfer: Appl. Phys. B **46**, 305 (1988)
9.17   Q. Zhao, S. Szatmári, F.P. Schäfer: Appl. Phys. B **47**, 325 (1988)
9.18   S. Szatmári, B. Rácz, F.P. Schäfer: Opt. Commun. **62**, 271 (1987)
9.19   Q. Zhao, F.P. Schäfer, S. Szatmári: Appl. Phys. B **46**, 1339 (1988)
9.20   S. Szatmári, F.P. Schäfer, E. Müller-Horsche, W. Mückenheim: Opt. Commun. **63**, 305 (1987)
9.21   S. Szatmári, F.P. Schäfer: Opt. Soc. Am. B **6**, 1877 (1989)
9.22   S. Szatmári, G. Kühnle: Opt. Commun. **69**, 60 (1988)
9.23   Zs. Bor: Opt. Lett. **14**, 119 (1988)
9.24   A.J. Taylor, R.B. Gibson, J.P. Roberts: Opt. Lett. **13**, 814 (1988)
9.25   P. Simon, H. Gerhardt, S. Szatmári: Opt. Lett. **14**, 1207 (1989)
9.26   S. Szatmári, G. Kühnle, P. Simon: Appl. Opt. **29**, 5371 (1990)
9.27   S. Szatmári, P. Simon, H. Gerhardt: Opt. Commun. **79**, 64 (1990)
9.28   G. Almási, S. Szatmári, P. Simon: to be published in Opt. Commun.
9.29   S. Szatmári, G. Almási, P. Simon: Appl. Phys. B**53**, 82 (1991)
9.30   S. Szatmári: to be published

Sandor Szatmári

10. Blue-Green Dye Laser Seeded Operation of a Terawatt Excimer Amplifier

F.K. Tittel, T. Hofmann, T.E. Sharp, P.J. Wisoff, W.L. Wilson and G. Szabó

With 5 Figures

A tunable, subpicosecond dye laser oscillator–amplifier system, generating 2-mJ pulses in the blue-green spectral region, has been developed. It consists of a hybrid synchronously pumped oscillator followed by a specially designed, high-performance, two-stage amplifier system. The pulse duration is either 800 fs or, by addition of a fiber-prism compressor, 250 fs. The ASE content of the dye oscillator–amplifier system is as low as 0.1%. The system is used to study the short pulse spectral and temporal gain characteristics of the $XeF(C \rightarrow A)$ amplifier. A saturation energy density of $50 \, mJ/cm^2$ for 250-fs pulses and a gain bandwidth of 60 nm were measured for the $XeF(C \rightarrow A)$ excimer transition. Generation of terawatt pulses (275 mJ, 250 fs) with the $XeF(C \rightarrow A)$ excimer amplifier was demonstrated.

10.1 Introduction

In recent years there has been an ever growing interest in the development of laboratory-scale, ultrahigh-power short-pulse laser systems [10.1–10]. However, an optimal solution to some of the fundamental problems encountered in high-power optical amplifiers still needs to be found. In laboratory-scale systems, terawatt levels can only be achieved by using short pulses. In the short-pulse regime, however, the extractable energy density is limited by the saturation energy density of the amplifying medium.

Up to now there are two fundamental approaches in solving this problem: one can either use active media with low saturation density (such as XeCl and KrF excimers with typical saturation densities in the mJ/cm^2 range) but with the potential of building large-cross-section modules (up to $1000 \, cm^2$), or one can choose an active medium with a large saturation density and relatively small cross section.

The first approach is successfully used by several groups [10.1–7], and peak powers as high as 4 TW have been generated [10.7]. For the second approach, the most promising candidates are solid-state media with saturation densities in the J/cm^2 range. This technique, however, originally suffered from the inherent large optical nonlinearities of solid-state media, which can cause catastrophic self-focusing at high intensities. However, a breakthrough was achieved by the

Topics in Applied Physics, Vol. 70
Dye Lasers: 25 Years Ed.: Dr. Michael Stuke
© Springer-Verlag Berlin Heidelberg 1992

introduction of chirped-pulse amplification (CPA) to these systems [10.8–10], resulting in the generation of 20 TW of output power [10.10].

This paper describes an alternative approach to high-power pulse amplification that employs a XeF(C–A) excimer amplifier. This amplifier has a saturation energy density of more than an order of magnitude larger than conventional UV excimers, but which is also scalable to large amplifier apertures. After 10 years of development, the XeF(C–A) technology has reached maturity, and a Joule-class amplifier module with nanosecond injection control has been demonstrated [10.11]. This makes it possible to study the gain characteristics and the optimization of energy extraction of a subpicosecond XeF(C–A) amplifier [10.12].

The highly repulsive lower state of the XeF(C–A) excimer transition results in two properties which are highly desirable for short-pulse, high-power amplification. The first one is that the saturation energy density, calculated from the stimulated emission cross section reported in [10.13], is expected to be $50\,mJ/cm^2$. The other, equally important property of the XeF(C–A) transition is that, compared to conventional excimers, it has an extremely broad (60 nm) gain bandwidth which theoretically is sufficient to support pulse durations as short as 10 fs.

Up to now, apart from our recent investigation [10.12], neither the saturation characteristic nor the gain bandwidth has been studied by direct short pulse measurements. Both for saturation studies and for energy extraction measurements, a subpicosecond, blue-green dye laser system is required. The requirements for such a laser system are rather stringent. For XeF(C–A) saturation measurements, because of the expected large saturation values, injection fluences in the range of $150\,mJ/cm^2$ are needed. This, however, cannot be achieved simply by tight focusing because the Rayleigh length of the probe beam should be larger than the 50-cm-long excimer amplifier cell. Therefore, the dye laser system should be able to generate subpicosecond pulses with energies in the mJ range. For the energy extraction experiments, because of the low single-pass gain, a similarly high-pulse energy is necessary. In addition, the level of the amplified spontaneous emission (ASE) from the injection laser should be extremely low. To meet these requirements a high-power subpicosecond dye laser system has been developed [10.14].

10.2 Dye Laser System

10.2.1 Oscillator Section

The schematic diagram of the dye laser system is depicted in Fig. 10.1. The oscillator was a modified blue-green Coherent 702 mode-locked dye laser, synchronously pumped by the third harmonic of a CW mode-locked Nd:YAG laser (Coherent Antares 76-s). The original pump system used KTP and BBO

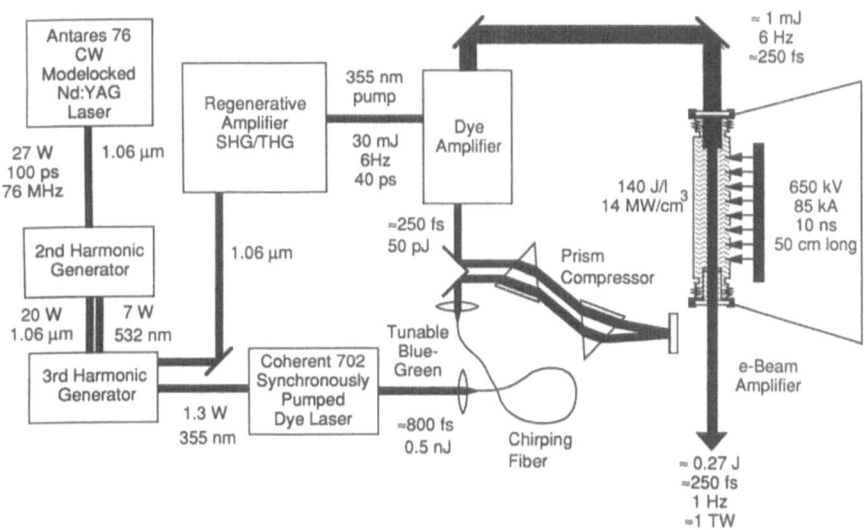

Fig. 10.1. Schematic diagram of the subpicosecond blue-green dye laser system

crystals for frequency doubling and tripling, respectively, and provided 1 W of output power at 355 nm. Since this is only slightly above the threshold (800 mW) of the dye laser, the stability was not satisfactory. By replacing both nonlinear crystals by LBO crystals, the system performance was improved significantly to yield a pump power of 1.3 W at 355 nm. The oscillator then was capable of generating stable 800-fs pulses, tunable from 465 to 515 nm when using coumarin 480 and DOCI as an amplifier and saturable absorber dyes, respectively. The average power of the dye laser oscillator at a repetition rate of 76 MHz was 20–40 mW, depending on wavelength.

The output from the oscillator was amplified either directly, or after being sent through a fiber and prism compressor. In the compressor section, the pulse coming from the oscillator was chirped by a 92.7-cm-long polarization-preserving fiber designed for the blue-green spectral region (Newport F-SPA), and compressed by a two-prism system. The prismatic compressor consisted of two 60° SF-14 prisms used in a double-pass arrangement. The optimum prism separation was found experimentally to be 95 cm. When operating under optimal conditions, the compressor produced 250-fs pulses. It is important to note, however, that the 250 fs should be considered as an upper bound for the pulse duration, because the BBO crystal, used in the multiple-shot background-free autocorrelator (Inrad 514-A), was 0.5 mm thick, which may lead to apparent pulse broadening in the autocorrelation measurements [10.15]. The tuning range was restricted to above 488 nm by the cutoff wavelength of the fiber. The overall transmission of the compressor (fiber + prisms) was 20%.

10.2.2 High-Performance Subpicosecond Dye Amplifier System

For pulse amplification to the millijoule level, we have designed a special dye amplifier system. In this system, contrary to previous millijoule subpicosecond amplifiers that have generally relied on a three-stage configuration [10.16], we used a two-stage design. In systems operating in the red, a three-stage design can be afforded because the ASE can be suppressed by insertion of saturable absorbers in the amplifier chain. In the blue-green region, where no good saturable absorbers are available, the only way for ASE suppression is spatial filtering. However, spatial filtering can effectively be used only once, because after the first spatial filter the divergence of the signal and the ASE is the same. Another conceptual point of our design was to use short pump pulses for the amplifiers because in this case the ASE can be further decreased by proper synchronism between the pump and signal pulses.

The two-stage amplifier system presented here can be used in any application provided that the input energy is high enough to drive the first amplifier into saturation. The minimum necessary energy can be estimated as follows: our experimental experience supports the assumption that the smallest practical diameter and highest achievable gain at the first cell is 0.1 mm and 10^5, respectively. Then, allowing for $10 \, \text{mJ/cm}^2$ output energy density, which is about 5 times higher than the saturation energy density of typical dyes and represents a good practical compromise between pulse broadening and energy extraction efficiency [10.17], the minimum input energy is 0.01 nJ. This obviously covers all kinds of practical dye lasers.

The dye amplifier was pumped with 40-ps-long, frequency-tripled pulses of a Nd:YAG regenerative amplifier which was seeded by the residual fundamental beam after frequency tripling of the CW mode-locked Nd:YAG laser. The regenerative amplifier (Continuum Corp. RGA-69) was configured with a closed cavity amplifier and one additional single-pass module. This system produced 250-mJ, 1064-nm pulses at 6 Hz repetition rate and pulses of 40 mJ energy at 355 nm after third-harmonic generation.

The experimental layout of the subpicosecond dye laser and amplifier can be seen in Fig. 10.2. The beam waist of the input dye laser was transformed by a 450-mm achromatic lens to match the cross section of the first amplifier stage. The first amplifier stage consisted of an 8-mm-long, dye flow cell pumped by 10% of the total pump energy via an adjustable delay line. The diameter of the dye laser beam at the first cell was 0.2 mm, corresponding to an injected energy density of $2 \times 10^{-7} \, \text{J/cm}^2$ (without pulse compressor). The output energy of the first stage was 25 μJ, representing an output energy density of $20 \, \text{mJ/cm}^2$. To minimize pulse broadening, care was taken to operate both amplifier stages at about the same level of saturation.

ASE suppression was accomplished by spatially filtering the output of the first stage with an achromatic telescope of 10 times magnification and a 50-μm pinhole as depicted in Fig. 10.2. The telescope expanded the dye beam to fill the 2-mm aperture of the dye flow cell of the second amplifier. The construction of

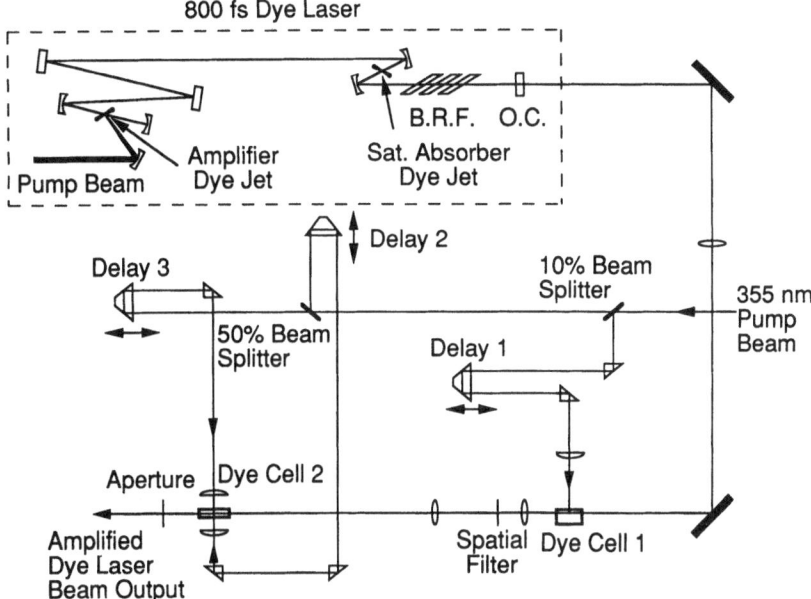

Fig. 10.2. Experimental layout of the dual jet dye laser and its two-stage dye amplifier. BRF: birefringence filter, OC: output coupler

the dye cell allowed symmetrical, double-sided transverse pumping to achieve a uniform energy deposition in the dye. The pump beam for the second stage was divided into two equal parts which could be adjusted by independent delay lines. Two independent delay lines were required, because the pump timing sensitivity was measured to be 10 ps. Hence, in this design the use of Bethune type prismatic dye cells [10.18] should be avoided. At the amplifier output the off-axis ASE was removed by means of an aperture.

The dye oscillator–amplifier system delivered more than 1 mJ pulse energy in the range from 470 to 510 nm, with a maximum of 2 mJ at around 505 nm. The output energy in the 470–480 nm region is due to the lower performance of the coumarin 480 dye, since in this range the oscillator provided 2 times higher input signals to the amplifier than around 500 nm.

While the pulse duration of the oscillator was continuously monitored by a multiple-shot autocorrelator, the amplified output was studied by a single-shot, phase-resolved autocorrelator [10.19]. The pulse duration measurements revealed that no significant pulse broadening in the dye amplifier occurs. The best fit for the autocorrelation trace was achieved by assuming an asymmetric exponential pulse shape with a rise/fall ratio of 1:5 and a pulse duration of 800 fs.

The spatial far field beam profile was close to Gaussian with a slightly elliptical shape in horizontal direction. The divergence was nearly diffraction-limited.

By insertion of the fiber and prism compressor stage, the input energy to the amplifier dropped to about 20%, but since the amplifier system was operating in saturation, this resulted only in a less than 50% decrease in output energy. The highest output energy with 250-fs pulses was 1.2 mJ.

As mentioned earlier, the ASE content of the dye laser pulses is a critical issue for the XeF(C–A) studies. For a measurement of the ASE level, the output of the dye laser was spectrally dispersed so that the spectrum filled the cathode of a photodiode. The injected dye laser beam was then separated from the ASE by a beam stop in front of the diode. Using this method, the ASE was found to be less than 0.1% of the output energy.

In order to measure the XeF(C → A) gain with picosecond pulses, a simple dye laser oscillator was designed that consists of a Littrow grating, a transverse-flow dye cell and an output coupler. The oscillator and a subsequent dye amplifier stage were pumped by the 40-ps-long, third-harmonic output of the regenerative amplifier used for the femtosecond dye laser system described above. This laser produced pulses of ∼ 100 ps duration and 2 mJ maximum energy and, by using different dyes, the entire XeF(C → A) spectrum from 450 to 530 nm could be probed.

10.3 XeF(C–A) Experiments

10.3.1 Experimental Arrangement

A detailed description of the XeF(C → A) excimer laser system is given in [10.11]. The XeF(C–A) excimer gas mixture consisted of 12 Torr Xe, 1 Torr F_2, 12 Torr NF_3, 750 Torr Kr and 4100 Torr Ar, which was transversely excited by energetic (650 keV), short duration (10 ns) electron beam pulses. All optical elements were made of fused silica to reduce induced absorption effects [10.13], except for the 0.375″ thick excimer chamber windows, for which MgF_2 was chosen to avoid self-focusing.

For gain measurements, the dye laser injection signal was sent in a single pass through the excimer gas cell, attenuated by neutral density filters to probe different energy densities. The laser input and output energies were monitored simultaneously by vacuum photodiodes, calibrated relative to each other before gain measurements. The photodiodes were also used to monitor the relative timing between the injection signal and the electron-beam excitation. Additionally, the energy of the amplified pulses was observed by a pyroelectric energy meter. The output beam profile at the gas cell window was imaged by a lens onto a CCD array and recorded for each pulse.

10.3.2 Gain Measurements

A spectrum of the free-running XeF(C → A) excimer laser [10.20] is shown in Fig. 10.3. The broadband laser spectrum is interrupted by several narrow-band

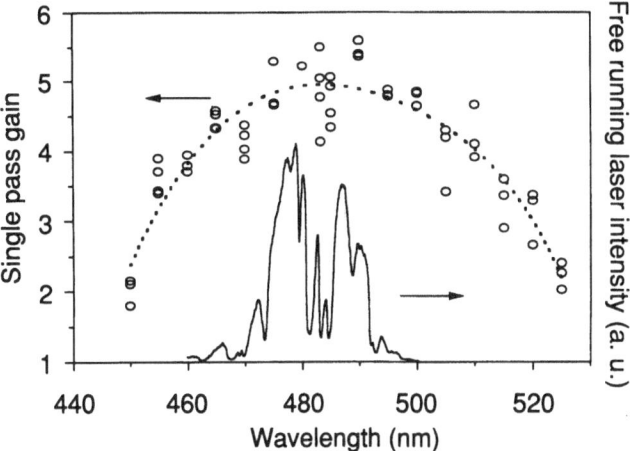

Fig. 10.3. Spectral dependence of the XeF(C → A) excimer single-pass gain for injection pulses of ~ 100 ps duration. The gain length is 50 cm, the injected energy density is 1 mJ/cm². The spectrum of the free running, nanosecond laser is shown for comparison. The gain spectrum shows a smooth profile with no apparent influence of narrow-band transient atomic absorbers, visible in the laser spectrum

absorption lines due to excited argon, krypton and xenon atoms created as by-products of the electron beam excitation [10.21]. These absorbers limit the gain bandwidth and therefore restrict amplification of ultrashort pulses. However, if it is possible to saturate these absorbers the full bandwidth is available for amplification. Most of the absorbing species have a larger cross section than the stimulated emission cross section of the XeF(C → A) transition and therefore saturate at smaller energy densities than the gain. This effect has been demonstrated in nanosecond gain measurements at a narrow-band absorber wavelength, where the gain actually increases with higher injected intensities [10.21].

In an attempt to characterize the bandwidth of the XeF(C → A) excimer with short pulse injection, which was investigated earlier for nanosecond pulses [10.20], the gain was measured with 100-ps pulses over a wide spectral region. The single pass gain for 100-ps, 1-mJ/cm² pulses, also shown in Fig. 10.3 is characterized by a smooth profile over a bandwidth of 60 nm. Similar to the results obtained with 800-fs pulses, it is also possible with 100-ps pulses to reach an energy density sufficiently high to saturate narrow-band absorbers and to make the entire XeF(C → A) bandwidth accessible to tunable high power generation or ultrafast, large bandwidth amplification.

The dependence of the gain for 250-fs, 800-fs and 100-ps pulses is shown in Fig. 10.4. The gain measurements were performed in a single pass through the 50-cm-long active gain medium at 490 nm, which corresponds to a maximum of the free-running XeF(C → A) laser spectrum. The single-pass gain for the 100-ps and 800-fs pulses was measured to be 5.7, which translates to a gain coefficient of $0.034\,\text{cm}^{-1}$. Using the Frantz–Nodvik model [10.22] the saturation energy

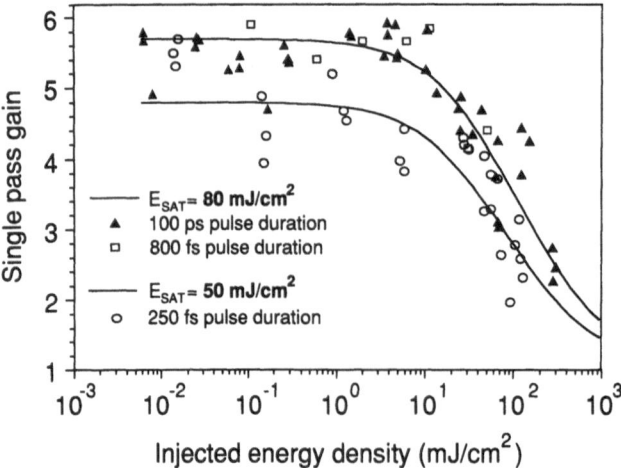

Fig. 10.4. The dependence of single-pass gain of a 50-cm-long XeF(C → A) excimer amplifier for 250 fs, 800 fs and ~ 100 ps pulse durations on the injected energy density at a wavelength of 490.5 nm. The Frantz–Nodvik curves were fitted taking into account beam-broadening effects

density was calculated to be $80 \, \text{mJ/cm}^2$ for both pulse durations. The identical saturation behavior for both pulse durations implies that within the experimental accuracy no significant repumping of the gain occurred on a 100 ps time scale.

Measurements with 250 fs probe pulses showed a slightly lower small-signal gain of 4.8, corresponding to a gain coefficient of $0.032 \, \text{cm}^{-1}$ and a saturation energy density of $50 \, \text{mJ/cm}^2$. A decrease of these gain parameters for 250-fs pulses was expected, since the rotational reorientation time of the XeF(C) molecule, estimated to be approximately 0.8 ps prevents complete utilization of the gain medium [10.12].

10.3.3 Energy Extraction

Energy extraction from the XeF(C → A) excimer amplifier for 250-fs, 490-nm pulses was investigated using a specially designed unstable resonator. The unstable resonator was of the positive-branch, confocal type with a magnification of $M = 4$. The injected beam made a total of five passes through the gain medium, resulting in a gain of ~ 1000. The dependence of the output energy of the XeF(C → A) amplifier upon the injected energy is shown in Fig. 10.5. A maximum output energy of 275 mJ was obtained with an injection energy of 0.5 mJ. The highest achieved energy was about one quarter of the maximum energy of 1 J observed for nanosecond injection [10.23], which represents a good efficiency for a femtosecond amplifier. The solid line in Fig. 10.5 depicts the output energy calculated by a numerical model, based on the gain saturation

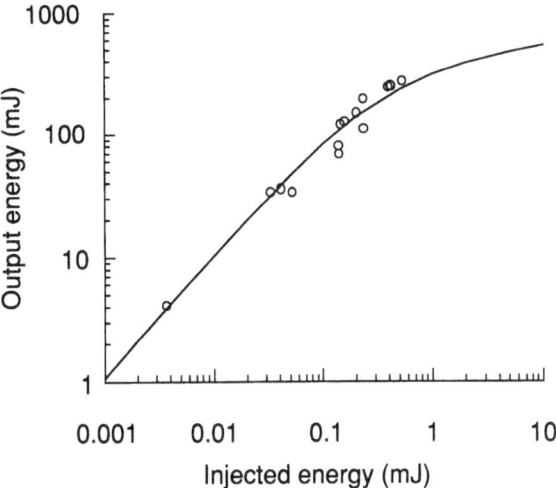

Fig. 10.5. The dependence of the output energy of the XeF(C → A) excimer amplifier on the injected energy for 250-fs pulses at a wavelength of 490.5 nm. The solid line is calculated from gain saturation measurements

and gain lifetime [10.13] measurements. It is apparent that the amplifier is driven into moderate saturation for maximum output, resulting in good energy extraction efficiency. Deep saturation, however, was avoided in order to prevent temporal pulse broadening.

An upper level for the ASE energy was measured at a distance of 4 m by blocking the injection dye laser. The detected energy was smaller than 1 mJ, corresponding to an ASE level of < 0.4%. Further reduction of the ASE level can be achieved by spatial and spectral filtering of the amplified beam and by the use of a saturable absorber. The autocorrelation measurements of the injected and amplified pulses showed large shot-to-shot fluctuations of the pulse duration. Minimum pulse durations of 250 fs after amplification were recorded, indicating no systematic temporal pulse broadening by the amplifier. The 250 fs pulse width should again be considered as an upper bound for the pulse duration because the BBO crystal used in the single-shot autocorrelator was 1 mm thick. Furthermore, contrary to the multiple-shot autocorrelators, the exact effect of the crystal length on the measured pulse duration has not yet been worked out for the single-shot case.

The focusability of the amplified beam was measured by observing the focal spot of a 4-m lens with a CCD camera. A $1/e^2$ spot diameter of 160 μm for the unamplified beam was measured, which corresponds to a diffraction-limited beam taking into account the top hat, torus-shaped beam profile. After amplification the beam focal spot increased to 200 μm or 1.3 times the diffraction limit. This increase is presumably due to the intensity variation in the near-field profile, created by the single-sided transverse electron-beam excitation geometry and the resulting gain gradient. Compared to a Gaussian beam of the same $1/e^2$ beam diameter of 4 cm, the amplified beam exhibits a 3 times larger focal spot diameter. Assuming the measured beam quality and an output power of 1 TW, the use of $f/1$ parabolic reflector would result in an intensity at the focal point of more than 10^{18} W/cm^2.

10.4 Conclusions

A 250-fs dye laser-amplifier system was constructed for the blue-green spectral region. A two-stage amplifier design, incorporating a spatial filter, was used to generate tunable 1.2 mJ laser pulses with less than 0.1% of ASE.

The gain of an electron-beam-pumped XeF(C $\rightarrow$ A) amplifier has been measured for 250-fs, 800-fs and 100-ps pulses. The small signal and saturation behavior for both the 800 fs and the 100 ps measurements was identical to within the measurement accuracy, giving no indication of gain repumping on a 100 ps time scale. The single-pass gain for both pulse durations was 5.7, corresponding to a gain coefficient of 3%/cm. The saturation energy density was 80 mJ/cm^2. For 250-fs pulses the gain coefficient was reduced by 10% and the saturation energy density decreased to 50 mJ/cm^2, believed to be related to the rotational reorientation time of the XeF(C $\rightarrow$ A) molecule of approximately 0.8 ps. The saturation energy density of 50 mJ/cm^2 for 250-fs pulses is about 30 times larger than for conventional excimer amplifiers, allowing the design of more efficient and compact high-power amplifiers. Narrow-band absorbers in the XeF(C $\rightarrow$ A) spectrum can be saturated, resulting in a smooth gain profile. A gain bandwidth of 60 nm centered at 480 nm was demonstrated, which can be used either for amplification of extremely short or broadly tunable laser pulses.

Amplification of 490-nm, 250-fs pulses yielded a maximum energy of 275 mJ and laser powers in the terawatt range. An upper limit of 0.4% was found for the ASE energy level. The beam quality of the amplified pulses was determined as 1.3 times diffraction-limited, considering the torus-shaped beam profile.

The amplification of 250-fs pulses with ~ 2 nm bandwidth made use of only a fraction of the XeF(C $\rightarrow$ A) gain bandwidth of 60 nm. It is expected that pulses of much shorter duration, such as the blue-green 10-fs pulses demonstrated by *Schoenlein* et al. [10.24] or frequency-doubled, mode-locked Ti:sapphire pulses can be amplified in this excimer system, possibly further increasing the peak output power. In fact, modeling of the XeF(C $\rightarrow$ A) transition suggests that both the gain and the saturation energy density do not change significantly for injection pulse durations as short as 50 fs [10.25]. Scaling of the XeF(C $\rightarrow$ A) excimer system has been demonstrated successfully for nanosecond systems [10.11], and therefore the design of electron-beam-pumped, large-aperture systems, as demonstrated for the KrF excimer [10.7] should also be applicable to the XeF(C $\rightarrow$ A) excimer amplifier, increasing the performance of this system even further.

Acknowledgements. The authors thank Jim Hooten at Rice University for his technical assistance. This work was supported by the Office of Naval Research, the Air Force Office of Scientific Research and by the Robert Welch Foundation.

References

10.1   P.B. Corkum, R.S. Taylor: Picosecond amplification and kinetic studies of XeCl, *IEEE J. Quantum Electron.* **18**, 1962 (1982)

10.2   J.H. Glownia, J. Misewich, P.P. Sorokin: 160-fsec XeCl excimer amplifier system, *J. Opt. Soc. Am.* **B4**, 1061 (1987)

10.3   S. Szatmári, F.P. Schäfer, E. Müller-Horsche, W. Mückenheim: Hybrid dye-excimer laser system for the generation of 80 fs, 900 GW pulses at 248 nm, *Opt. Commun.* **63**, 305 (1987)

10.4   A.J. Taylor, C.R. Tallman, J.P. Roberts, C.S. Lester, T.R. Gosnell, P.H.Y. Lee, G.A. Kyrala: High-intensity subpicosecond XeCl laser system, *Opt. Lett.* **15**, 39 (1990)

10.5   T.S. Luk, A. McPherson, G.N. Gibson, K. Boyer, C.K. Rhodes: Ultrahigh-intensity KrF* laser system, *Opt. Lett.* **14**, 1113 (1989)

10.6   J.R.M. Barr, N.J. Everall, C.J. Hooker, I.N. Ross, M.J. Shaw, T.W. Toner: High energy amplification of picosecond pulses at 248 nm, *Opt. Commun.* **66**, 127 (1988)

10.7   S. Watanabe, A. Endoh, M. Watanabe, N. Sarukura, K. Hata: Multiterawatt excimer-laser system, *J. Opt. Soc. Am. B* **6**, 1870 (1989)

10.8   P. Maine, D. Strickland, P. Bado, M. Pessot, G. Mourou: Generation of ultrahigh peak power pulses by chirped pulse amplification, *IEEE J. Quantum Electron.* **24**, 398 (1988)

10.9   A. Sullivan, H. Hamsler, H.C. Kapteyn, S. Gordon, W. White, H. Nathel, R.J. Blair and R.W. Falcone: "Multiterawatt, 100fs laser" *Opt. Lett.* **16**, 1406 (1991)

10.10   C. Sauteret, D. Husson, G. Theill, S. Seznec, S. Gary, A. Migus: Generation of 20-TW pulses of picosecond duration using chirped-pulse amplification in a Nd:glass power chain, *Opt. Lett.* **16**, 238 (1991)

10.11   C.B. Dane, G.J. Hirst, S. Yamaguchi, Th. Hofmann, W.L. Wilson, R. Sauerbrey, F.K. Tittel, W.L. Nighan, M.C. Fowler: Scaling characteristics of the XeF(C → A) excimer laser, *IEEE J. Quantum Electron.* **26**, 1559 (1990)

10.12   T.E. Sharp, Th. Hofmann, C.B. Dane, W.L. Wilson, F.K. Tittel, P.J. Wisoff, G. Szabó: Ultrashort-laser-pulse amplification in a XeF(C → A) excimer amplifier, *Opt. Lett.* **15**, 1461 (1990); Th. Hofmann, T.E. Sharp, C.B. Dane, P.J. Wisoff, F.K. Tittel and G. Szabó: "Characterization of an ultra high power XeF(C → A) Excimer Laser System" to appear in *IEEE J. Quantum Electron* 1992.

10.13   F.K. Tittel, G. Marowsky, W.L. Wilson, Jr., M. Smayling: Electron beam pumped broad-band diatomic and triatomic excimer lasers, *IEEE J. Quantum Electron.* **17**, 2268 (1981)

10.14   T.E. Sharp, C.B. Dane, F.K. Tittel, P.J. Wisoff, G. Szabó: A tunable, high power, sub-picosecond blue-green dye laser system with a two-stage dye amplifier, in *IEEE J. Quantum Electron.* **27**, 1221 (1991)

10.15   A.M. Weiner: Effect of group velocity mismatch on the measurement of ultrashort optical pulses via second harmonic generation, *IEEE J. Quantum Electron.* **19**, 1276 (1983)

10.16   R.L. Fork, C.V. Shank, R.T. Yen: Amplification of 70-fs optical pulses to gigawatt powers, *Appl. Phys. Lett.* **41**, 223 (1982)

10.17   Z. Bor, B. Rácz, F.P. Schäfer: Nitrogen-laser-pumped ultrashort pulse amplifier, *Sov. J. Quantum Electron.* **12**, 1050 (1983)

10.18   D.S. Bethune: Dye cell design for high-power low-divergence excimer-pumped dye lasers, *Appl. Opt.* **20**, 1897 (1981)

10.19   G. Szabó, Z. Bor, A. Müller: Phase-sensitive single-pulse autocorrelator for ultrashort laser pulses, *Opt. Lett.* **13**, 746 (1988)

10.20   C.B. Dane, S. Yamaguchi, Th. Hofmann, R. Sauerbrey, W.L. Wilson, F.K. Tittel: Spectral characteristics of an injection-controlled XeF(C → A) excimer laser, *Appl. Phys. Lett.* **56**, 2604 (1990)

10.21   N. Hamada, R. Sauerbrey, F.K. Tittel, W.L. Wilson, W.L. Nighan: Performance character-istics of an injection-controlled electron-beam pumped XeF(C → A) laser system, *IEEE. J. Quantum Electron.* **24**, 1571 (1988)

10.22   L.M. Frantz, J.S. Nodvik: Theory of pulse propagation in a laser amplifier, *J. Appl. Phys.* **34**, 2346 *(1963)*

10.23   S. Yamaguchi, Th. Hofmann, C.B. Dane, R. Sauerbrey, W.L. Wilson, F.K. Tittel: Repetitively pulsed operation of an injection controlled high power XeF(C → A) excimer laser, *IEEE. J. Quantum Electron.* **27**, 259 (1991)

10.24   R.W. Schoenlein, J.-Y. Bigot, M.T. Portella, C.V. Shank: Generation of blue-green 10 fs pulses using an excimer pumped dye amplifier, *Appl. Phys. Lett.* **58**, 801 (1991)

10.25   F. Kannari: Theoretical study of subpicosecond pulse amplification in XeF(C → A), submitted to *IEEE J. Quantum Electron.*

Frank K. Tittel

11. Atomic Optics with Tunable Dye Lasers

V.S. Letokhov

With 9 Figures

11.1 Introduction and History

Many new effects in optical physics and spectroscopy owe their origin to tunable dye lasers. Even if more practical and efficient tunable lasers are developed in the future, the remarkable role of the dye lasers in the advancement of science and technology cannot be overestimated. I would like to illustrate this statement with the example of the discovery of the methods for manipulating atomic motion by means of laser radiation, particularly the methods of the laser-induced optics of atomic beams. Incidentally, this area of laser-atomic physics will also be twenty five in two years.

My interest in this problem was aroused in 1968 in connection with the search for new Doppler-free laser spectroscopic techniques. In the sixties, new Doppler-free laser absorption saturation spectroscopic methods were developed on the basis of the *Lamb dip* [11.1] and the *inverted Lamp dip* [11.2]. A shortcoming of this powerful technique is that it is capable of eliminating the linear Doppler broadening only in the quantum transition being saturated and in those coupled to it. I was aware of the fact that a principally different method existed to suppress the Doppler broadening, based on the restriction of the particle's motion within an area less than about λ (radiation wavelength) across. This effect, referred to as the *Lamp-Dicke regime* [11.3], was successfully realized in the microwave band in the Ramsey hydrogen maser (storage bulb method) [11.4]. To implement such an approach in the optical region of the spectrum, I suggested [11.5] *trapping* atoms or molecules in small regions of space, less than λ_{opt} in size, by means of the so-called spatially periodic gradient force in a standing light wave (Fig. 11.1a). Thanks to the restriction (localization) of the atomic motion, there should have taken place a severe distortion of the Doppler profile with a narrow peak in its center, of all spectral lines observed along the axis of the standing light wave.

I believed this method to hold much promise for high-resolution laser spectroscopy (in addition to *saturation spectroscopy* [11.1, 2] and *two-photon spectroscopy* in counter-propagating waves [11.6]) and called it the *spectroscopy of trapped particles* [11.7].

However, this idea, attractive as it was, could not be carried into effect without tunable lasers. I remember how, at the beginning of my research work at the Institute of Spectroscopy, Dr. O. Kompanets and myself made an attempt to

Topics in Applied Physics, Vol. 70
Dye Lasers: 25 Years Ed.: Dr. Michael Stuke
© Springer-Verlag Berlin Heidelberg 1992

observe the channeling and localization of molecular motion in a standing light wave produced by a powerful continuous-wave CO_2 laser. Unfortunately, the low vibrational polarizability of the molecules in the IR region and their distribution among a great many rotational–vibrational levels rendered this observation impossible. Nowadays it is probably worth returning to such an experiment with the use of the cooled molecular jet technique.

In the middle seventies, when dye lasers became available and pioneering ideas were suggested on the laser cooling of atoms [11.8] and laser cooling of trapped ions [11.9] (Fig. 11.1b)., Dr. V. Balykin and myself went back to these problems.

I would like to mention here the proposals concernig the levitation of dielectric microparticles in the focus of a nontunable laser [11.10], which have now become very useful in controlling the motion of microscopic particles [11.11], specifically microbiological ones.

The invention of lasers, and particularly tunable dye lasers, has armed the scientists with an entirely new source of coherent light possessing high spectral brightness, monochromaticity, tunability and directivity. Light pressure, formerly a barely perceptible phenomenon, has become, with the advent of lasers, a means for controlling atomic motion. The first experiments were soon conducted on the action of laser light upon atomic motion: deflection of atoms by laser radiation [11.12], focusing of atoms by light-pressure potential-gradient force [11.13], deceleration of longitudinal atomic motion in a laser beam [11.14], velocity monochromatization of longitudinal atomic motion and lowering of its temperature to 1.5 K [11.15], further lowering of the temperature to 0.07 K

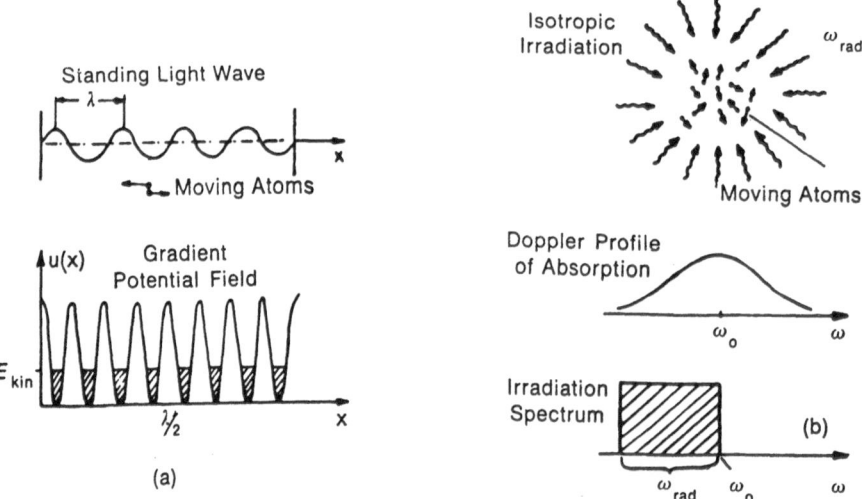

Fig. 11.1a,b. Illustrating the idea of (a) localization of atoms [11.5] and (b) cooling of atoms or ions [11.8, 9] by laser radiation

[11.16], reaching of the minimum temperature of 2.5×10^{-4} K theoretically predicted for two-level atoms [11.17, 18], resonant collimation of an atomic beam by its two-dimensional cooling [11.19], stopping of an atomic beam [11.20] and, finally, localization of cooled atoms in a magnetic trap [11.21] and an optical trap [11.22]. Of special interest are the drecent results on the ultradeep cooling of atoms [11.23] and their trapping in a three-dimensional standing light wave [11.24], which open up new possibilities. All these ex-periments have demonstrated that physicists now have at their disposal a new and sufficiently powerful tool to exert influence upon atomic motion, which one could, in principle, try to employ to develop one more type of optics, namely, the optics of neutral atomic beams.

The present report discusses the experiments on the collimation, focusing and mirror reflection of neutral atomic beams that have already been performed and can be considered as the first steps towards the development of optical elements for handling such beams. To make the actual physical material presented here easier to grasp, we will first introduce the principal concepts relating to the action of light on atomic motion.

11.2 Radiation Force Acting on Atoms in a Resonant Light Field

By radiation force we mean the total force arising upon interaction between laser light and an atom. Depending on the spatial and temporal structure of the light field, its strength and wavelength, radiation force may be an extemely complex function of the atomic position and velocity. Since all the known studies on the application of light pressure forces have been carried out using three types of light fields, namely a plane wave, a Gaussian laser beam and a standing light wave, or their combinations, we will restrict ourselves to consid-eration of these types of fields only. To date, the theory of atomic motion in such fields has been developed quite well (see reviews [11.25] and monograph [11.26]). We will therefore consider the behavior of the radiation pressure force in them only qualitatively.

a) Plane Wave

Consider a plane wave directed along the Z axis and having its frequency tuned to resonance with the absorption frequency of an atom placed in it. The atom absorbs the laser photons directed along the Z axis and re-emits them spontan-eously in all directions (Fig. 11.2a, b). As a result, the atom is acted upon in the direction of the wave by a radiation force whose maximum magnitude is given by the product of the photon momentum $\hbar k$ by the photon scattering rate γ, i.e. $F_{max} = \hbar k \gamma$, where $k = 2\pi/\lambda$ and $\hbar$ is Planck's constant. If the atom is not exactly in resonance with the laser frequency ω_1 and has a velocity projection of V_z on the Z axis in the direction of the light wave, the radiation pressure force

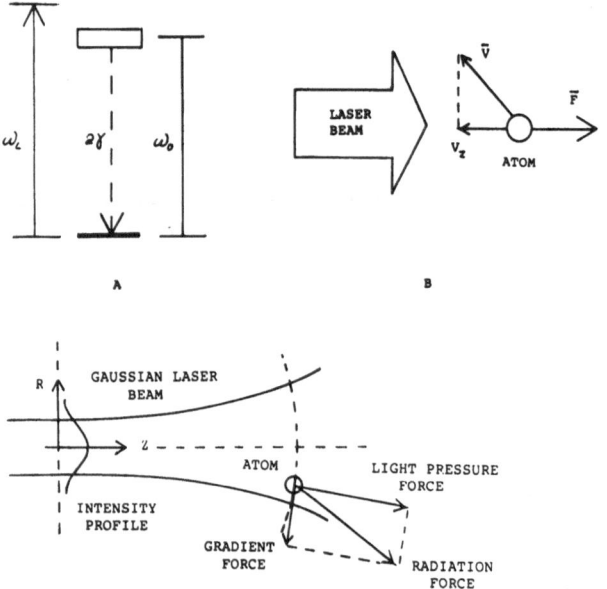

Fig. 11.2a,b. Radiation pressure force acting on an atom in a traveling light wave: (**a**) two-level scheme of resonant interaction between an atom and a plane traveling light wave; (**b**) illustrating the action of a plane traveling light wave on an atom; (**c**) radiation pressure force in a Gaussian laser beam

will be expressed in terms of the atomic velocity projection V_z and the detuning Ω of the field frequency ω_1 with respect to the atomic absorption frequency ω_0 as

$$F = \hbar k \gamma G / [1 + G + (\Omega - k v_z^2) / \gamma^2], \tag{11.1}$$

where $G = I/I_s$, I is the light wave intensity, I_s the atomic transition saturation intensity and $\Omega = \omega_1 - \omega_0$. The acceleration of the atom under the action of this force reaches 10^8 cm/s^2, which is a hundred thousand times the acceleration due to gravity.

b) Gaussian Laser Beam

An atom placed in a laser beam restricted on the diameter is also acted upon, in addition to the above force directed along the beam, by a gradient force due to the transverse inhomogeneity of the light field (Fig. 11.2c). The origin of this force can be understood most simply from the classical consideration of the interaction between an atom and an inhomogeneous field. An atom in a laser field acquires a dipole moment, and the inhomogeneous field acts on this dipole moment. Depending on the detuning of the laser radiation frequency with respect to the atomic transition frequency, the field either expels the atom out of the beam ($\Omega > 0$) or draws it towards the beam center ($\Omega < 0$). In terms of the atomic and laser parameters, the gradient force is expressed as

$$F_{\mathrm{gr}} = \frac{\hbar(\rho/\rho_0^2)(\Omega - kv_z)G}{1 + \gamma + (\Omega - kv_z)^2/\gamma^2} \; , \tag{11.2}$$

where ρ_0 is the laser beam radius and ρ the distance between the atom and the laser beam axis.

c) Plane Standing Light Wave

This wave is formed from two counterpropagating plane traveling waves. When the radiation intensity is low $(G \ll 1)$, the radiation force in the standing wave is determined by the sum of the forces due to each traveling wave. As the radiation intensity is increased, there comes into play the modulation of the laser field intensity with a period of $\lambda/2$, which in turn causes the gradient force to become manifest. What is more, when the radiation intensity is high $(G \gg 1)$, the atomic motion is affected by the processes of stimulated re-emission by the atom of photons between the two traveling waves, which, in the presence of a spatial inhomogeneity of the field, give rise to what is known as the retarding radiation pressure force component [11.27–29]. This force arises when an atom moves in an intense and highly inhomogeneous field, whose strength varies perceptibly during the time it takes for the atom to decay spontaneously. It exerts influence on atoms moving with slow velocities along the z axis: $v_z k/\gamma$. Unlike the radiation pressure force in a traveling wave, the retarding force is not restricted in magnitude and grows with the intensity of light.

Let us now consider how all the above types of fields and forces produced by them can be used to control atomic beams.

11.3 Collimation

The need to collimate a beam of particles arises whenever it is required to increase their phase density, i.e. to compress them in both velocity distribution and space. The collimation (compression) of particle beams increases the efficiency of their application in experimental studies and improves the accuracy of these experiments. Depending on the particular particle species, different methods may be employed to collimate them. Common to all these methods is the use of dissipative processes. For example, light charged paricles are collimated by radiative friction, protons and antiprotons, and electron cooling, and heavy partcles by ionization loss.

The dissipative force most effective for neutral atomic particles is the laser radiation pressure. The atomic beam collimation experiments performed to date have used either the radiation pressure force [11.30] or the retarding force [11.31]. Each of these schemes has advantages and disadvantages. The use of radiation pressure force (11.1) allows for obtaining narrower angular divergences of atomic beams and, accordingly, higher phase densities of the beams. The merit of utilizing the retariding force is a faster beam collimation, but the ultimate collimation angle is in this case much larger.

Let us consider collimation by radiation pressure force (11.1) in more detail. This collimation scheme was first suggested by *Balykin* et al. [11.32]. The atomic beam (see Fig. 11.3) is in this case irradiated on all sides with an axisymmetric light field whose frequency ω_1 is shifted toward the red end relative to the atomic transition frequency ω_0. The axisymmetric field is formed by a laser radiation reflecting from the inner surface of a conical reflecting axicon. In the plane of the figure, this field consists of two counter-propagating light waves, the intensities of which are equal at any point of the axicon space.

An atom moving with a transverse velocity of v_ρ in the axisymmetric light field is acted upon by a radiation pressure force which at $\omega_1 < \omega_0$ is directed against the radial velocity vector v_ρ, and at $\omega_1 > \omega_0$ with the vector v_ρ. Such a direction of the radiation pressure force is due to the fact that the atom, because of the Doppler shift of the atomic transition frequency, absorbs more effectively those photons which come from the light wave propagating counter to the velocity vector v_ρ. This in turn means that at $\omega_1 > \omega_0$ the total force from the two light waves is directed against the radial atomic velocity vector. Owing to the action of this force, at $\omega_1 < \omega_0$ in the inner region of the axicon there occurs a rapid narrowing of the transverse velocity distribution of the atomic beam, which reduces the angular divergence of the beam and increases its density, i.e. improves the beam collimation. The joint effect of radiative friction and momentum diffusion leads to the establishment of a stationary atomic velocity distribution which determines the ultimate collimation angle

$$\Delta\phi_{\min} = (1/v_z)\,(\hbar\gamma/M)^{1/2} \; , \tag{11.3}$$

where M is the atomic mass and v_z the atomic velocity along the z axis. For a thermal atomic beam, the ultimate collimation angle is of the order of 10^{-3}–10^{-4} rad. If the initial beam divergence is 0.1 rad, the on-axis atomic beam intensity can be expected to rise 10^4–10^6 times.

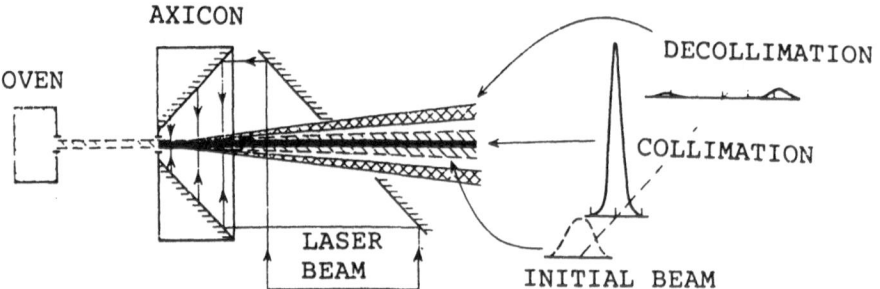

Fig. 11.3a,b. Collimation and decollimation of an atomic beam by laser radiation: (**a**) schmematic of a radiative atomic beam collimator; (**b**) atomic beam profiles

The collimation experiment [11.30] was conducted with a beam of sodium atoms. Figure 11.1b presents beam profiles prior to and after interaction with the laser field. The collimation process is very sensitive to the position of the laser radiation frequency. To effect collimation, the laser radiation frequency was detuned toward the red side ($\omega_1 - \omega_0 = -13$ MHz). The longitudinal atomic velocity v_z was 7.3×10^4 cm/s. Comparison between the above beam profiles shows, firstly, a substantial increase (by a factor of 5) in the on-axis atomic beam intensity and secondly, a considerable narrowing (collimation) of the beam. Changing over to positive detunings caused a material broadening (decollimation) of the beam. The on-axis beam intensity in that case was observed to change by more than three orders of magnitude.

11.4 Focusing an Atomic Beam; Imaging an Atomic Source

The principal element in any of the existing types of optics is a lens. It is therefore quite obvious that creating laser field configurations capable of focusing neutral atomic beams is essential to the development of optics for such beams. There are at present two possibilities for focusing an atomic beam: by means of the gradient force and using the spontaneous radiation pressure force.

The possibility of focusing an atomic beam by means of the gradient force was first considered by *Bjorkholm* et al. [11.33, 34] at the Bell Telephone Laboratories. In their scheme, the atomic beam propagated along and inside a narrow laser beam. The laser radiation frequency was tuned below the atomic transition frequency. In that case, the gradient force was directed towards the laser beam center. With the atomic path length inside the laser beam being great enough, the atoms were drawn transversely toward the laser beam axis so that a maximum atomic beam compression was observed to take place at some point. The minimum observed atomic beam diameter was 26 μm.

Balykin et al. [11.35] used a different laser field configuration (Fig. 11.4). In the general case, it comprises four divergent Gaussian beams propagating towards each other. The waists of the beams are located at equal distances from

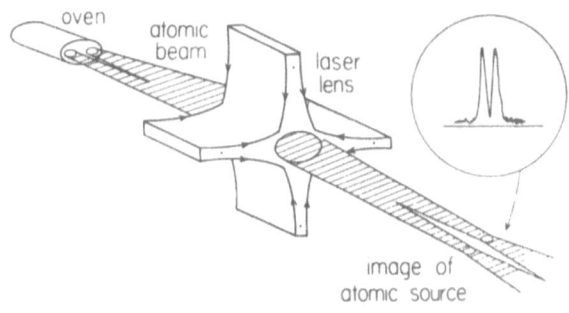

Fig. 11.4. Focusing of an atomic beam and "imaging" of the atomic source. Inset: atomic beam profile in the image plane

the center of the configuration. The laser radiation frequencies are tuned to resonance with the atomic absorption frequency. Under such conditions, an atom moving away from the atomic beam axis is acted upon by spontaneous radiation pressure force (11.1) which tends to bring the atom back to the beam axis.

The effect of the gradient force on the atomic motion is in this case insignificant. Such a configuration is really a "laser lens" for a beam of neutral atoms, for it can be demonstrated that an atomic beam issuing from a point (an analog of a point source in light optics) is focused also into a point after interacting with the laser field of this configuration. What is more, one may express the focal length of such a lens as

$$f = (v^2/w^2 d)[1 + 4G_0 + (\Omega^2/\gamma^2)/G_0] \tag{11.4}$$

and derive the laser lens formula

$$\frac{1}{S} + \frac{1}{L} = \frac{1}{f}\left(1 - \frac{d}{2L}\right), \tag{11.5}$$

where d is the thickness of the laser lens, S the distance from the atomic source to the lens, L the distance from the lens to the image of the source, G_0 the saturation parameter on the lens axis, $w = (8\hbar k\gamma l/M)\,(q_{0z}^2 l/b_z^2 q_z^2)$, l the distance from the lens axis to the laser beam waist, q_z and q_{0z} are the beam radii along the z axis on the lens axis and at the beam waist, respectively, and $b_z = 2kq_{0z}^2$. The laser lens is sufficiently localized in space: its characteristic size d along the atomic beam is much less than the distances S and L from the atomic source to the lens and from the lens to the image. Under these conditions ($d \ll S, L$), lens formula (11.5) coincides with the optical lens formula.

The first successful experiment on the focusing of an atomic beam by means of a laser lens [11.35] was conducted with a beam of sodium atoms. The laser lens was formed by two counter-propagating Gaussian beams. Such a laser lens is analogous to a cylindrical lens in light optics. Figure 11.4 illustrates the imaging of a "two-point" atomic source–an atomic oven with two 0.5-mm diameter holes spaced 2 mm apart. Without the laser lens, the atomic beam profile is a wide diffuse spot. With the laser radiation on and the lens formed in the path of the atoms, the atomic trajectories are observed to obey the laws of neutral particle beam optics. In the image plane, the beam profile consists of two well-defined peaks, each corresponding to one of the atomic source holes.

An essential drawback of the above laser lens is its insufficiently high resolution (the minimum focal spot diameter is about 50 μm). As in [11.33], the limit on resolution is due to momentum diffusion which smears the trajectory of the atom interacting with the laser lens field.

Resolution can be materially improved by resorting to the idea of using the gradient force–a potential force by nature, but now with a different laser field configuration and a different atom-field interaction geometry. *Balykin* and *Letokhov* [11.36] have suggested an "atomic objective lens" with a resolution of a few Å units. Such an objective lens is a strictly focused laser field of the TEM$_{01}^*$

mode configuration, the frequency of which is tuned far enough from the atomic transition frequency (Fig. 11.5). The atoms propagate close to the axis of the atomic objective lens. The rigid focusing of the laser radiation forming the lens allows for its being localized enough in space (thin lens), and the choice of the TEM_{01}^* mode and the large detuning of the field frequency permit the chracteristics of the atomic objective lens to be made close to those of an ideal lens.

To determine the maximum resolution of such an atomic objective lens, it is necessary to consider the atomic beam in the form of the de Broglie waves $\psi = \psi_0 \exp[(-i/\hbar)p(z)\,dz]$ and the laser field, as a transparency whole phase transmission function is $T(x, y)$. These waves acquire a certain phase advance in the laser field which leads to their subsequent focusing in the focal plane of the atomic objective lens. The minimum focal spot diameter is determined first of all by the diffraction of the de Broglie waves by the aperture of the lens. Apart from this principal restriction on the resolution of the atomic objective lens, there are also limitations due to various aberrations, such as spherical and chromatic aberrations and, in addition, diffusion aberration associated with the atomic momentum diffusion in the laser field. Nevertheless, calculations show that the laser field and atomic beam parameters can be selected so that the resolution of the atomic objective lens will differ only slightly from the diffraction-limited resolution. For instance, if the transverse dimension of the lens is taken to be

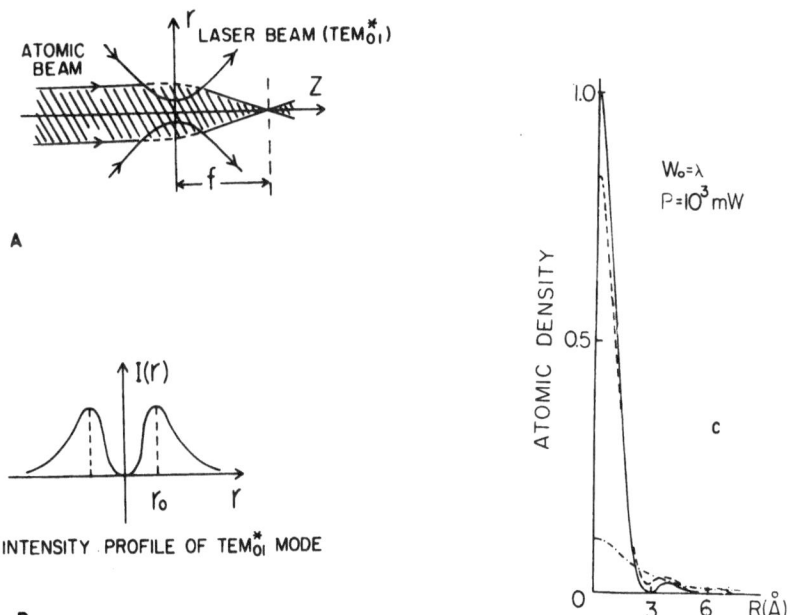

Fig. 11.5a–c. Deep focusing of an atomic beam into an Å-size spot: (a) laser field and atomic beam arrangement; (b) transverse intensity profile of the TEM_{01}^* laser mode field; (c) transverse atomic density distribution in the focal plane of the "atomic objective lens" ($p = 1$ W)

equal to a few wavelengths and the laser power is around one watt, the focal spot diameter of the atomic beam should not be much greater than a few Å units (Fig. 11.5c).

11.5 Atomic Beam Reflection

Besides the lens, a no less important element of any optics is the mirror. It can be used for both focusing and reflection purposes. Focusing by means of a concave mirror has certain merits compared to lens focusing, there being no chromatic abberation in the former case. This property is particularly important in optics handling particle beams instead of light because chromatic aberration is especially severe for particles.

The atomic mirror idea was suggested by *Cook* and *Hill* [11.37]. Figure 11.6 shows a schematic diagram of such an atomic mirror. It is formed by a very thin surface wave originating upon the total internal reflection of a laser beam at a dielectric–vacuum interface. The thickness of the surface wave may vary between a few fractions of a wavelength and a few wavelengths. The laser radiation intensity on the surface of the dielectric is equal to the intensity of the initial laser wave in the dielectric, whereas in the vacuum it drops sharply practically to zero at a distance of a few wavelengths from the surface.

In such a surface wave, there arises an enormous light intensity gradient, the maximum possible in optics. An atom placed in a surface wave whose frequency is higher than the atomic transition frequency is acted upon by a gradient force tending to expel it from the wave and into the vacuum. If the atom comes up against the surface wave from the vacuum side, it moves along a straight trajectory in the vacuum, slows sharply down to zero normal velocity in the surface wave, accelerates in the wave in the reverse direction away from the

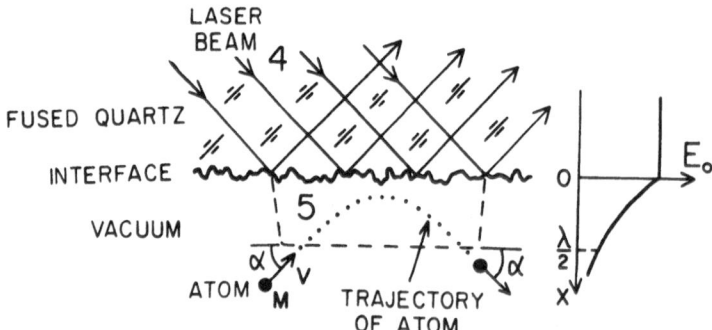

Fig. 11.6. Schematic diagram of an "atomic mirror"

surface, and then follows a straight trajectory in the vacuum once more, its angle of incidence being equal to the angle of reflection if its transverse velocity does not exceed some maximum value. The maximum transverse atomic velocity is determined from the condition of equality between the kinetic energy of the transverse atomic motion and the barrier height of the surface wave as $v_{max} = [2U_{grad}(0)/M]^{1/2}$, where $U_{grad}(0)$ is the barrier height of the surface wave. If the transverse atomic velocity is higher than v_{max}, the atom will reach the dielectric surface and reflect from it in a diffuse manner. For a typical laser radiation with a power of $P = 1$ W and, for example, a sodium atom with a thermal velocity of $v = 6 \times 10^4$ cm/s, the maximum transverse velocity $v_{max} \approx 5 \times 10^2$ cm/s and, correspondingly, the maxium angle of reflection $\alpha_{max} \approx 10^{-2}$ rad. When using laser-cooled atomic beams or pulsed dye laser radiation, there is no limitation on the maximum angle of incidence: reflection is possible even if the atomic beam is normal to the atomic mirror.

Reflection from an atomic mirror was observed for the first time by *Balykin* and coworkers [11.38]. The experiment was conducted with sodium atoms. The atomic mirror was a plane-parallel plate of fused quartz with a bevelled lateral face through which laser radiation was passed. To increase the surface area of the atomic mirror, use was made of repeated total internal reflection of the laser beam. Figure 11.8b shows the arrangement of the reference, incident, and reflected atomic beams relative to the atomic mirror, the respective beam profiles registered being presented in Fig. 11.7a. When the atomic mirror is parallel to the atomic beam axis so that the atoms fly past it, only the reference and incident beams are detected (top of Fig. 11.7a). If the mirror is tilted so that it shuts off the atomic beam, the latter is observed to reflect (peak 3). As the tilt

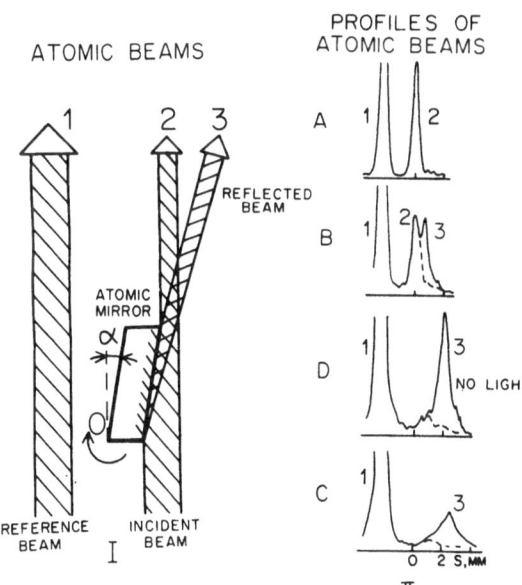

Fig. 11.7a,b. Mirror reflection of atoms by a laser field: (a) reference, incident, and reflected beams (AM: atomic mirror); (b) atomic beam profiles: 1 – reference; 2 – incident; 3 – reflected. Dashed line: laser field off

angle of the mirror is increased, the angle of reflection of the atomic beam also increases, but the number of atoms undergoing mirror reflection decreases because some atoms reach the mirror surface and suffer diffuse reflection. The maximum angle of reflection observed in the experiment amounted 0.4°. The mirror reflectivity in that case was close to 100%.

Another remarkable property of the atomic mirror is its ability to reflect atoms in a quantum-state-selective fashion. Such a selectivity arises from the dispersive character of the relationship between the gradient force acting on the atoms and the laser radiation frequency: when the detuning of the laser field frequency with respect to the atomic absorption frequency is positive, the gradient force drives the atoms away from the surface, and when it is negative, the force draws the atoms to the surface, whereupon they undergo diffuse scattering. Assume that a beam of atoms distributed among several sublevels of the ground state impinges upon the atomic mirror. The atoms at a sublevel for which frequency of a transition to an excited state is lower than the laser frequency, will then be reflected at the mirror, whereas the rest of the atoms will scatter in a diffuse manner. The reflected beam will thus contain only particles in one and the same quantum state. The above experimenters [11.39] observed selective reflection of atoms at two hyperfine splitting sublevels of the ground state in sodium. The ratio between the mirror reflectivities for atoms at the sublevels $F = 2$ and $F = 1$ (i.e. the selectivity of reflection) was no less than 100.

Note that such a selective reflection can also be expected for molecules. In that case, possibilities will be opened up for the production and spectroscopy of beams of molecules in one and the same vibrational–rotational state.

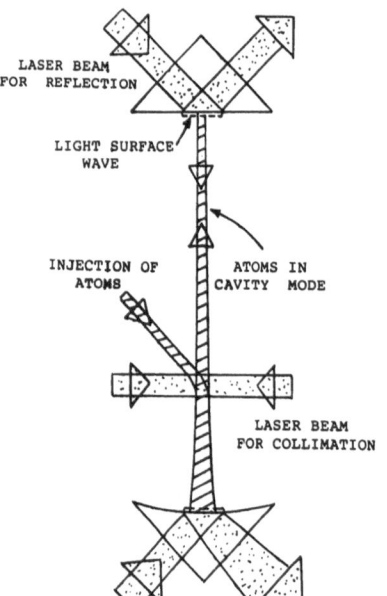

LASER BEAM
FOR REFLECTION

LIGHT SURFACE
WAVE

INJECTION OF
ATOMS

ATOMS IN
CAVITY MODE

LASER BEAM
FOR COLLIMATION

Fig. 11.8. Atomic-mirror-based cavity for de Broglie waves. An atomic mirror is a surface electromagnetic wave originating upon total internal reflection of a laser beam at a dielectric–vaccum interface

It should also be noted that it is possible to create spherical atomic mirrors and cavities on their basis for atomic de Broglie waves [11.40, 41]. Figure 11.8 shows one of the possible atomic cavity configurations. It is similar to an optical cavity with its physical mirrors replaced by light-induced mirrors. A possible mechanism by which atoms can be injected into such a cavity is the laser collimation of an external atomic beam (Fig. 11.8). The maximum steady-state density of atoms in the cavity will be determined by the rate of their injection into the cavity and their lifetime therein. One of the parameters characterizing the light field in an ordinary optical cavity is its degeneracy equal to the number of photons per cavity mode. For laser radiation, the degeneracy parameter is fairly high. The estimates made in [11.41] show that the realization of a pseudooptical cavity for atomic de Broglie waves allows, in principle, a high wave degeneracy to be achieved at a relatively low atomic density. This will result from atomic velocity monochromatization and collimation of the atomic waves.

11.6 Conclusion and Prospects

Despite the fact that only the very first experiments have been conducted on exerting influence upon atomic motion, we can already say that investigators now have an entirely new method for controlling various atomic beam parameters. As laser technology develops further, it will become possible not only to perfect the main elements of neutral atomic beam optics considered above, but also to pass to molecular beams. Some possible applications of neutral atomic beam optics are already clear. Collimation of atomic beams allows shaping of their spatial parameters and improvement of their divergence and phase density in an isotope-selective fashion. Apart from increasing atomic beam density, atomic beam focusing can help to create an atomic microscope. It is not very difficult to envisage a scheme for such a microscope. It may be similar to a scanning transmission or reflection electron microscope. The scattered or reflected atoms can be registered by the well-developed single-atom detection techniques.

Atomic mirrors can be used as high-speed deflectors, modulators and shutters for neutral atomic beams. They can also be used to create traps for ultracold atoms. Concave atomic mirrors can be employed as optical elements in an atomic microscope because they can focus atomic beams into a spot with a diameter comparable with the de Broglie wavelength, the atom–laser field interaction time being too short for momentum diffusion to become a limiting factor. Sharply focused atomic beams with a focal spot diameter of a few Å units will enable one to observe atomic collisions and study the scattering of atoms on one another under strictly controlled conditions. It would be of interest to extend this technique to molecular beams, for this would allow better insight into molecular interaction dynamics.

NEW TYPE of AMPLIFYING (ABSORBING)

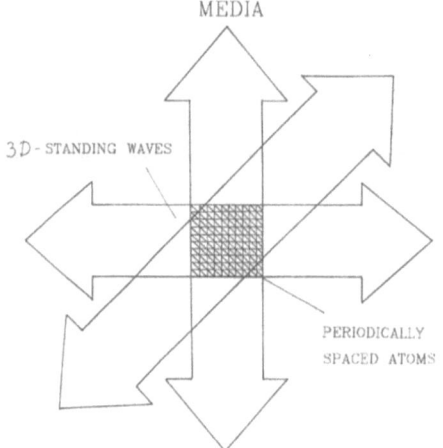

Fig. 11.9. Illustrating a new type of amplifying (absorbing) medium in the form of a regular array of atoms strongly localized by a light field

It seems very interesting to study the absorption and amplification properties of an atomic ensemble subject to three-dimensional trapping in a three-dimensional standing light wave, as suggested by *Letokhov* et al. [11.42]. and experimentally demonstrated by *Westbrook* and coworkers [11.24]. In that case, one can achieve a spatially periodic arrangement of the atoms, which can be localized quite well within regions $(50-100 \text{ Å})^3$ in volume in the nodes or loops of the three-dimensional standing light wave by a strong visible light field Obviously it will be thus possible not only to narrow the Doppler width of all spectral lines, UVU lines included, but also to effect the Bragg diffraction like interaction between the spatially periodic atomic ensemble (Fig. 11.9) and another light wave. Such amplifying or absorbing media, which should feature substantially enhanced amplification or absorption properties in certain diffraction directions, are of great interest in laser physics. They are also interesting from the standpoint of suppressing or increasing their coupling to the radiation modes of the vacuum.

Acknowledgements. I express my deep gratitude to Drs. V. Balykin, Yu. Ovchinnikov, and A. Sidorov for many fruitful discussions. I am especially grateful to Dr. V. Balykin for his invaluable help in preparing some sections of this report.

References

11.1   W.E. Lamb, Jr.: Phys. Rev. **134A**, 580 (1964); R.A. McFarlane, W.R. Bennett, W.E. Lamb Jr.: Appl. Phys. Lett. **2**, 189 (1963) A. Szoke, A. Javan: Phys. Rev. Lett. **10**, 521 (1963)
11.2   V.S. Letokhov: Pis'ma Zh. Eksp. Teor. Fiz. (Russian) **6**, 597 (1967) P.H. Lee, M.L. Skolnick: Appl. Phys. Lett. **10**, 303 (1967) V.N. Lisitsyn, V.P. Chebotayev: Zh. Eksp. Teor. Fiz. (Russian) **54**, 419 (1968); R.L. Barger, J.L. Hall: Phys. Rev. Lett. **22**, 4 (1969)

11.3   R. Dicke: Phys. Rev. **82**, 472 (1953)
11.4   H.M. Goldenberg, D. Kleppner, N.F. Ramsey: Phys. Rev. Lett. **8**, 361 (1960)
11.5   V.S. Letokhov: Pis'ma Zh. Eksp. Teor. Fiz. (Russian) **7**, 348 (1968)
11.6   L.S. Vasilenko, V.P. Chebotayev, A.B. Shishayev: Pis'ma Zh. Eksp. Teor. Fiz. (Russian) **12**, 161 (1970)
11.7   V.S. Letokhov: Science **190**, 344 (1975)
11.8   T.W. Hansch, A.L. Schawlow: Opt. Commun. **13**, 68 (1975)
11.9   D.J. Wineland, H.G. Dehmelt: Bull. Am. Phys. Soc. **20**, 637 (1975) H.G. Dehmelt: Nature **262**, 777 (1976)
11.10 A. Ashkin: Phys. Rev. Lett. **24**, 156 (1970); **25**, 1321 (1970); A. Ashkin, J.M. Dziedzik: Appl. Phys. Lett. **19**, 283 (1971)
11.11 A. Ashkin, J.M. Dziedzik: Science **235**, 1517 (1985)
11.12 R. Schieder, H. Wather, L. Wöste: Opt. Commun. **5**, 337 (1972)
11.13 J.E. Bjorkholm, R.E. Freeman, A. Ashkin, D.B. Pearson: Phys. Rev. Lett. **41**, 1361 (1978)
11.14 V.I. Balykin, V.S. Letokhov, V.I. Mishin: Pis'ma Zh. Eskp. Teor. Fiz. (Russian) **29**, 614 (1979)
11.15 S.A. Andreev, V.I. Balykin, V.S. Letokhov, V.G. Minogin: Pis'ma Zh. Eksp. Teor. Fiz. (Russian) **34**, 463 (1981); Sh. Eksp. Teor. Fiz. (Russian) **82**, 1429 (1982)
11.16 W.P. Phillips, H.J. Metcalf: Phys. Rev. Lett. **48**, 596 (1982)
11.17 V.S. Letokhov, V.G. Minogin, B.D. Pavlik: Zh. Eksp. Teor. Fiz. (Russian) **72**, 1328 (1977)
11.18 S. Chu, L. Hollberg, J. Bjorkholm, A. Cable, A. Ashkin: Phys. Rev. Lett. **55**, 48 (1985)
11.19 V.I. Balykin, V.S. Letokhov, A.I. Sidorov: Pis'ma Zh. Eksp. Teor. Fiz. (Russian) **40**, 251 (1984)
11.20 W. Ertmer, R. Blatt, J. Hall, M. Zhu: Phys. Rev. Lett. **55**, 996 (1985)
11.21 A.L. Migdall, J.V. Prodan, W.P. Phillips, T.H. Bergeman, H.J. Metcalf: Phys. Rev. Lett. **54**, 2596 (1985)
11.22 S. Chu, J. Bjorkholm, A. Ashkin, A. Cable: Phys. Rev. Lett. **57**, 314 (1986)
11.23 C. Cohen-Tannoudji, W.D. Phillips: Physics Today, October, No. 10, 33 (1990)
11.24 C.I. Westbrook, R.N. Watts, C.E. Taner, S.L. Rolston, W.D. Phillips, P.D. Lett, P.L. Gould: Phys. Rev. Lett. **65**, 33 (1990)
11.25 A. Ashkin: Science **210**, 1081 (1980); V.S. Letokhov, V.G. Minogin: Phys. Rep. **73**, 1 (1981); A.P. Kazantsev, G.A. Ryabenko, G.I. Surdutovich, V.P. Yakovlev: Phys. Rep. **129**, 75 (1985); Mechanical Effects of Light, Special Issue of J. Opt. Soc. Am. **B2**, (1985); S. Stenholm: Rev. Mod. Phys. **58**, 699 (1986)
11.26 V.G. Minogin, V.S. Letokhov: *Laser Light Pressure on Atoms* (Gordon and Breach, New York 1987)
11.27 A.P. Kazantsev: Zh. Eksp. Teor. Fiz. (Russian) 66, 1599 (1974)
11.28 V.G. Minogin, O.T. Serimaa: Opt. Commun. **30**, 373 (1979)
11.29 J. Dalibard, C. Cohen-Tannoudji: J. Opt. Soc. Am. **B2**, 1707 (1985)
11.30 V.I. Balykin, V.S. Letokhov, A.I. Sidorov: Pis'ma Zh. Eksp. Teor. Fiz. (Russian) **40**, 251 (1984); V.I. Balykin, A.I. Sidorov: Appl. Phys. **42B**, 51 (1987)
11.31 C. Salamon, J. Dalibard, A. Aspect, H. Metcalf, C. Cohen-Tannoudji: *Phys. Rev. Lett.* **59**, 1659 (1987)
11.32 V.I. Balykin, V.S. Letokhov, V.G. Minogin, T.V. Zueva: Appl. Phys. **35B**, 149 (1984)
11.33 J.E. Bjorkholm, R.R. Freeman, A. Ashkin, D.P. Pearson: Phys. Rev. Lett. **41**, 1361 (1978); Appl. Phys. Lett. **36**, 99 (1980)
11.34 J.E. Bjorkholm, R.R. Freeman, A. Ashkin, D.B. Pearson: Opt. Lett. **5**, 111 (1980)
11.35 V.I. Balykin, V.S. Letokhov, A.I. Sidorov: Pis'ma Zh. Eksp. Teor. Fiz. (Russian) **43**, 172 (1986); V.I. Balykin, V.S. Letokhov, A.I. Sidorov, Yu.B. Ovchinnikov: J. Mod. Opt. **35**, 17 (1988)
11.36 V.I. Blaykin, V.S. Letokhov: Opt. Commun. **64**, 151 (1987); V.I. Blaykin, V.S. Letokhov: Zh. Eksp. Teor. Fiz. (Russian) **94**, 140 (1988)
11.37 R.J. Cook, R.K. Hill: Opt. Commun. **43**, 258 (1982)
11.38 V.I. Blaykin, V.S. Letokhov, Yu.B. Ovchinnikov, A.I. Sidorov: Pis'ma Zh. Eksp. Teor. Fiz. (Russian) **35**, 282 (1987)

11.39   V.I. Blaykin, V.S. Letokhov, Yu.B. Ovchinnikov, A.I. Sidorov: Phys. Rev. Lett. **23,** 2137 (1988)

11.40   V.I. Blaykin, V.S. Letokhov, V.G. Minogin: Proc. Intern. Symp. Physics of Cooling and Traping of Low-Energy Particles, 1987, Stockholm, Physica Spectra **T22,** 119 (1988)

11.41   V.I. Blaykin, V.S. Letokhov: J. Opt. Soc. Am. **B48,** 517 (1989)

11.42   V.S. Letokhov, V.G. Minogin, B.D. Pavlik: Opt. Commun. **19,** 72 (1976); Zh. Eksp. Teor. Fiz. (Russian) **72,** 1328 (1977); V.S. Letokhov, B.D. Pavlik: Appl. Phys. **9,** 229 (1976)

V.S. Letokhov

12. Laser Spectroscopy of Small Molecules

V. Beutel, G. Bhale, W. Demtröder, H.-A. Eckel, J. Gress and M. Kuhn

With 10 Figures

Recent progress in high-resolution spectroscopy of molecules resulting from a combination of laser spectroscopy in cold molecular beams and mass spectroscopy is illustrated by some isotope selective spectra of Ag_2 dimers and mass selective spectroscopy of Na_3 in a broad mass distribution of Na_n clusters.

12.1 Introduction

Tunable dye lasers, either in their single-mode cw version or as narrow-band high peak-power pulsed lasers have become indispensable tools for high-resolution molecular spectroscopy. Optical frequency doubling, tripling or mixing of visible and near UV dye lasers allow selective excitation of high-lying molecular levels and open new possibilities for investigating photon-induced molecular reactions, such as isomerization or fragmentation, to study molecular Rydberg states and their autoionization channels or to follow up the molecular dynamics of intra- or inter-molecular energy transfer.

Spectroscopic measurements of such processes can be correctly interpreted only if the assignment of the excited levels is known. For complex molecular spectra this assignment is by no means trivial. Here the combination of different experimental techniques from various fields of physics and chemistry has brought about considerable progress for the analysis of molecular spectra and a detailed understanding of molecular dynamics. These techniques are:

a) Sensitive detection techniques, such as resonant two- or three-photon ionization, laser-induced fluorescence with improved collection efficiency of the photons [12.1] or optothermal detection with superconducting bolometers [12.2].

b) Sub-Doppler resolution achieved with collimated molecular beams or with nonlinear techniques in vapor cells or beams [12.3]

c) Adiabatic cooling of molecules in supersonic seeded beams which leads to a compression of the thermal population distribution into the lowest vibrational–rotational levels of the electronic ground state, resulting in a drastic simplification of otherwise complex molecular absorption spectra [12.4, 5]

d) Photoionization of selectively excited molecular levels in combination with

Topics in Applied Physics, Vol. 70
Dye Lasers: 25 Years Ed.: Dr. Michael Stuke
© Springer-Verlag Berlin Heidelberg 1992

mass selective detection of the ions which allows a more reliable assignment of spectra taken in samples with several molecular species or mixtures of different isotopes. This technique can be realized with pulsed lasers and time of flight mass spectrometers [12.6] or with cw lasers and quadrupole mass spectrometers.

The present paper illustrates the combination of these techniques by some examples, taken from recent experiments in our laboratories in Kaiserslautern.

12.2 Determination of Singlet–Triplet Potentials of Alkali Molecules

If alkali atoms have been cooled down by optical cooling techniques into the milli- to micro-Kelvin range, they can be trapped within a defined volume by an inhomogeneous magnetic field B [12.7]. These trapped atoms are spin-aligned and the force acting upon them is $F = -\mu \cdot \text{grad } B$. The interaction potential between two spin-aligned Na atoms, for example, corresponds to the $^3\Sigma_u$ potential curve of Na_2, while the interaction between atoms with opposite electron spins corresponds to the $^1\Sigma_g$ potential.

At large internuclear separations R the energy difference $\Delta E(R) = E(^3\Sigma_u)$ $- E(^1\Sigma_g^+)$ becomes comparable to the hyperfine splitting of the atomic ground states. Under these conditions the nuclear spin–electron spin coupling can cause an electron spin flip. This means that trapped atoms can escape out of the trapping volume after spin-flip collisions. In the potential diagram, such a spin-flip collision would correspond to a transition from the $^3\Sigma_u$ to the $^1\Sigma_g$ potential (hyperfine-induced g–u mixing [12.8]).

In order to determine the cross sections for such spin-flip collisions, the potential curves of the $^3\Sigma_g$ states have to be measured up to large internuclear distances R [12.9]. This can be done in the following way, illustrated by the example of the Cs_2 molecule (Fig. 12.1). A single-mode dye laser L1 is tuned to a transition $X^1\Sigma_g^+(v'', J'') \rightarrow D^1\Sigma_u (v', J')$ from a level (v'', J'') in the electronic ground state to a high-lying rovibrational level (v', J') in the $D^1\Sigma_u$ state. The Franck–Condon factors for transitions from the outer turning point in the upper level allow one to reach very high vibrational levels of $v'' > 130$ of the $X^1\Sigma_g$ ground state through stimulated emission induced by a second tunable single-mode dye laser. These transitions can be detected either through the decrease of the fluorescence from the $D^1\Sigma (v', J')$ levels or, in the case of photoionization of the excited level by an argon laser, through the corresponding decrease of the ionization rate.

In Fig. 12.2 the stimulated emission signals, which are obtained with a pump laser L1, kept on the transition $X^1\Sigma_g^+(v'' = 0, J'' = 47) \rightarrow D^1\Sigma_u (v' = 50, J = 48)$, and the stimulating probe laser L2 tuned through the bands $v' = 50 \rightarrow v'' = 120, 130, 136$ are shown. With increasing values of v'' the coupling between

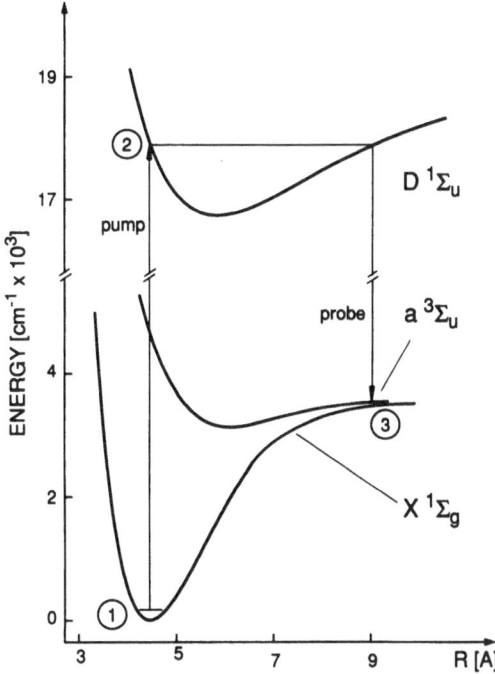

Fig. 12.1. Level scheme for stimulated emission pumping of high vibrational levels v_3 of the $Cs_2X^1\Sigma_g$ ground state, close to the dissociation energy

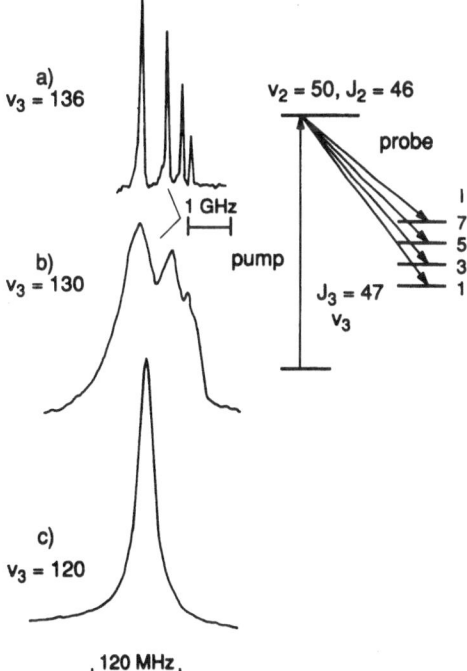

Fig. 12.2. Hyperfine-induced mixing of levels (v_3, J_3) of the $X^1\Sigma_g^+$ and $^3\Sigma_u$ states. The four components correspond to the four possible quantum numbers I of the total nuclear spin of Cs_2

the $X^1\Sigma_g^+$ levels and the $^3\Sigma_u$ levels become stronger resulting in an increasing splitting of the lines into four components.

From the intensity ratios of these components, which differ for even rotational levels and odd levels and which correspond to the four possible values of the total nuclear spin $I = 7, 5, 3, 1$ for ortho levels and $I = 6, 4, 2, 0$ for para levels, the coupling matrix elements can be obtained [12.8] which allow determination of the spin-flip cross sections.

12.3 Isotope-Selective Spectroscopy of Ag_2 Dimers

When a silver surface is bombarded with 10-KeV argon ions, not only Ag atoms are sputtered but also Ag_2 dimers and larger clusters of Ag_n with $n \geq 3$. The question now arises as to whether these dimers and clusters are directly emitted from the surface of the solid or whether they are emitted as atoms but recombine within the dense vapor cloud close above the surface. A way of answering this question is the spectroscopic determination of the internal energy distribution of the emitted particles. If they are formed by recombination within the vapor clouds they should have a large internal energy, i.e. a vibrational population distribution reaching up to high vibrational levels.

This population distribution $N(v'', J'')$ of Ag_2 dimers can be measured by exciting the sputtered molecules by a pulsed dye laser into an excited electronic state which is then ionized by a second laser (Fig. 12.3). The Ag_2^+ ions are detected by a time of flight mass spectrometer, which also yields information on the translation energy distribution.

For the correct assignment of the absorption lines $(v'', J'') \rightarrow (v', J')$ the molecular constants of Ag_2 have to be known. Although several papers have appeared on the spectroscopy of Ag_2 [12.10], no rotationally resolved spectrum

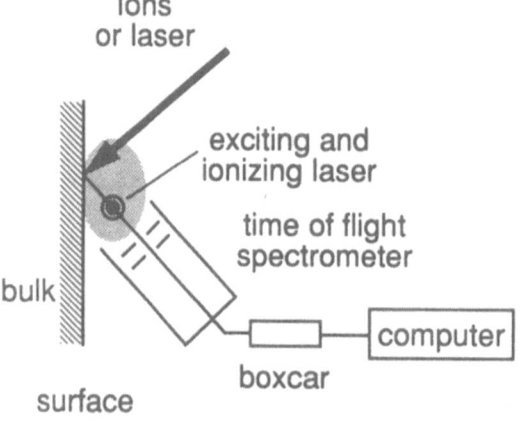

Fig. 12.3. Schematic experimental arrangement for measuring the internal state distribution of sputtered Ag_2 molecules by state-selective resonant two-photon ionization with time-of-flight mass selective detection

has been measured so far. We report here on high-resolution isotope-selective rotationally resolved spectroscopy of Ag_2 dimers, formed in a cold supersonic argon beam seeded with silver vapor. The measurements yield molecular constants in different electronic states, the ionization energy of Ag_2 and extrapolations of the dissociation energy.

The experimental setup is shown in Fig. 12.4. The Ag_2 molecules are formed during the adiabatic expansion of an argon–silver vapor mixture ($P_{Ar} \approx 3$ bar, $T \approx 2000$ K) from a high-temperature oven through a 70-µm nozzle into the vacuum. The rotational temperature of the Ag_2 molecules is reduced below 250 K by adiabatic cooling. The molecular beam is collimated by a skimmer and is then crossed perpendicularly by two superimposed laser beams. The first, narrow-band pulsed dye laser is tuned through the wanted electronic band system of Ag_2, thus populating levels (v', J') in an excited electronic state. For the higher states B, C, D, E frequency doubling of the dye laser is necessary. These levels are then ionized by a second laser (either an excimer laser or another dye laser pumped by the same excimer laser). The ions are extracted by an electric field, enter a time of flight mass spectrometer and are recorded after their mass specific flight time by a gated box car integrator.

Figure 12.5 shows a section of the $2 \leftarrow 0$ band in the $A^1\Sigma_u \leftarrow X^1\Sigma_g^+$ system recorded without mass selection. One clearly recognizes the three band heads of the three isotopes $^{107}Ag^{107}Ag$, $^{107}Ag^{109}Ag$ and $^{109}Ag^{109}Ag$, which are shifted against each other. However, the P and R branches of the three isotopes overlap, which makes an umambiguous assignment difficult, because the "lines" in

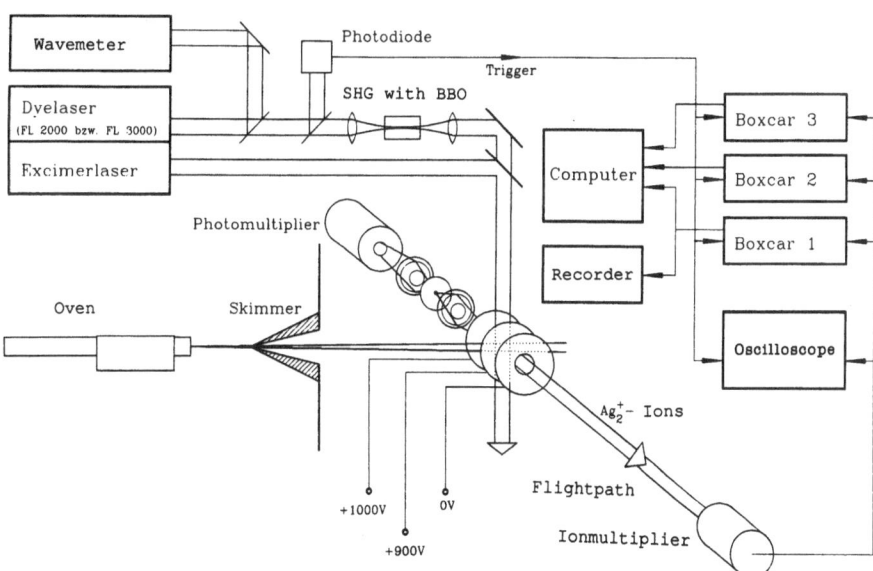

Fig. 12.4. Experimental setup for isotope selective spectroscopy of Ag_2 dimers in a supersonic molecular beam using a time of flight spectrometer and three gated box cars

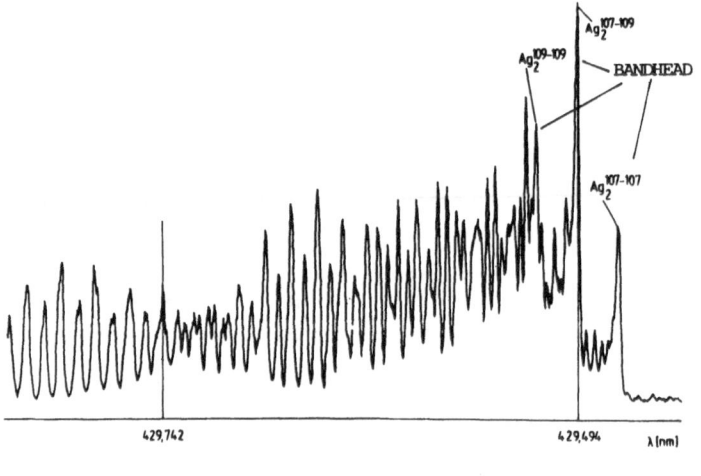

Ag$_2^{107-109}$

Ag$_2^{109-109}$

BANDHEAD

Ag$_2^{107-107}$

429,742   429,494   λ [nm]

F

Fig. 12.5. Section of the $2 \leftarrow 0$ band in the $A \leftarrow X$ system of Ag$_2$. The isotope shift of the band heads is clearly recognized, while for the larger wavelengths the rotational lines of the three isotopes overlap

Fig. 12.3 are in fact overlapping rotational transitions with different J values of the three isotopes.

With the time of flight spectrometer the three isotopes can be separately but simultaneously recorded by three boxcar integrators with appropriate gate times (Fig. 12.4). The calculated isotope shifts of corresponding rotational lines in the three spectra greatly facilitates the rotational assignments.

Of particular interest is the ionization potential IP of Ag$_2$. A comparison of the values IP(n) for Ag$_n$ clusters gives information about the change from localized electron orbitals for small n to a delocalized electron distribution for large n. From the dissociation energy D of the Ag$_2$ ground state and the ionization potential IP, the dissociation energy $D(\text{Ag}_2^+)$ of the ion ground state can be deduced according to

$$D(\text{Ag}_2^+) = \text{IP}(\text{Ag}) - \text{IP}(\text{Ag}_2) + D(\text{Ag}_2).$$

This gives insight into the binding characteristics of the electron which is removed upon photoionization

The IP can be determined in different ways: One method is based on the direct one-photon ionization starting from thermally populated levels (v'', J'') in the $X^1\Sigma_g$ ground state. Since in vapor cells or furnaces at high temperatures the thermal population distribution $N(v'', J'')$ is very broad, the ion yield, i.e. the number of produced ions as a function of the ionizing laser wavelength, follows a more or less flat curve determined by the convolution of population densities $N(v'', J'')$ and ionization probability. In such cases, the onset of this curve, which yields the ionization potential, cannot be determined very accurately. Here cold molecular beams which reduce the energy spread of the thermal distribution, are really advantageous.

A more selective ionization can be achieved by resonant two-photon two-color ionization with two tunable lasers. The first frequency-doubled laser is tuned to a selected transition $(v', J') \leftarrow (v'', J'')$ from the ground state to a level (v', J') in an intermediate electronic state and a second tunable laser ionizes the excited molecules in level (v', J').

The main problem of both methods is the uncertianty of the vibrational level v^+, reached in the ion ground state. If the potential curve of the ion ground state is shifted against that of the initial level of the ionizing transition, vertical transitions generally end in excited vibrational levels of the ion ground state and the apparent ionization threshold is too high. The second method has the advantage that different vibrational levels of the intermediate state can be selectively populated, thus probing different vertical transitions. A comparison of the measured onsets of the ionization yield $N_I(\lambda_L)$ as a function of laser wavelength λ_L generally allows the correct assignment of v^+ and therefore the adiabatic ionization potential.

Since the ions or photoelectrons are extracted by an electric field E, the appearance potential AP of the ions is lowered to

$$AP = IP - \left(\frac{e^3}{\pi\varepsilon_0}E\right)^{1/2}$$

and shows a shift $[(e^3/\pi\varepsilon_0)E]^{1/2}$ against the true ionizational potential IP. One, therefore, has to measure AP at different electric fields and extrapolate to $E \rightarrow 0$.

The most accurate method for the determination of the ionization potential is based on the measurement of a series of autoionizing Rydberg levels with term values $T(n, v^*, J^*)$ which converge for $n \rightarrow \infty$ against the level $(v^+ = v^*, J^+ = J^*)$ of the ion ground state [12.11]. If several series $T(n, v^*, J^*)$ for different vibrational levels can be assigned even the vibrational and rotational constants of the molecular ion ground state can be determined [12.12]. Figure 12.6 shows such a spectrum of Rydberg series $T(n, v^*, J^* \approx 10)$ which converge for $n \rightarrow \infty$ towards the vibrational levels $v^+ = 3, 4, ..., 9$.

With the last two methods the IP of Ag_2 has been determined as

$$IP(Ag_2) = (61\ 747 \pm 4)\ cm^{-1}.$$

The difference between the extrapolated appearance potential and the Rydberg convergence limits is about $3\ cm^{-1}$. The uncertainty of $\pm 5\ cm^{-1}$ is partly due to the wavelength measurement of the two lasers (performed with a monochromator) and the uncertainty of the extrapolation from the lowest measured vibrational level $v^* = 2$ to the level $v^* = 0$. A small contribution comes from the unresolved rotational distribution in the band heads of the vibrational bands, which were selected for the two-step excitation because of intensity reasons.

Isotope-specific rotationally resolved spectra, such as that shown in Fig. 12.4, have been measured for several bands in the B, C, D and E electronic states of Ag_2. The analysis of these spectra, which is not yet completed, yields the

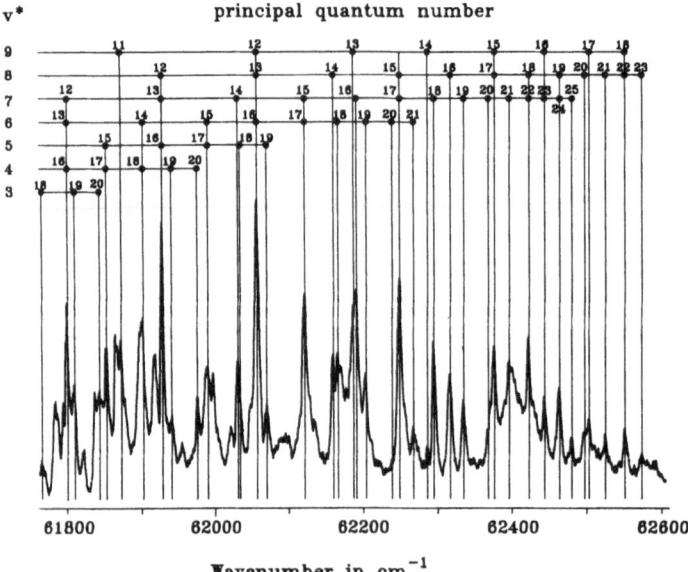

Fig. 12.6. Rydberg series $T(n, v^*, J^* \approx 10)$ of Ag_2, converging for $n \to \infty$ towards the vibrational levels $v + = v^*$ in the Ag_2 ground state

rotational and vibrational constants, the potential curves and the transition probabilities.

With the constants obtained so far the analysis of the population distribution sputtered from the silver surface under argon ion bombardment yields vibrationally populated levels up to 1 eV energy [12.13]. This proves that the dimers Ag_2 are formed by collision between sputtered atoms in the vapor cloud outside the solid.

12.4 Sub–Doppler Spectroscopy of the Sodium Cluster

Sodium clusters belong to the most thoroughly investigated metal clusters [12.14]. They are very good candidates for detailed studies of the transition region between molecules, small liquid drops and the solid metal. There are three main questions:

a) How is the geometrical structure changing from the well-defined potential curves of Na_2 and more or less floppy clusters of Na_n $(n \geq 3)$ to the rigid fcc structure of solid sodium? At which values of n is the transition to the solid phase occurring?

b) How is the electron distribution and the electronic structure changing from the well-known molecular orbital concept in Na_2 to the electronic band structure of the metal ? At which values of n does the delocalization of the electron distribution in a Na_n cluster start?

c) What is the dynamics of the nuclear motions in a cluster and how is it determined by the electron distribution and electron correlation?

Many laboratories have worked on these problems and the present paper on sub-Doppler laser spectroscopy of Na_3 can only give selected contributions to their solution.

The first studies of the Na_3 spectra with pulsed dye lasers have been performed by *Wöste* and coworkers [12.15, 16]. The spectral resolution of these pulsed experiments was limited by the laser bandwidth and was not sufficient to resolve the rotational structure. On the other hand, information on the geometry of Na_3 based on the knowledge of rotational constants, is essential not only to test ab initio calculations [12.17] but also to gain more detailed insight into the structure of a molecule without rigid geometry. The Na_3 cluster represents a good example of a small "floppy" molecule with a shallow potential, Jahn–Teller distortion and pseudo-rotation [12.18].

In order to resolve the rotational structure, sub-Doppler spectroscopy with single-mode cw dye lasers is necessary. Such measurements have been performed of Na_3 in our group [12.19]. The experimental arrangement is shown in Fig. 12.7. Cold sodium clusters Na_x are formed in a supersonic argon/sodium beam during the· adiabatic expansion from a heated stainless steel oven ($T = 900$–970 K, $P_{Ar} \leq 10$ bar). The molecular beam is skimmed and is crossed perpendicularly by the beam of a cw dye laser tuned to the A ← X system of Na_3. The laser-induced fluorescence is viewed by a photomultiplier PM1 or can be spectrally resolved by a monochromator at a fixed excitation wavelength. In a second crossing point, located in the ion source of a quadrupole mass spectrometer collinearly arranged with the laser beam, the dye laser excites levels in the A state which are then ionized by an argon ion laser.

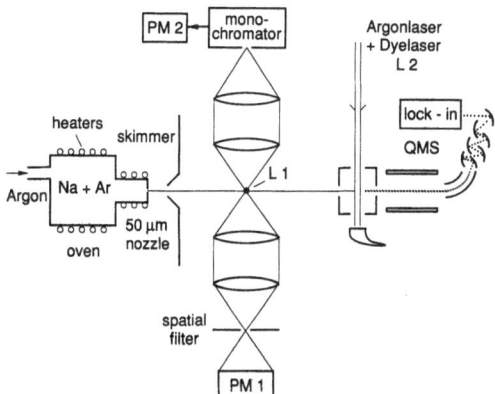

Fig. 12.7. Experimental arrangement for sub-Doppler spectroscopy of Na_x clusters. At the first crossing point LIF can be observed; at the second crossing point in the ion course of the quadrupole mass filter, resonant two-photon ionization is performed

Efficient resonant two-photon ionization with cw lasers in a molecular beam requires a special experimental arrangement. The ionization step must take place before the excited level decays. Which decay times 10^{-8}s the molecules travel at $\bar{v} = 10^3$ m a distance of 10^{-5}m $\approx$ 10 µm. This implies that the two laser beams must overlap within 10 µm. The experimental realization uses a combination of spherical and cylindrical lenses to form a rectangular focal spot of dimensions $(20 \times 1000 \ \mu\text{m}^2)$ in the molecular beam so that all molecules in the beam pass through both laser beams [12.1].

A section of the sub-Doppler spectrum is shown in Fig. 12.8. The high density of lines, which, even at a resolution of 30 MHz, are only partly resolved, is due to the rotational structure of a nonrigid asymmetric top with fine and hyperfine structure. The spectrum represents perpendicular transitions with $\Delta K = \pm 1$, a very narrow Q branch (because the geometry of upper and lower state is very similar), and weaker P and R branches.

Although the general structure of the spectrum can be reproduced using ab initio calculations of molecular constants [12.17, 20], the umambiguous assignment of individual lines turns out to be very difficult without further information.

This information is provided by optical–optical double resonance techniques (Fig.12.9). A dye laser L1 is tuned to a selected transition $|k\rangle \leftarrow |i\rangle$ of the A $\leftarrow$ X system, and depletes the population N_i due to partial saturation of the transition . If L1 is chopped at a frequency f_1, the population N_i is accordingly modulated. In the second crossing point the resonant two-photon ionization with a second dye laser and an argon laser takes place. When the ion signal of $M(Na_3^+)$ is measured after a lock-in amplifier tuned to the chopping frequency f_1, only those transitions are monitered which start from the labeled level $|i\rangle$. In

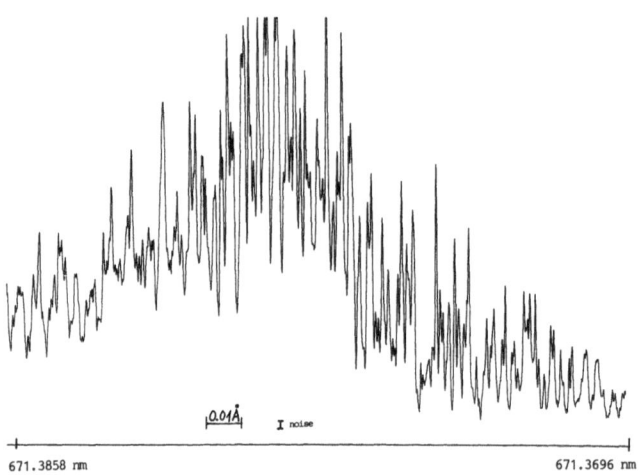

671.3858 nm   671.3696 nm

Fig. 12.8. Section of a sub-Doppler excitation spectrum of Na_3 in the A $\leftarrow$ X system monitored with resonant two-photon ionization. The noise level is indicated by the bar

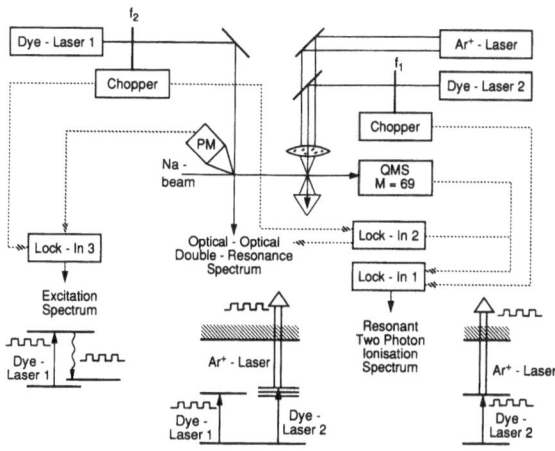

Fig. 12.9. Arrangement for optical–optical double resonance spectroscopy where the OODR signals are monitored by resonant two-photon ionization

order to discriminate against ionization signals caused by two argon laser photons, the second dye laser L2 may be chopped at a frequency f_2 and the OODR signal is then recorded at the sum frequency $f_1 + f_2$.

For a single lower labeled level one expects three OODR transitions with $\Delta J = 0, \pm 1$ from this level into three rotational levels in the upper state. Figure 12.10 shows, however, that three *groups* of transitions are observed which represent the hyperfine structure of the rotational transitions.

Ab initio calculations yield rotational constants sufficiently accurate that at least for transitions with low rotational quantum numbers, the rotational assignments can be performed unambiguously from the measured positions and spacings of the OODR signals. For the example of Fig. 12.10 the ratios $[\bar{v}(R) - \bar{v}(Q)]/[\bar{v}(Q) - \bar{v}(P)]$ yield $N''_{K_a K_c} = 2_{0,2}$ for the lower labeled level. Measurements of several of such OODR signals obtained with the pump laser kept on different absorption lines of Fig. 12.8 finally allow the assignment of the whole rotational structure of the A ← X spectrum [12.21].

When the probe lasers are kept on one of the OODR signals while the pump laser is tuned around the pump transition, the lower curve of Fig. 12.10 is obtained. This shows the degree of saturation (about 50%) and it also demonstrates that several different pump lines contribute to the OODR signal. This is due to the partial overlap of lines in the dense spectrum of Fig. 12.8.

The explanation of the hyperfine structure of the OODR signals is more complex [12.19] and can be traced to the slow pseudo-rotation of the Na_3 molecule in the ^2X ground state, which mixes hfs components during the time between pumping and probing the molecule. The anomalous intensity ratios of P, Q and R OODR transitions can also be explained by pseudo-rotation [12.21]

In the literature the question has been discussed of whether the higher vibronic levels in the Na_3 ground state may exhibit chaotic behavior [12.22]. This may be experimentally proved by measuring the statistics of the nearest neighbor distribution of vibronic levels [12.23], which are populated by stimulated emission pumping. Such experiments are under way in our laboratory.

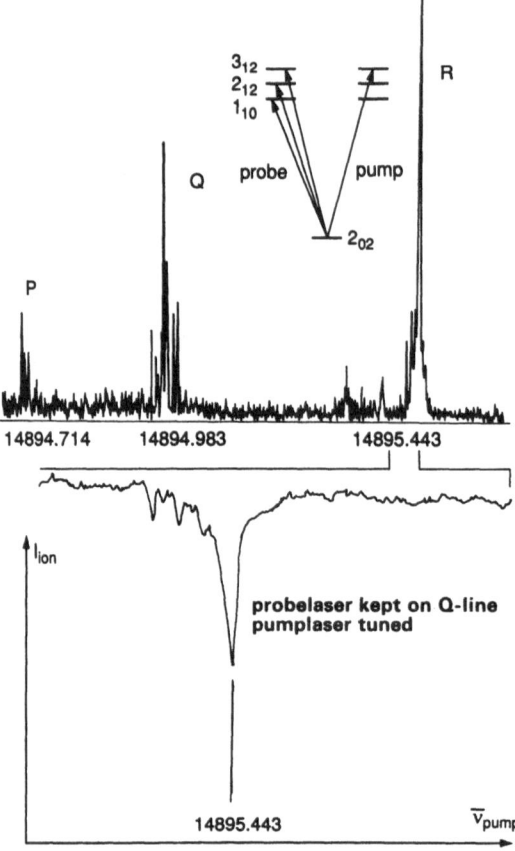

Fig. 12.10. P, Q and R OODR signals obtained when the dye laser L1 is kept on the R transition $3_{1,2} \leftarrow 2_{0,2}$ in the $A \leftarrow X$ system of Na_3. The lower trace is obtained when the probe laser wavelength is kept constant while the pump laser L1 is tuned. It shows a depletion of the level $N''_{K_aK_c} = 2_{0,2}$ by about 50%

Acknowledgement This work has been supported by the Deutsche Forschungsgemeinschaft within the Sonderforschungsbereich SFB 91.

References

12.1   W. Demtröder: Fresenius, J. Anal. Chem. **337**, 830 (1990)

12.2   T.E. Gough, R.E. Miller, G. Scoles: Appl. Phys. Lett. **30**, 338 (1977); D. Bassi, A. Boschetti, M. Scotoni: In Applied Laser Spectroscopy, ed. by W. Demtröder, M. Inguscio, NATO ASI-Series B, Vol. 241 (Plenum, New York 1991)

12.3   F. Bylicki, G. Persch, E. Mehdizadeh, W. Demtröder: Chem. Phys. **135**, 225 (1989)

12.4   D.H. Levy, L. Wharton, R.E. Smalley: Laser Spectroscopy in **Supersonic Jets**: In *Chemical and Biochemical Applications of Lasers Vol. II*, ed. by C.B. Moore (Academic, New York 1977)

12.5   G. Scoles (ed): *Atomic and Molecular Beam Methods Vols. I and II* (Oxford University Press, Oxford 1988 and 1991)

12.6   H.J. Neusser, V. Boesl, R. Weinkauf, E.W. Schlag: Int. J. Mass Spectrosc. and Ion Processes **60**, 147 (1984)

12.7   E. Tiesinger, S.J.M. Kuppens, B.J. Verhaar, H.T.C. Stoof: Phys. Rev. Lett. in press; D. Sesko, T. Walker, C. Monroe, A. Gallagher, C. Wieman: Phys. Rev. Lett. **63**, 961 (1989)

12.8   H. Weickenmeier, U. Diemer, W. Demtröder, M. Broyer: Chem. Phys. Lett. **124**, 470 (1986)

12.9   H. Weickenmeier, U. Diemer, M. Wahl, M. Raab, W. Demtröder, W. Müller: J. Chem. Phys. **82**, 5354 (1985)

12.10   D.J. Pesic, B.R. Vujisic: J. Mol. Spectrosc. **146**, 561 (1991) and references therein.

12.11   R.E. Stebbings, F.B. Dunning: *Rydberg states of atoms and molecules* (Cambridge University Press, Cambridge MA. 1983)

12.12   M. Schwarz, R. Duchowicz, W. Demtröder, Chr. Jungen: J. Chem. Phys. **89**, 5460 (1988)

12.13   A. Wucher, K. Franzreb, H. Oechsner: To be published

12.14   C. Brechignac, Ph. Cahuzac, F. Carlier, M. de Frutos, J. Leygnier: J. Chem. Soc. Farad. Trans. **86**, 2525 (1990); M. Kappes, M. Schär, U. Röthlisberger, C. Yeretzian, E. Schumacher: Chem. Phys. Lett. **143**, 251 (1988)

12.15   M. Broyer, G. Delacretaz, P. Labastie, J.P. Wolf, L. Wöste: J. Phys. Chem. **91**, 2620 (1987)

12.16   M. Broyer, G. Delecretaz, G.Q. Ni, R.L. Whetten, J.P. Wolf, L. Wöste: J. Chem. Phys. **90**, 843 (1988)

12.17   F. Cocchini, W. Andreoni, T.H. Upton: J. Chem. Phys. **88**, 6068 (1988)

12.18   G. Delecretaz, E.R. Grant, R.L. Whetten, L. Wöste, J.W. Zwanziger: Phys. Rev. Lett. **56**, 2598 (1986)

12.19   H.-J. Foth, J.-M. Gress, Chr. Hertzler, W. Demtröder: Z. Phys. **D18**, 257 (1991)

12.20   W. Meyer: FB Chemie, Univ. Kaiserslautern, Private communication

12.21   H.-A. Eckel, J.-M. Gress, W. Demtröder: To be published

12.22   J.M. Llorente, H.S. Taylor: J. Chem. Phys. **91**, 953 (1989)

12.23   G. Persch, E. Mehdizadeh, W. Demtröder, Th. Zimmermann, H. Köppel: Ber. Bunsengesellschaft Phys. Chem. **92**, 312 (1988)

Wolfgang Demtröder

13. In Situ Gas-Phase Diagnostics by Coherent Anti-Stokes Raman Scattering

W. Richter

With 9 Figures

This chapter reviews the characteristics of coherent anti-Stokes Raman scattering (CARS) with respect to gas-phase diagnostics. The basic features of the CARS process and its spectroscopic application for diagnostic purposes are discussed. This is illustrated by examples taken from the area of vapour-phase epitaxial growth of semiconductor layers.

13.1 Introduction

The goal of gas-phase diagnostics is to determine all relevant parameters of a gas such as temperature, the kind of species present and their number densities. Quite often this information has to be obtained with spatial or temporal resolution because the system under study is not homogeneous. Gradients may occur and/or stationary conditions may not be reached. Examples of such systems are plasma processes, gas-phase processes stimulated by pulsed laser radiation and the large area of chemical vapour deposition.

Diagnostics in such systems with total pressures typically in the range from 1 mbar to 1 bar is essentially limited to optical methods. In order to get specific information for each molecular species, interaction of the electromagnetic (EM) radiation with either electronic, vibrational or rotational excitations of the molecule can be utilized. Electronic transitions are conveniently observed in the UV/VIS spectral range by absorption spectroscopy or laser-induced fluorescence. The latter is an especially sensitive technique, with high spatial and temporal resolution. However, sharp spectral features usable for identification of molecular species are only obtained for atoms or very simple molecules. Thus, the application is limited to specific situations. First-order interactions of EM radiation with vibrational or rotational excitations, on the other hand, are generally located in the infrared spectral region. There, in comparison to the visible spectral range, spectroscopic work over a larger spectral range is in general more difficult and usually less sensitive. Thus, for the measurement of vibrational/rotational excitations the higher-order, Raman type, EM interactions are quite often utilized which transfer the experimental range by frequency sum or difference processes into the visible spectrum.

One of these higher-order Raman type processes is coherent anti-Stokes Raman scattering (CARS), which was first observed in 1965 [13.1, 2]. It is a four-

Topics in Applied Physics, Vol. 70
Dye Lasers: 25 Years Ed.: Dr. Michael Stuke
© Springer-Verlag Berlin Heidelberg 1992

wave-mixing process usually for spectroscopic purposes performed with two laser frequencies (at least one tunable) and provides quite good spectral (< 0.1 cm^{-1}), spatial (< 1 mm) and temporal (ns) resolution. Furthermore, its sensitivity as a nonlinear optical interaction can be made quite high (detection limit below 0.1 mbar) and it is not limited to a certain class of molecules. Thus, a wide range of molecular species can be accessed. Of course single atoms are outside the detection range of CARS. For them, however, laser-induced fluorescence is an excellent technique.

13.2 Coherent Anti-Stokes Raman Scattering

The principle of CARS is displayed in the energy transition diagram of Fig. 13.1. Strong contributions to the CARS matrix elements arise when the process becomes resonant with the vibrational–rotational (or just rotational) molecular transitions, i.e. when $\omega_L - \omega_S = \omega_{ij}$. Under such circumstances the first two transitions may be interpreted as preparing the molecules into coherently excited vibrational–rotational states while the last two transitions correspond to an anti-Stokes Raman process. Electronic resonances ($n = 1$ to $n = 2$) could, in principle, be exploited too in order to increase the vibration–rotation resonant signal. However, this in general requires an additional laser providing a third frequency and complicates the experimental set-up considerably.

The CARS intensity may be written as [13.2–6]

$$I_{\text{CARS}} \propto |\chi^{(3)}|^2 I_L^2 I_S L^2 \left(\frac{\sin\left(\frac{1}{2}\Delta k L\right)}{\frac{1}{2}\Delta k L} \right)^2 , \tag{13.1}$$

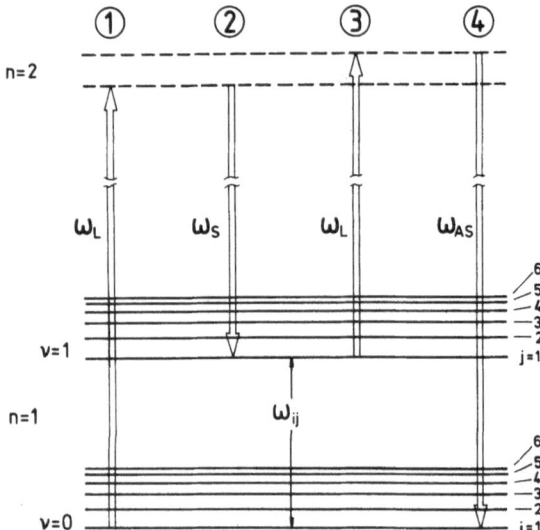

Fig. 13.1. Schematic diagram of transitions between molecular energy levels for a CARS process. Quantum numbers are i, j for rotational, v for vibrational and n for electronic states

where L is the interaction length, $I_{L,S}$ are the laser intensities for the scheme displayed in Fig. 13.1 (two transitions with ω_L) and Δk is the phase mismatch. The third-order susceptibility $\chi^{(3)}_{CARS}$ is obtained from perturbation theory as [13.3]

$$\chi^{(3)}_{CARS} \propto N \sum_{ij} \frac{\Delta\rho_{ij}}{\omega_{ij} - \omega_L + \omega_S - i\Gamma_{ij}} \left| \sum_n \left(\frac{\mu_{jn}\mu_{ni}}{\omega_{ni} - \omega_L} + \frac{\mu_{jn}\mu_{ni}}{\omega_{nj} + \omega_L} \right) \right|^2 , \qquad (13.2)$$

with N being the molecular number density, $\Delta\rho_{ij}$ the difference in occupation between states i and j and Γ_{ij} the transitional linewidth. The summation over all electronic states n contains the electronic dipole matrix elements μ in the numerator, while the denominator expresses the possible electronic resonances discussed above. Together with the perturbation expression for the spontaneous Raman cross-section [13.3],

$$\left(\frac{d\sigma}{d\Omega} \right)_{ij} \propto \omega_L \omega_S^3 \left| \sum_n \left(\frac{\mu_{jn}\mu_{ni}}{\omega_{ni} - \omega_L} + \frac{\mu_{jn}\mu_{ni}}{\omega_{jn} + \omega_L} \right) \right|^2 , \qquad (13.3)$$

equation (13.1) can then be rewritten in the form [for $\Delta k = 0$ and by inserting (13.2, 3)]

$$P_{CARS} \propto \left| N \sum_{ij} \left(\frac{d\sigma}{d\Omega} \right)_{ij} \frac{\Delta\rho_{ij}}{\omega_{ij} - \omega_L + \omega_S - i\Gamma_{ij}} + \chi^{(3)}_{nr} \right|^2 P_L^2 P_S , \qquad (13.4)$$

where in addition a nonresonant ($\omega_{ij} \neq \omega_L - \omega_S$) contribution $\chi^{(3)}_{nr}$ has been added to $\chi^{(3)}_{CARS}$ and the use of laser powers $P_{L,S}$ has eliminated the interaction length L under the assumption of Gaussian beam profiles.

Equation (13.4) shows that the following information can be obtained from a CARS measurement:

i) identification of the molecule through determination of ω_{ij};
ii) relative number densities (N) which might be calibrated with external standards;
iii) temperature via the difference in occupation number $\Delta\rho_{ij}$; and
iv) possibly total pressure via its influence on Γ_{ij}.

Examples of these possibilities will be discussed in Sect. 13.4.

13.3 Experimental Considerations

In order to achieve resonance with the rotational–vibrational transitions at least one tunable laser is needed. Commonly either two dye lasers pumped by an excimer laser or one dye laser and a frequency-doubled Nd:YAG laser are used. Detailed information concerning the CARS experiment can be found in [13.2, 7, 8]. Here only three topics which are important for gas-phase diagnostics will be discussed. These concern the operational mode of the dye laser, the beam configuration to achieve phase matching and the polarisation of the EM fields involved.

As explained in Fig. 13.2 CARS spectra can be obtained either with a narrow band dye laser by scanning (scanning CARS) or with a broad band operating dye laser in a "one shot" experiment by employing an optical multichannel analyser (broadband CARS). While in scanning CARS the spectral resolution is determined by the linewidth of the two lasers providing the frequencies $\omega_{L,S}$, in broadband CARS the resolution is in general determined by the optical multichannel detection system. Thus, in broadband CARS rotational structures, if present, are in general obscured, while in scanning CARS the rotational transitions may be resolved by tuning the laser to the appropriate linewidth. This is an important feature for the identification of molecules and also simplifies temperature measurements. Only in cases where the origin of the spectral feature is well-known should broadband CARS with its inherent time advantage be utilized.

The second remark concerns the phase matching. In transparent gases because of their negligible dispersion phase matching, i.e. $\Delta k = 0$ in (13.1), can be achieved in the simple collinear beam arrangement (Fig. 13.3a). However, the planar so-called BOXCARS (Fig. 13.3b) or the folded BOXCARS arrangement (Fig. 13.3b) have additional advantages. First of all, the spatial resolution is considerably increased since the beams have less overlap. Secondly, it is easier to separate the anti-Stokes radiation from the input laser beams since their

a) Scanning-CARS:

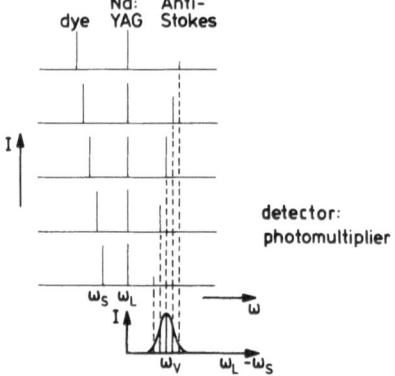

b) Broadband-CARS:

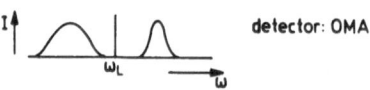

Fig. 13.2a, b. Experimental procedures for obtaining CARS spectra by either (a) scanning the dye laser or (b) using a broad band operating dye laser and spectrally analysing the CARS signal with an optical multichannel analyser (OMA)

a) Collinear CARS:

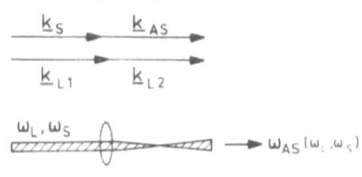

b) BOXCARS:

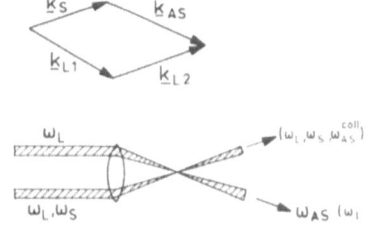

c) Folded BOXCARS:

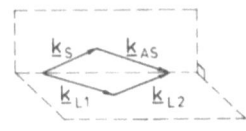

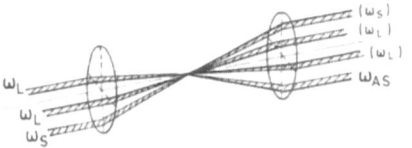

Fig. 13.3. Different possibilities to achieve phase matching in CARS. The wavevector diagram and the corresponding experimental configurations are shown in each case

directions are different. The result is a much more efficient reduction of background in the spectra.

The last point to be made in relation to the experiment is connected with the tensor properties of $\chi^{(3)}_{CARS}$. This fourth rank tensor connects the electric field polarisations of the input fields to that of the anti-Stokes field. In all cases where the depolarisation ratio,

$$\rho = \chi^{(3)}_{1221}/\chi^{(3)}_{1111} \; , \tag{13.5}$$

is different for the non-resonant background χ_{nr} and the vibration under study, selection of certain polarisations can be utilized to suppress the non-resonant background. Figure 13.4 gives an example for the A_1 vibration of CH_4 ($\rho = 0$). While in Fig. 13.4d only the background ($\rho = 1/3$) is seen, in Fig. 13.4b the vibrational signal appears with almost no background, by using the polarisations as indicated in Fig. 13.4b. A considerable increase in signal-to-background ratio can be achieved this way. Intermediate situations are displayed in Fig. 13.4a, c, where both coherent signals are present and the lineshape is strongly affected by interference effects. In such cases the integrated signal is difficult to obtain and determination of the number densities becomes quite inaccurate. Besides this large increase of signal-to-background ratio possible with appropriate polarisations, one notices, however, that in line with the

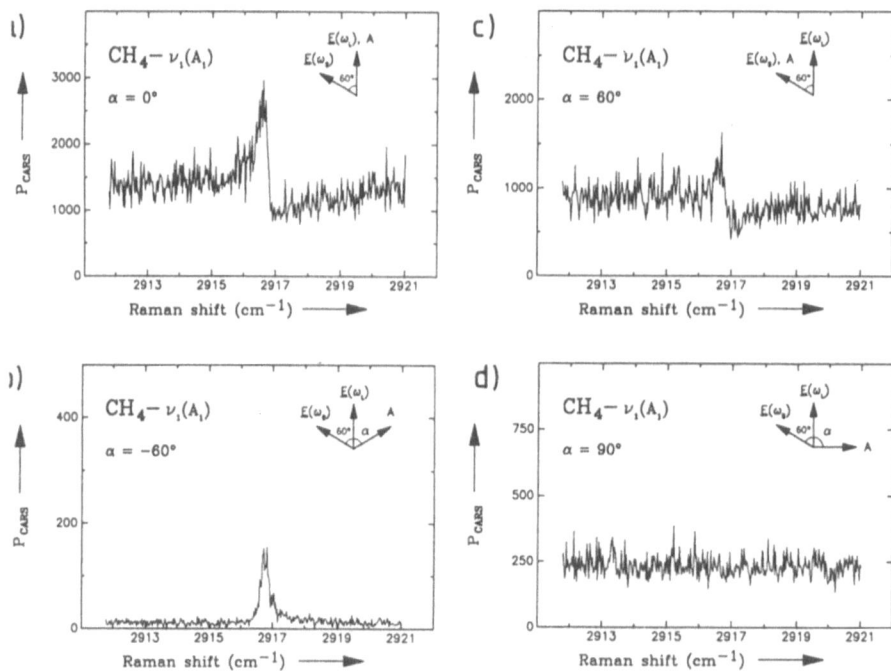

Fig. 13.4. CARS spectra of the $\nu_1(A_1)$ vibration of CH_4 taken with different polarisations (as indicated in the inserts) for the pump laser (index L), the Stokes laser (index S) and the anti-Stokes signal (A). α denotes the angle between anti-Stokes and pump laser polarisation

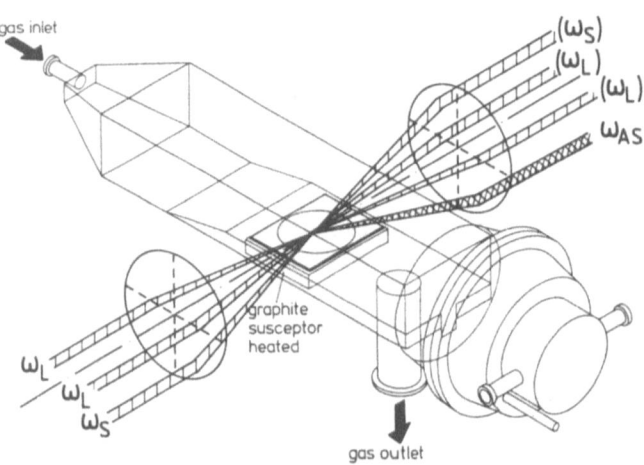

Fig. 13.5. Experimental arrangement to perform folded BOXCARS in a horizontal reactor used for chemical vapour deposition on a heated substrate. Working distance between the two lenses may be up to 1 m

background reduction the signal is also strongly attenuated (by a factor of 8 in Fig. 13.4). This signal reduction may limit the detection of molecular species in situations where the background originates not from χ_{nr} but from other sources (e.g. stray light, noise).

Finally, a typical experimental arrangement is shown in Fig. 13.5 with a folded BOXCARS configuration in connection with a typical horizontal reactor used for chemical vapour deposition. Focusing distances of up to 0.5 m can be applied and thus access to many different "reaction cells" like motors or plasma environments is possible as long as windows can be attached to the experimental chamber.

13.4 Experimental Results

A number of papers have been published which review the CARS results in different fields such as combustion [13.9] or CVD [13.10]. We will present here more recent results obtained in studies of the metal-organic vapour-phase epitaxy (MOVPE) and which illustrate nicely the possible information that can be obtained by CARS as discussed in Sect. 13.2.

Identification of molecular species may be performed by comparing the measured CARS spectra with spectra or line frequencies quoted in the literature. This is a straightforward method which may be also applied to molecular subgroups. Figure 13.6 gives an example [13.11]. The rotational substructure of vibrational modes is an additional feature used for identification. For a given molecular structure the moments of inertia can be calculated with high accuracy. Together with the degeneracy of rotational states and the thermal occupation function (Bose–Einstein) the spectral shape can be calculated. Figure 13.7 shows the comparison of such calculations (Fig. 13.7c, d) with corresponding experiments (Fig. 13.7a, b) [13.12]. In many cases the high spectral resolution of the CARS signal results in an even more accurate determination of the molecular structure by regarding the moments of inertia as fit parameters.

From the rotational structure the *temperature* can also be determined, as can be seen from Fig. 13.7. This can be done either by calculations, where the temperature enters through the thermal occupation function, or by comparing with a catalogue of spectra taken at different temperatures and calibrated by other means (thermocouple). The former method even provides an absolute temperature determination. Temperature measurements have been performed with a time resolution of nanoseconds in non-stationary photoreactions [13.13]. Care has to be taken, however, not to perform these measurements at too high laser powers. In this case the occupation function may deviate considerably from thermal equilibrium. The CARS experiment then still measures the occupation difference between two states but this can no longer be related to the temperature. Experimentally this is easy to check, however, since deviations in

tertiary-butyl-phosphine

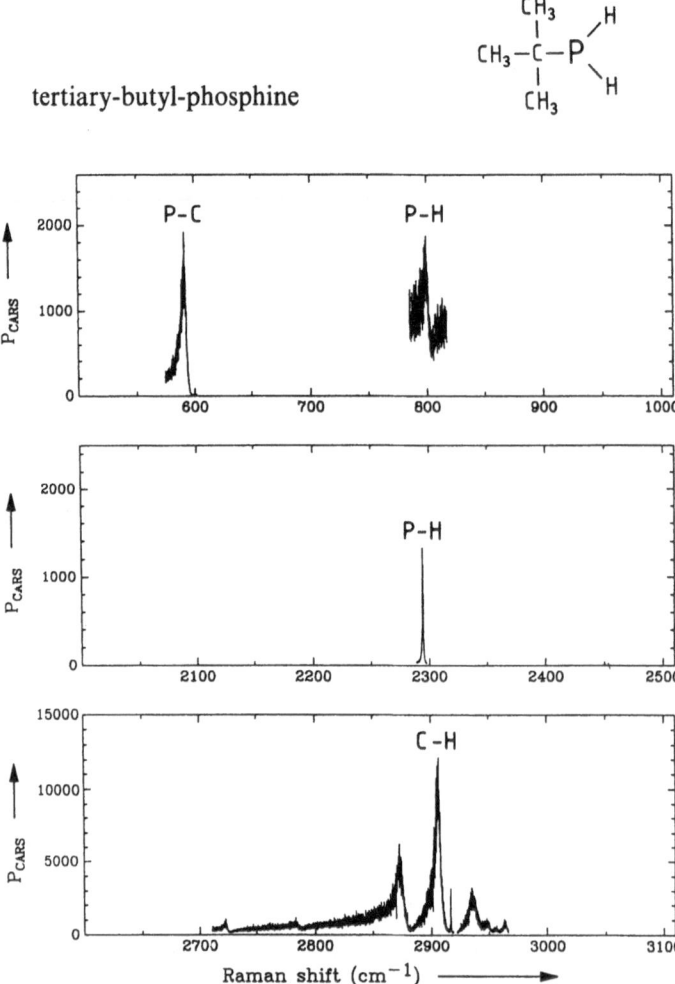

Fig. 13.6. CARS spectrum of tertiary butyl-phosphine (t-BPH$_2$) shown in parts for different spectral regions. The bonds indicated above the peaks are those which are mainly involved in the corresponding vibration

the spectral shape with increasing laser power signal the onset of non-thermal distributions.

The determination of *number densities* of molecular species constitutes a more difficult task than the two previous ones. It is, nevertheless, relatively simple if the molecule under study is also available in larger quantities as a pure gas or liquid with sufficient vapour pressure. In this case CARS measurements with a well-controlled partial pressure can be performed as a "standard" and used for calibration. H$_2$ or CH$_4$ quite often produced in gas-phase reactions

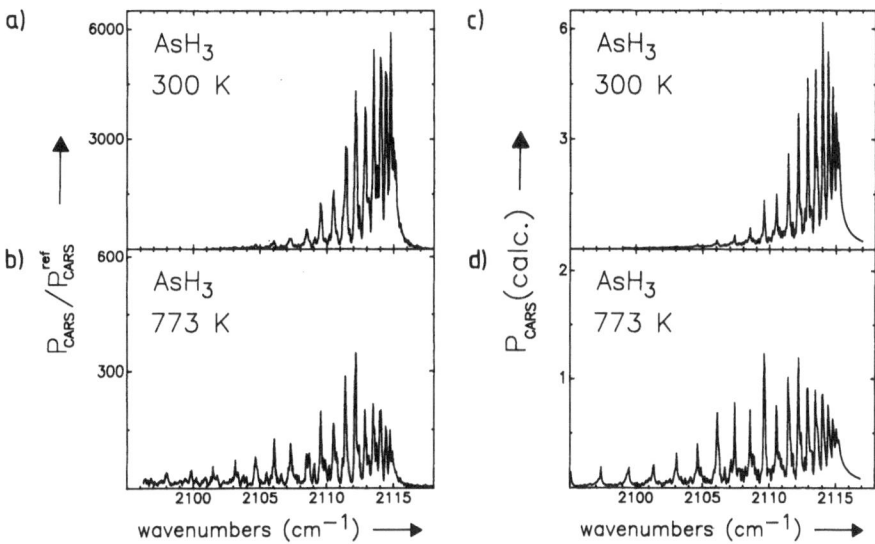

Fig. 13.7a–d. Measured (**a, b**) and calculated (**c, d**) CARS spectra of the $v_1(A_1)$ vibration of AsH_3 at two different temperatures. From [13.12]

offer examples of this determination of concentrations. In other cases it is often possible to perform a relative calibration to some known value. This procedure can be performed with results such as those presented in Fig. 13.8 [13.11]. There the thermal decomposition of t-butylphosphine, a new precursor for the epitaxial growth of InP, is measured as a function of temperature for constant input partial pressure. The strong decrease of the peak at 2293 cm^{-1} signals the decomposition of the molecules while the new structure appearing at higher temperatures can be assigned to the generation of PH_3 in the decomposition reaction.

The square root of the integrated peak areas is usually assumed to be proportional to the number density according to (13.4). This, of course, introduces some uncertainty because destructive interference terms from transitions with the same ω_{ij} but different quantum numbers i, j may require a complete simulation of the spectra. However, it turns out that this error is at most 10–20% and can be tolerated in most cases. At room temperature t-butylphosphine is known not to decompose noticeably. The 300-K spectrum is therefore taken as a reference spectrum to normalize the higher temperature spectra, taking into account the obvious decrease of number density due to thermal expansion.

The result of such a procedure for the thermal decomposition of AsH_3 (relevant for the growth of GaAs) is shown in Fig. 13.9. With increasing temperature, AsH_3 is decomposed and H_2 is produced. The decomposition and production rates, however, are different when measured over different surfaces.

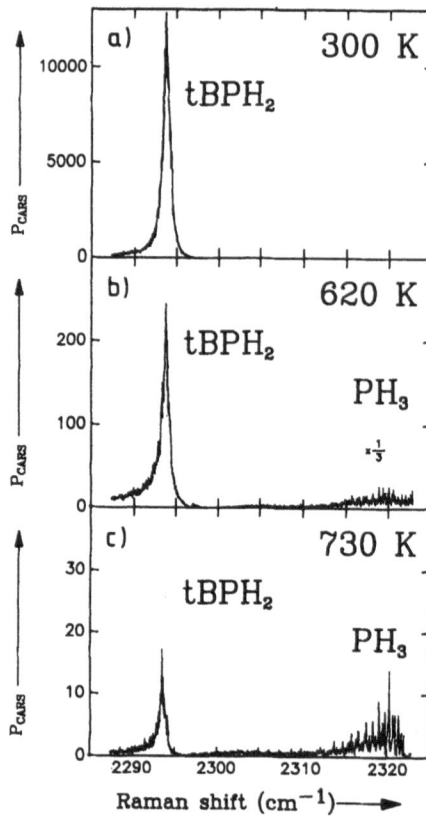

Fig. 13.8. CARS spectra of tertiary butyl-phosphine (t-BPH$_2$) in the spectral region of the P–H stretching vibration for different temperatures. Note the different scales for P_{CARS}. At the highest temperature t-BPH$_2$ is mainly decomposed, while simultaneously PH$_3$ is produced. From [13.13]

Therefore, one must conclude from these CARS data that surface reactions are involved in the chemical transformations. Thus, the decomposition of AsH$_3$ cannot be only a homogeneous reaction, but at lower temperatures (relevant for epitaxial growth) heterogeneous catalytic reactions at the GaAs surface must dominate.

Finally, the evaluation of *total pressures* from the linewidth Γ_{ij} is a difficult task since the measured CARS spectra are obtained from (13.4) by convoluting with the laser linewidths. However, at higher pressures (1 bar or more) the influence is significant enough that, for the simulation of spectra at least, the change of Γ_{ij} with total pressure has to be taken into account [13.14].

13.5 Summary and Conclusions

The previous section illustrates that quite detailed information about gas-phase parameters and reactions can be obtained by CARS. Many more examples

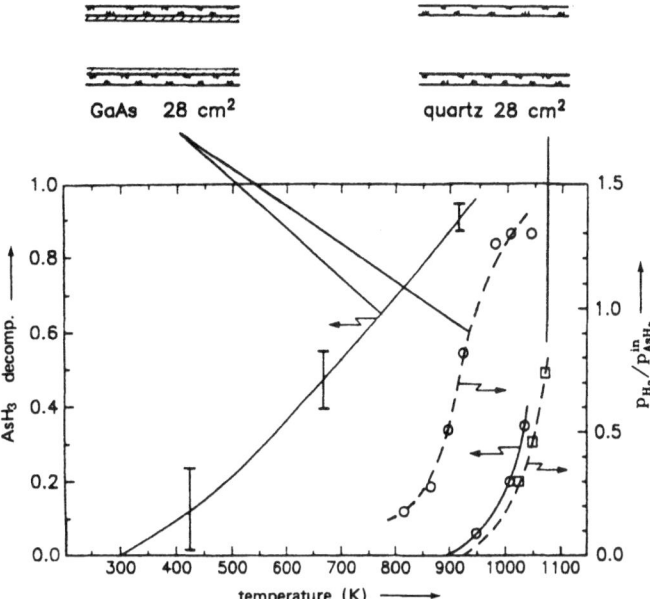

Fig. 13.9. Rate of gas-phase decomposition of AsH_3 (solid line) and production of H_2 (dashed line) during the thermal decomposition of AsH_3 above a quartz and a GaAs surface. The difference in the data for the two surfaces indicates surface reactions. From [13.13]

could have been added. The main advantage of CARS in all these cases lies in its excellent diagnostic properties, i.e. spectral ($< 0.1 \ cm^{-1}$), spatial ($< 1 \ mm$) and temporal (ns) resolution as well as sensitivity (detection limit $< 0.1 \ mbar$). The other quite important experimental advantages are the large working distances combined with the properties of coherent beams. They allow diagnostics to be performed in quite hostile environments without the need of large optical apertures.

The CARS technique itself as a spectroscopic tool has become possible only through the tunable dye laser. The development of dye lasers from a difficult scientific instrument to a "one button" machines has made it possible nowadays to perform CARS spectroscopy as a routine measurement. Consequently, even complete spectroscopic set-ups are commercially available today.

Acknowledgements. This work was supported by the Stiftung Volkswagenwerk (AZ I/64061) and in part also by the Bundes-Ministerium für Forschung und Technologie (AZ TK452).

References

13.1   P.D. Maker, R.W. Terhune: Phys. Rev. **A801**, 137 (1965)
13.2   D.A. Greenhalgh: Quantitative CARS Spectroscopy, Chap. 5 in *Advances in Non-Linear Spectroscopy* (Advances in Spectroscopy Vol. 15) ed. by R.J.H. Clark, R.E. Hester (Wiley, Chichester 1988)
13.3   S.A.J. Druet, J.P.E. Taran: Prog. Quantum Electron. **7**, 1 (1981)
13.4   R.J. Hall, A.C. Eckbreth: CARS: Application to Combustion Diagnostics in *Laser Applications* (Vol. V), ed. by Erf, Ready (Academic, Orlando 1984)
13.5   J.P. Taran: In *Nonlinear Raman Spectroscopy and Its Application*, ed. by W. Kiefer, D.A. Long (Reidel, Dordrecht 1982)
13.6   J.W. Nibler, G.V. Knighten: In *Raman Spectroscopy of Gases and Liquids*, ed. by A. Weber, Topics Curr. Phys. Vol. 11 (Springer, Berlin, Heidelberg 1979) Chap. 7
13.7   M.D. Levenson, J.J. Song: In *Coherent Nonlinear Optics*, ed. by M.S. Feld, V. S. Letokhov, Topics Curr. Phys., Vol. 21 (Springer, Berlin, Heidelberg 1980)
13.8   G.L. Eesley: *Coherent Raman Spectroscopy* (Pergamon, Oxford 1981)
13.9   A.C. Eckbreth, P.W. Schreiber: CARS: Application to Combustion and Gas-Phase Diagnostics, in *Chemical Applications of Nonlinear Raman Spectroscopy*, ed. by A.B. Harvey (Academic, New York 1981)
13.10   R. Devonshire: Chemtronics **2**, 183 (1987)
13.11   P. Kurpas, M. Motzkus, W. Richter: To be published
13.12   R. Lückerath, P. Tommack, A. Hertling, H.J. Koss, P. Balk, K.F. Jensen, W. Richter: J. Crystal Growth **93**, 151 (1988)
13.13   W. Richter, P. Kurpas, R. Lückerath, M. Motzkus, M. Waschbüsch: J. Crystal Growth **107**, 13 (1991)
13.14   L.A. Rahn, R.E. Palmer: J. Opt. Soc. Am. B3, 1164 (1986)

Wolfgang Richter

14. High-Resolution Spectroscopy at Short Wavelengths Using Pulsed Dye Lasers

S. Svanberg

With 5 Figures

The invention of the dye laser [14.1, 2] 25 years ago has brought about an enormous development of optical spectroscopy (for reviews see, e.g., [14.3, 4]). The tunability and high spectral intensity of these sources in combination with the possibility to achieve extremely narrow bandwidths for cw lasers, or pulses of extremely short duration, have greatly facilitated many spectroscopic investigations and made entirely new types of experiments possible.

In the present chapter we will discuss high-resolution laser spectroscopy of free atoms and especially focus on the UV and VUV wavelength region. (For a more detailed discussion of this topic we refer to [14.5]). Dye lasers do not operate at such wavelengths, but their radiation can be frequency converted to this spectral region using nonlinear optics techniques. Such conversion can most conveniently be performed with pulsed lasers, that, however, necessarily have a much larger linewidth than the single-mode cw systems available at longer wavelengths.

The topic of this chapter is primarily to discuss how a resolution, limited only by the Heisenberg uncertainty relation, can still be obtained by combining intense, broadband excitation with "classical" high-resolution techniques such as optical double-resonance (ODR), level-crossing (LC) and quantum-beat (QB) spectroscopy. Such techniques proved very useful at an earlier stage of development of dye lasers. At the beginning of the 1970s only multi-mode cw lasers were available and in a very limited wavelength range. Such lasers, used for stepwise excitations and combined with ODR and LC spectroscopy, allowed a substantial extension of high-resolution alkali-atom spectroscopy (see e.g. [14.6, 7]). At the same time, QB spectroscopy was shown to effectively extend the wavelength range of high-resolution laser spectroscopy [14.8].

While awaiting the development of broadly tunable, narrow-band cw laser sources allowing the full utilization of Doppler-free laser spectroscopy methods using collimated atomic or ionic beams, spectral holeburning, two-photon absorption and cooled atoms or ions in traps, the ODR, LC and QB methods can provide precision spectroscopic data on atoms and ions. However, it must be remembered that these resonance and coherence methods are only useful for the measurement of level *splittings* due to, e.g., fine and hyperfine interactions, Zeeman and Stark effects. Measurements of isotopic shifts, scalar Stark interaction, and of course investigations of the absolute wavelength/frequency of the optical transition, still require the narrow-band laser sources. For selected wavelengths such radiation can be achieved, e.g., for Lamb-shift/Rydberg

Topics in Applied Physics, Vol. 70
Dye Lasers: 25 Years Ed.: Dr. Michael Stuke
© Springer-Verlag Berlin Heidelberg 1992

constant measurements or for optical frequency standards (see [14.5] for references).

14.1 Generation of UV/VUV Radiation

A variety of methods is available for the generation of short wavelength coherent radiation. Most commonly tunable radiation in the UV or VUV spectral region is generated using nonlinear optics frequency shifting of dye laser radiation.

14.1.1 Sum-Frequency Generation in Crystals

Frequency doubling or mixing in non-linear crystals is an efficient way to generate radiation in regions where the crystals are transparent and phase-matching conditions can be obtained. A review of modern materials and techniques can be found in [14.9]. Of particular importance is β-barium borate (BBO), which provides frequency doubling down to 205 nm. Excimer-pumped blue dye lasers can very efficiently be doubled in this way. Frequency tripling of red dye-laser radiation, by first doubling in KDP and subsequent mixing with the residual fundamental in BBO, can be performed with high efficiency with a frequency cut-off at 197 nm [14.10]. We have been able to generate almost 10 mJ of tunable radiation in the region around 200 nm using the radiation from a Nd:YAG pumped dye laser.

14.1.2 Frequency Conversion in Gases

Gases are transparent at shorter wavelengths than crystals and different non-linear techniques can be used to attain VUV radiation. A simple technique not requiring phase matching is stimulated Raman scattering in high-pressure H_2 gas [14.11]. By the generation of successively higher anti-Stokes components, a photon energy increase of 4155 cm^{-1} or $\approx 0.5 \text{ eV}$ can be gained in each step. This means that, using a primary laser wavelength of 200 nm, the first anti-Stokes Raman component will be at 185 nm and the second anti-Stokes component at 170 nm.

Sum and difference frequency mixing can be performed in noble gases such as Kr and Xe, and also in metal vapours such as Hg and Mg. The efficiency is greatly improved if resonance enhancement through two-photon resonances in the gas can be obtained. For sum frequency mixing, phase matching must be achieved. The index of refraction of the gas can be manipulated by adding noble gas to the converting metal vapour. Most of the range 100–200 nm can be covered in this way. Even shorter wavelengths can be achieved by high (odd) harmonic generation in the noble gases. The field of non-linear frequency conversion in gases is covered in [14.12–14].

Recently, very high odd overtones have been generated in the interaction of intense laser radiation with noble gases [14.15, 16]. The highest overtone generated so far is the 97th (8.3 nm) of a titanium–sapphire solid-state laser operating at about 807 nm [14.17]. In Lund, a tunable titanium–sapphire terawatt laser system suitable for high overtone generation is now being assembled.

14.2 High-Resolution Laser Spectroscopy Using Pulsed Dye Lasers

We will now illustrate how pulsed dye laser radiation at short wavelengths can be combined with ODR, LC and QB spectroscopy to produce precision data on radiative lifetimes, Landé g_J factors and hyperfine interaction constants. The three methods will be illustrated by recent work at the Lund Institute of Technology, with examples from silver and ytterbium. Finally, some hybrid experiments on nitrogen atoms that are produced by photodissociation of parent molecules using the resonant spectroscopic laser pulse are described.

14.2.1 Optical Double Resonance Experiments

In the ODR method [14.18–20] polarised light is used to excite the different substates of the excited atomic level, and the decay exhibits a spatial anisotropy and certain polarisation properties. By inducing radio-frequency transitions between the sublevels the population differences are leveled out and the angular and polarisation pattern changes. A sufficiently high rf amplitude is needed to bring about the transfer during the limited time available in the short-lived excited state. The signal linewidth directly reflects the level broadening due to the uncertainty relation. A useful experimental set up for pulsed ODR experiments is described in [14.21], discussing measurements on the 6p ^{2}P states of Ag reachable with 206 nm radiation. A diagram of the magnetic sublevels of the 6p $^2P_{3/2}$ state is shown in the upper part of Fig. 14.1. Silver has two stable isotopes, ^{107}Ag and ^{109}Ag, both with nuclear spin $I = 1/2$. In the Paschen–Back region $2I + 1$ rf signals are expected, symmetrically around the position for the spin-zero case. Such signals are shown in the lower left part of Fig. 14.1, also featuring the narrowing down of the signals when detection is restricted to "old" atoms, i.e. when the detection system is switched on after a certain delay. Recordings of this kind yielded $a = -9.05(25)$ MHz for the magnetic dipole interaction constant and $g_J = 1.336(2)$ for the Landé factor [14.21].

14.2.2 Level-Crossing Measurements

The LC method [14.20] originates in *Hanle*'s early work on magnetic depolarisation [14.22] which was extended to the case of high-field crossings by *Colgrove* et al. [14.23]. Using σ-light, pairs of crossing magnetic sublevels can be

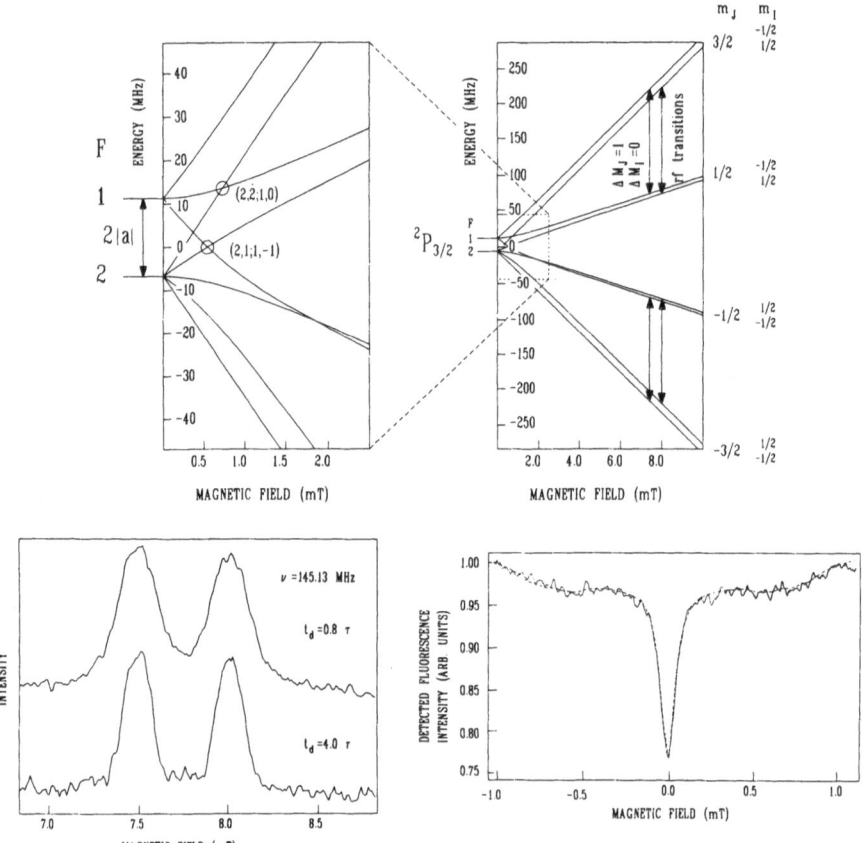

Fig. 14.1. Magnetic sublevel diagram for the 6p $^2P_{3/2}$ state of Ag (top); ODR signals (lower left); LC signal (lower right). From [14.21]

coherently excited and a redistribution in the fluorescence light can be detected that is related to the magnetic fields of level crossing. In Fig. 14.1 (top), detectable level crossings for the silver 6p $^2P_{3/2}$ state are indicated. A recording of the fluorescence light intensity as a function of an external magnetic field is included in this figure. Apart from the zero-field level crossing (the Hanle effect) the unresolved structure due to two level crossings for each of the two silver isotopes is shown. This recording was again taken for 206 nm excitation and basically yields similar information on hyperfine structure as the ODR recording in the same figure. In a previous investigation, resolved level-crossing signals were recorded for the 5p $^2P_{3/2}$ state of copper [14.24] that was excited at 202 nm.

14.2.3 Quantum-Beat Measurements

Quantum-beat experiments [14.25, 26] can be considered as time-resolved LC investigations. Following short-pulse coherent excitation the quantum-mechanical interference between pairs of sublevels is manifested as oscillations superimposed on the temporal decay of the fluorescence light. An experimental set-up for QB experiments with VUV laser light excitation is shown in Fig. 14.2 [14.27]. A hyperfine structure QB recording for the 7p ^{2}P$_{3/2}$ state of silver, excited at 185 nm is shown in Fig. 14.3. By frequency tripling of the output from

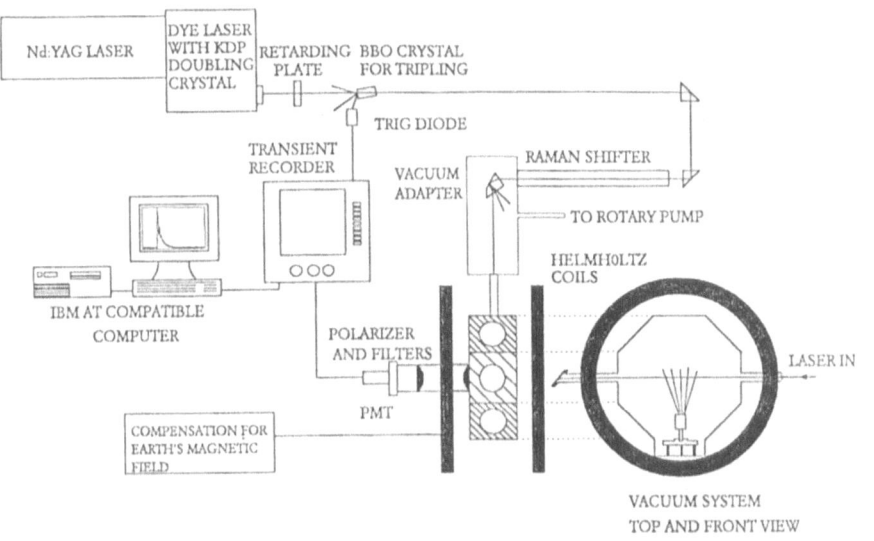

Fig. 14.2. Experimental set-up for QB spectroscopy in the YUV region. From [14.27]

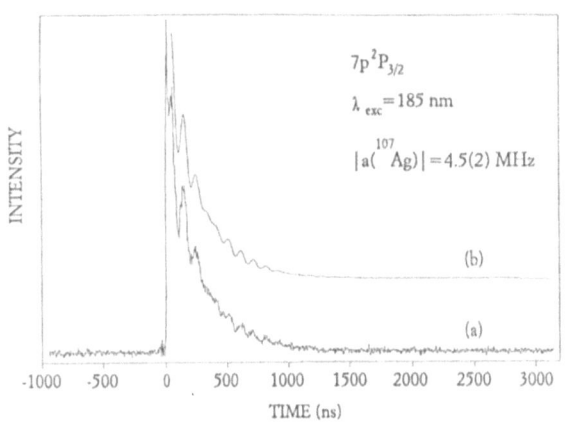

Fig. 14.3. Experimental QB recording of the structure of the 7p ^{2}P$_{3/2}$ state of Ag. A theoretical fit to the data is also included. From [14.27]

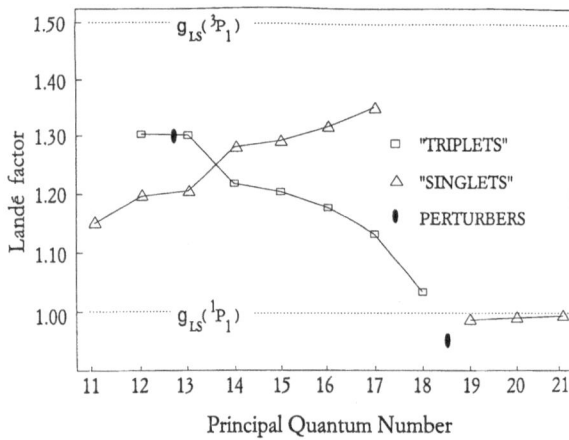

Fig. 14.4 Values of g_J for np $^{1,3}P_1$ sequences in Yb, obtained using Zeeman quantum-beat experiments. From [14.28]

a red dye laser, 200 nm was first achieved. By anti-Stokes Raman shifting the required 185 nm radiation was then generated. Because of the oxygen Schumann–Runge band absorption the Raman shifter assembly is evacuated and is directly attached to the atomic-beam vacuum chamber. In the figure a theoretically generated QB curve using a magnetic dipole interaction constant a (^{107}Ag) $= -4.5$ MHz is included. This recording provides a good example on how a very high resolution can be achieved in the VUV region using broadband pulsed lasers in combination with resonance/coherence techniques.

As a further example of QB spectroscopy at short wavelengths, we choose Zeeman quantum beat measurements in highly excited ytterbium Rydberg state. From the beat frequency and the applied magnetic field value, the Landé g_J factor of the state investigated can be evaluated. A systematic study of the Zeeman effect in the np $^{1,3}P_1$ Rydberg sequences of ytterbium has been performed ($\lambda_{exc} = 207–200$ nm) and results are given in Fig. 14.4 [14.28]. In this way it is possible to study the influence of perturbing states in a similar way as previously done for the barium atom [14.29].

14.2.4 Time-Resolved Studies on Atoms Formed by Short-Wavelength Dissociation

Many of the light non-metallic elements occur naturally only as molecules. Furthermore, the resonance lines frequently fall in the deep VUV region. Both these problems can be overcome by photodissociation of molecules by short-wavelength laser radiation that is two-photon resonant with transitions in the produced free atoms. Following a study of excited-state lifetimes in oxygen atoms produced from NO_2 molecules [14.30], we have just finished corresponding studies for nitrogen atoms formed from N_2O [14.31, 32]. In Fig. 14.5 a

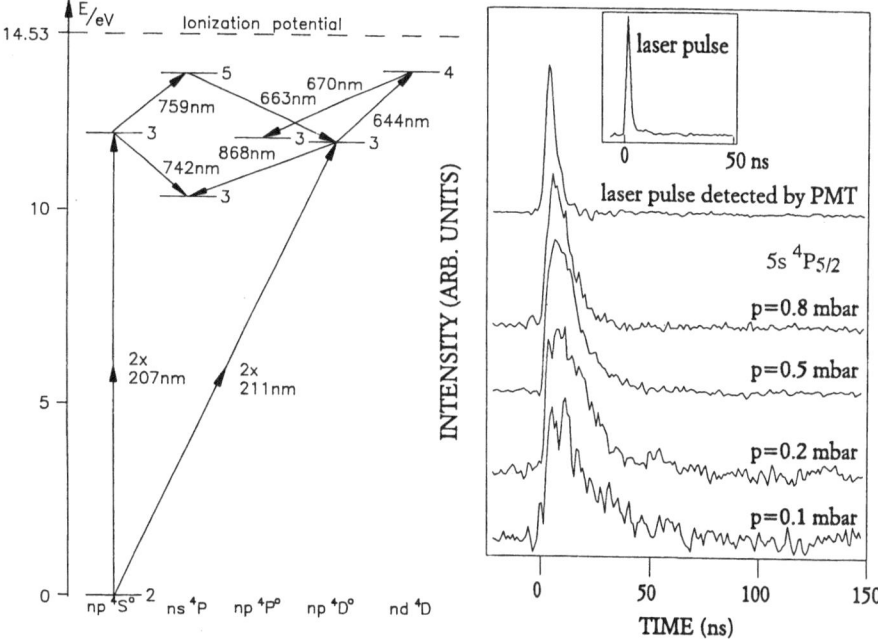

Fig. 14.5. Partial energy-level diagram for nitrogen and fluorescence decay of nitrogen atoms following step-wise excitations. From 14.32]

partial energy-level diagram for nitrogen and recordings of fluorescence light at different gas pressures following stepwise laser excitations are shown. A similar scheme can be used for many other light atoms such as C, F, Cl, P and S. Using resonance cells with molecular species enriched in odd isotopes high-resolution data on hyperfine structure can also be obtained.

14.3 Conclusion

As illustrated in the present chapter, high-resolution laser spectroscopy can readily be performed at "difficult" wavelengths combining pulsed laser radiation with resonance/coherence techniques. Basically, the ODR, LC and QB techniques provide the same type of information and the choice of method will be determined by practical aspects. QB spectroscopy requires a high time resolution on both the laser and the detection electronics. In the ODR technique, rf fields of suitable frequency must be available and at sufficient strength, which can be a limitation for short-lived states. In many respects the level-crossing method is the least demanding technique.

Acknowledgements. The author acknowledges a very fruitful cooperation with a large number of past and present coworkers and graduate students in the field of basic atomic laser spectroscopy. The examples discussed in the present paper are due to work by G.J. Bengtsson, Jiang Zhankui, P. Jönsson, J. Larsson, C.-G. Wahlström and D.D. Wang. This work was supported by the Swedish Natural Science Research Council.

References

14.1   P.P. Sorokin, J.R. Lankard: IBM J. Res. Dev. **10**, 162 (1966)
14.2   F.P. Schäfer, W. Schmidt, J. Volze: Appl. Phys. Lett. **9**, 306 (1966)
14.3   W. Demtröder: *Laser Spectroscopy*, 3rd corrected printing, Springer Ser. Chem. Phys. Vol. 5 (Springer, Berlin, Heidelberg 1988)
14.4   S. Svanberg: *Atomic and Molecular Spectroscopy—Basic Aspects and Practical Applications*, Springer Ser. Atoms and Plasmas, Vol. 6 (Springer, Berlin, Heidelberg 1991)
14.5   S. Svanberg: In *Applied Laser Spectroscopy*, ed. by W. Demtröder, M. Inguscio (Plenum, New York 1990) p. 149
14.6   S. Svanberg, P. Tsekeris, W. Happer: Phys. Rev. Lett. **30**, 817 (1973)
14.7   S. Svanberg: In *Laser Spectroscopy III*, ed. by J.L. Hall, J.L. Carlsten, Springer Ser. Opt. Sci. Vol. 7 (Springer, Berlin, Heidelberg 1977) p. 187
14.8   S. Haroche, J.A. Paisner, A.L. Schawlow: Phys. Rev. Lett. **30**, 948 (1973)
14.9   R.C. Eckart, X.Y. Fan, M.M. Fejer, W.J. Kozlovsky, C.N. Nabors, R.L. Byer, R.K. Route, R.S. Feigelson: In *Laser Spectroscopy VIII*, ed. by W. Persson, S. Svanberg (Springer, Berlin, Heidelberg 1987)
14.10   W.L. Glab, J.P. Hessler: Appl. Opt. **26** (16), 3181 (1987)
14.11   A.P. Hickman, J.A. Paisner, W.K. Bischel: Phys. Rev. **A33**, 1788 (1986)
14.12   W. Jamroz, B.P. Stoicheff: Progress in Optics **XX**, 325 (1983)
14.13   R. Hilbig, G. Hilber, A. Lago, B. Wolf, R. Wallenstein: Comments Atom. Molec. Phys. **18**, 157 (1986)
14.14   A. Borzutsky, R. Brünger, R. Wallenstein: In *Applied Laser Spectroscopy*, ed. by W. Demtröder, M. Inguscio (Plenum, New York 1990) p. 63
14.15   X.F. Li, A. L'Huillier, M. Ferray, L.A. Lompré, G. Mainfray: Phys. Rev. **A39**, 5751 (1989)
14.16   N. Sarukura, K. Hata, T. Adachi, R. Nodomi, M. Watanabe, S. Watanabe: Phys. Rev. **A43**, 1669 (1991)
14.17   J.D. Kmetec, J.J. Macklin, S.E. Harris: Private communication (1991)
14.18   A. Kastler, J. Brossel: Comp. Rend. **229**, 1213 (1949)
14.19   J. Brossel, F. Bitter: Phys. Rev. **86**, 368 (1952)
14.20   W. Happer, R. Gupta: In *Progress in Atomic Spectroscopy*, Part A, ed. by W. Hanle, H. Kleinpoppen (Plenum, New York 1979) p. 391
14.21   J. Bengtsson, J. Larsson, S. Svanberg: Phys. Rev. **A42**, 5457 (1990)
14.22   W. Hanle: Z. Phys. **30**, 93 (1924)
14.23   F.D. Colgrove, P.A. Franken, R.R. Lewis, R.H. Sands: Phys. Rev. Lett. **3**, 420 (1957)
14.24   J. Bengtsson, J. Larsson, S. Svanberg, C.-G. Wahlström: Phys. Rev. **A41**, 233 (1990)
14.25   J.N. Dodd, G.W. Series: In *Progress in Atomic Spectroscopy*, Part A, ed. by W. Hanle, H. Kleinpoppen (Plenum, New York 1978) p. 639
14.26   S. Haroche: In *High Resolution Laser Spectroscopy*, ed. by K. Shimoda, Topics Appl. Phys. Vol. 13 (Springer, Berlin, Heidelberg 1976) p. 253
14.27   G.J. Bengtsson, J. Larsson, S. Svanberg: Z. Phys. in press
14.28   Jiang Zhankui, J. Larsson: Z. Phys. in press
14.29   P. Grafström, C. Levinson, H. Lundberg, S. Svanberg, P. Grundevik, L. Nilsson, M. Aymar: Z. Phys. **A308**, 95 (1982)
14.30   S. Kröll, H. Lundberg, A. Persson, S. Svanberg: Phys. Rev. Lett. **55**, 284 (1985)

14.31   G.J. Bengtsson, J. Larsson, S. Svanberg, D.D. Wang: Phys. Rev. **A,** in press
14.32   G.J. Bengtsson, K. Hansen, J. Larsson, W. Schade, S. Svanberg: Z. Physik, in press

Sune Svanberg

15. Doppler-Free Spectroscopy of Large Polyatomic Molecules and van der Waals Complexes

H.J. Neusser, E. Riedle, T. Weber and E.W. Schlag

With 5 Figures

It is shown that high-resolution spectroscopy with tunable dye lasers leads to rotationally resolved electronic spectra of large molecular systems. Two-photon absorption of narrow-band cw light in an external cavity eliminates Doppler broadening in the $S_1 \leftarrow S_0$ transition of the prototype organic molecule benzene. Several thousands of rotational lines in the room temperature spectrum are analyzed, providing spectroscopic constants with a hitherto inaccessible precision. Investigations of the homogeneous linewidth of individual rotational transitions reveals that Coriolis coupling plays an important role in the intramolecular energy redistribution process in this molecule. Aided by the reduced Doppler broadening in a skimmed cooled supersonic beam, rotationally resolved UV spectra of benzene–noble-gas van der Waals clusters were measured. These measurements yield precise information on the van der Waals bond lengths and structures of these complexes.

15.1 Introduction

The advent of narrow-band tunable dye lasers [15.1, 2] has caused a breakthrough in the precision of atomic spectroscopy and, more recently, has allowed high-resolution molecular spectroscopy of systems with many vibrational degrees of freedom to be performed. Several techniques have been developed that push gas-phase spectral resolution below the natural barrier set by Doppler broadening [15.3]. Spectroscopy in collimated beams, saturation spectroscopy, polarization spectroscopy and Doppler-free two-photon spectroscopy were originally demonstrated for atoms to reveal the underlying line structure of a Doppler-broadened transition. In this contribution we would like to summarize some of our important new results achieved for isolated organic molecules. Two different types of experiments are described leading to Doppler-free spectra of either a large polyatomic molecule at room temperature or of van der Waals complexes at low temperature. Typical examples for the spectra and the new information revealed are presented.

15.2 Doppler-Free Two-Photon Spectroscopy of Benzene

Following the successful application of Doppler-free methods in atomic physics, these techniques have rapidly gained importance in the spectroscopy of poly-

Topics in Applied Physics, Vol. 70
Dye Lasers: 25 Years Ed.: Dr. Michael Stuke
© Springer-Verlag Berlin Heidelberg 1992

atomic molecules. Of the various methods mentioned above, Doppler-free two-photon absorption was the technique first successfully applied to large molecules [15.4]. One advantage of two-photon excitation is that narrow-band visible laser light can be used to observe electronic transitions occurring at UV photon energies. Indeed most polyatomic molecules, and in particular organic systems, begin to absorb in the UV part of the spectrum. Other important advantages are that all molecules within the interaction volume contribute to the Doppler-free signal independent of their velocity and that the Doppler-broadened background (due to the absorption of two photons from one laser beam) can often be suppressed by proper choice of laser polarization [15.5, 6].

The typical experimental set-up for a Doppler-free two-photon experiment consists of a tunable laser light source with a frequency width narrower than the Doppler width (typically a frequency stabilized single-mode cw dye laser) and a gas cell that contains the low-pressure molecular gas under investigation. Elimination of the Doppler broadening is achieved by the absorption of two photons propagating in opposite directions. This is accomplished by placing the gas cell within a standing light field, e.g., in an external resonator as originally suggested by *Vasilenko* et al. [15.7] or by back reflection of the laser beam onto itself. For molecular spectroscopy, the sensitivity of the external resonator arrangement is an important advantage. The oscillator strength of a molecular electronic transition is distributed over many vibrational transitions according to the Franck–Condon principle. Furthermore, even at room temperature, tens of thousands of rovibrational ground state levels are populated and the effective population density of a single rovibronic state is smaller by four orders of magnitude than the total gas density. Finally, in many polyatomic molecules the fluorescence quantum efficiency is smaller than unity due to fast nonradiative processes occurring within the isolated molecule. The greater sensitivity of the external resonator with its light intensity enhancement of one order of magnitude is often the only way to raise the two-photon signal above the detection threshold in a cw experiment.

The scheme of a Doppler-free two-photon experiment with external cavity as used in our laboratory [15.8] is shown in Fig. 15.1. The external cavity is locked to the cw dye laser frequency with a technique developed by *Hänsch* and *Couillaud* [15.9]. Two-photon absorption is monitored by detection of UV fluorescence from the excited level. At room temperature the Doppler-free spectrum of a polyatomic molecule consists of many thousands of rotational lines which arise from the large number of thermally populated ground state levels. For demonstration a small portion (10%) of the 14^1_0 vibronic band of benzene with a resolution of 10 MHz is shown in Fig. 15.2. Every line represents a single rotational transition. Typically the line density in the spectrum is more than 150 lines/cm^{-1}. In benzene, on an average, about 8 lines are located within the Doppler width of 1.6 GHz. For this reason every laser scan yields a tremendous number of data points which have to be rapidly transferred to and stored in a computer with sufficient memory. The complete spectrum is assembled by many individual laser scans of typically 30 GHz (1 cm^{-1}).

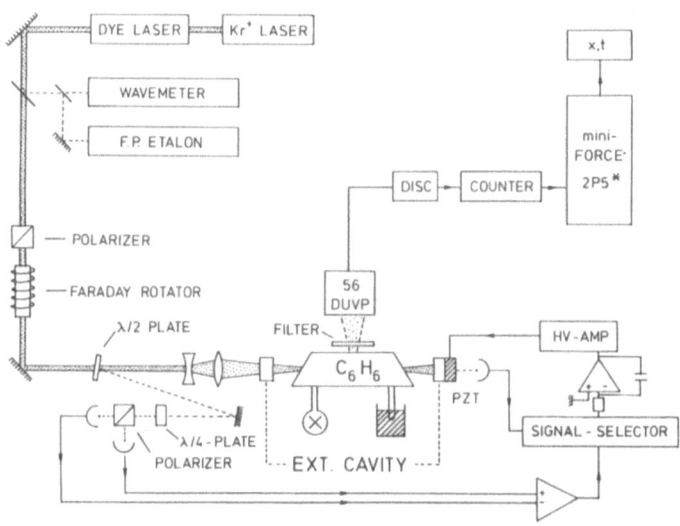

Fig. 15.1. Scheme of the experimental set-up for Doppler-free two-photon fluorescence excitation spectroscopy of polyatomic molecules. The cell containing the molecular gas is placed in an external concentric cavity, which creates the standing wave field for the two-photon absorption from counterpropagating light beams

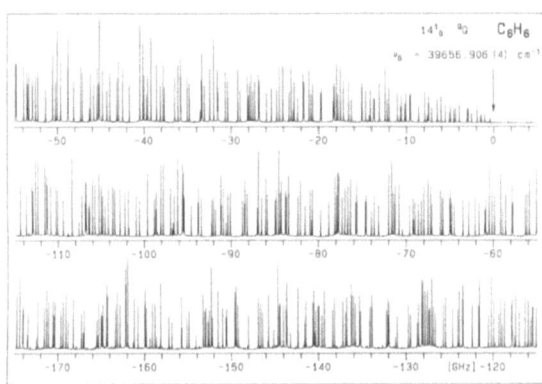

Fig. 15.2. Doppler-free two-photon spectrum of a small portion of the 14_0^1 vibronic band in benzene, C_6H_6. Every line corresponds to an individual rotational transition and has been assigned

The rotational line spectrum can be described by a semirigid symmetric top Hamiltonian including quartic centrifugal distortion constants. A fit of the theoretical spectrum to the accurate line positions yields precise rotational constants and centrifugal distortion constants of the ground and excited electronic state. The precision is higher by two orders of magnitude than the constants obtained from conventional high-resolution but Doppler-limited UV spectroscopy [15.10]. A typical result for the rotational constants in the electronically excited 14^1 state is $B'_{14} = 0.181284\,\text{cm}^{-1}$ and $C'_{14} = 0.090711\,\text{cm}^{-1}$ with an accuracy of $10^{-6}\,\text{cm}^{-1}$ [15.11].

In addition to the precise spectroscopic studies of molecular structure, Doppler-free two-photon absorption provides important information on molecular dynamics. After the inhomogeneous Doppler broadening is eliminated, the homogeneous linewidths of molecular transitions can be observed. At gas pressures of several mbar, the homogeneous linewidth is governed by collisional broadening, but at lower gas densities the linewidth is determined by dynamic processes taking place within the isolated molecule. In recent years there has been a growing interest in the study of intramolecular dynamics within the dense bound level structure typical of large molecules. Dynamics in a molecule with discrete level structure is of particular interest in photochemistry in order to understand the photophysical primary processes which occur before a polyatomic molecule undergoes a chemical reaction. Nonradiative processes such as internal conversion (IC), intersystem crossing (ISC) and intramolecular vibrational energy redistribution (IVR) are internal energy redistribution processes important for every chemist.

It is the goal of high-resolution spectroscopy to elucidate the above-mentioned dynamic processes. The precise investigation of these processes is only possible if the excitation is selective enough to lead to an excitation of single defined quantum states.

In Fig. 15.3 the measured homogeneous linewidths of three individual rotational transitions in the $14^1 1^2$ vibrational state at a high excess energy of

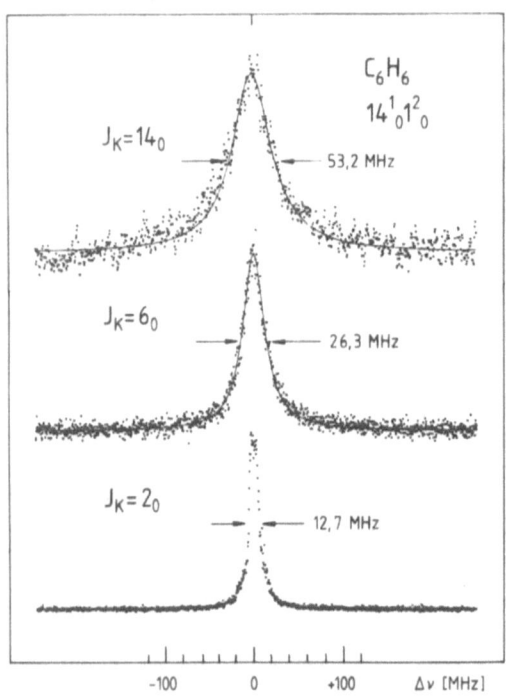

Fig. 15.3. Measured homogeneous linewidth of three rotational lines of the $14_0^1 1_0^2$ band at an excess energy of 3412 cm^{-1}. The solid lines represent a Lorentzian curve fitted to the experimental points. The linewidth increases with increasing J due to Coriolis coupling [15.8]

3412 cm^{-1} in the electronically excited state of benzene are shown [15.8]. The transitions lead to the defined states with rotational quantum numbers $J'_{K'}$ = $2_0, 6_0$, and 14_0, respectively. It is clearly recognized that the Lorentzian shape (solid line) of the lines increases from 12.7 to 53.2 MHz with increasing J' quantum number. This situation resembles the rotational predissociation phenomenon typical for small molecules [15.12]. Here, in the large molecule benzene the rotational dependence of the linewidth clearly points to parallel Coriolis coupling. Thus, it is seen that Coriolis coupling, i.e. coupling of vibrational states through the rotation of the molecules, is an important coupling mechanism leading to intramolecular redistribution processes (IVR) of energy in large molecules.

15.3 Doppler-Free UV Spectroscopy of Benzene–Noble-gas van der Waals Clusters

In the past decade there has been growing interest in the spectroscopy of van der Waals complexes. The production of isolated atomic and molecular clusters has become feasible in cooled supersonic jet expansions [15.13]. The cold molecular beam provides another important advantage for high-resolution spectroscopy, i.e. the reduction of the transversal velocity distribution and consequently of Doppler broadening by selecting the central part of the beam with a skimmer. Thus, elimination of Doppler broadening can be obtained without using nonlinear techniques such as saturation spectroscopy or Doppler-free two-photon absorption.

In a supersonic jet expansion it is not possible to produce a single-cluster species, but rather, in addition to the monomer constituents of the gas mixture, a variety of clusters of differing composition and size is produced. The spectra of the different species may overlap and it is often difficult to distinguish them in the fluorescence excitation spectra. For this reason we combined high-resolution spectroscopy with resonance-enhanced two-photon ionization [15.14]. The produced ions can be mass-analyzed and integrated for a selected mass. This leads to highly resolved (Δv = 130 MHz) mass-selected two-photon ionization spectra reflecting the UV intermediate state spectrum of the selected cluster species.

Pulsed excitation with high spectral resolution is necessary for an efficient two-photon ionization. Light having these properties is provided by amplification of the single-mode cw laser in a three-stage amplifier system pumped by an excimer laser. The light pulses are Fourier-transform-limited with a bandwidth of 80 MHz and a peak power of nearly 1 MW. Absorption of the narrow-band frequency-doubled light pulses leads·to an excitation of the S_1 electronically excited state of the clusters within the molecular beam. A second broadband laser pulse ionizes the excited clusters. Both light beams interact perpendicularly with the molecular beam sð that Doppler broadening is reduced to some

40 MHz and sharp intermediate state spectra can be obtained when the first laser is scanned. The ions are mass selected and detected in a linear time of flight mass spectrometer with a resolution of $m/\Delta m \approx 250$.

With this technique we were able to resolve the rotational structure in the vibronic bands of benzene–noble-gas dimers and trimers. This allows us to determine the structure and the exact bond lengths of these complexes.

The high spectral and mass selectivity achievable is demonstrated in Fig. 15.4 by the Doppler-free spectra of the strongest one-photon vibronic band, 6_0^1, of two isotopic benzene–noble-gas dimers $C_6H_6 \cdot {}^{84}Kr$ and $C_6H_6 \cdot {}^{86}Kr$[15.15]. By separate integration of the ion current at the different isotopic masses, both spectra are measured simultaneously within the natural isotopic mixture of Kr. The complexes are produced by expansion of 6-mbar benzene seeded in 5-bar Kr through a pulsed 300-μm nozzle. Since the natural abundance of ^{86}Kr (17.3%) is only one third of ^{84}Kr (57.0%), the vertical scale of the upper spectrum in Fig. 15.4 is expanded by a factor of 3.

Both hands display the rotational structure of a prolate symmetric top with a rotational temperature $T_{rot} \approx 1.5$ K. On the low- and high-energy side of the spectrum, well-resolved strong lines appear which are assigned to the P and R branch, respectively. The seven strong features in the center of the band are subbranches of the Q branch with partly overlapping rotational lines. The blue-shaded wings of the subbranch originate from lines with constant K' and ΔK but varying J' and indicate an increase of the B rotational constant in the S_1 state. Due to the small relative mass change between the two clusters of only 1.2%, the differences in band structure and line positions cannot be seen by naked eye, but only by detailed analysis of the data.

For a precise determination of the rotational constants a two-stage computer fit to the line positions was performed, according to a rigid symmetric top energy formula. In the first step, the ground state constant B_0'' is evaluated by combination differences $[B_0'' = 0.026562(13) \text{ cm}^{-1}$ for $C_6H_6 \cdot {}^{84}Kr$ and $B_0'' = 0.026263(21) \text{ cm}^{-1}$ for $C_6H_6 \cdot {}^{86}Kr]$, i.e. frequency differences of transitions

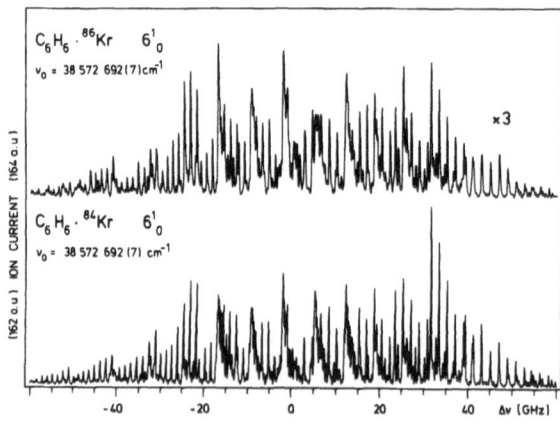

Fig. 15.4. Mass-selected rotationally resolved UV spectrum of the 6_0^1 band of the $C_6H_6 \cdot {}^{86}Kr$ (upper trace) and $C_6H_6 \cdot {}^{84}Kr$ (lower trace) van der Waals clusters. The two spectra were measured for Kr in natural isotopic abundance. Note the increased vertical scale of the less abundant $C_6H_6 \cdot {}^{86}Kr$ isotopic cluster

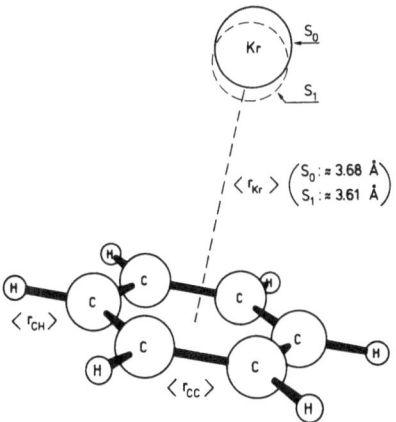

Fig. 15.5 Structure of the $C_6H_6 \cdot {}^{86}Kr$ van der Waals cluster. The Kr atom is located on the C_6 symmetry axis of the benzene molecule. The bond length of 3.68 Å in the electronic ground state decreases by 70 m Å upon electronic excitation of benzene

starting at the same ground state but leading to different excited states. In a second step the excited state constants A_v' [0.090838(5) cm^{-1}], B_v' [0.027266(15) cm^{-1} for $C_6H_6 \cdot {}^{84}Kr$ and 0.026962(24) cm^{-1} for $C_6H_6 \cdot {}^{86}Kr$], and ζ_{eff}' [$-0.5860(3)$] are fitted to about 100 unblended rotational lines of the experimental spectrum. By means of a simultaneously measured iodine absorption spectrum, it is possible to fix the band origin, v_0, to an accuracy of 0.007 cm^{-1} (see Fig. 15.4). The standard deviation of the fit is 23 MHz, which is less than a fifth of the spectral linewidth.

The low standard deviation and the absence of an asymmetry splitting of the lines with low K values within the experimental resolution demonstrate that the $C_6H_6 \cdot Kr$ cluster is a prolate symmetric top, with the Kr atom located above the benzene ring on the C_6 rotational axis (Fig. 15.5). From the vibrationally averaged rotational constants B_0'' and B_v', accurate values for the effective bond distance $\langle r_{Kr} \rangle$ in the S_0 and the S_1 state are calculated. They are identical for both isotopic species and are given in Fig. 15.5. Due to a slightly larger ring size and a higher polarizability of benzene in the S_1 state, the bond distance *decreases* upon electronic excitation.

Recently, similar experiments on benzene–Ar_2 have shown that the second Ar atom is located in the symmetrical position on the other side of the benzene ring [15.16]. No change in the benzene–Ar distance is found when the second Ar atom is adsorbed to the benzene surface. This points to a vanishing three-body interaction between the two Ar atoms in the cluster through the benzene plane. Such data represent basic information necessary for a microscopic understanding of the van der Waals interaction in organic molecules.

15.4 Summary and Conclusion

In this note we have shown that Doppler-free spectroscopy is now feasible for large molecular systems. Using narrow-band dye lasers, Doppler-free spectra were measured either by two-photon absorption from counterpropagating light beams in a room temperature gas sample or by expansion and cooling of a high-pressure gas sample through a nozzle, reducing the Doppler broadening. Analysis of these spectra provides hitherto inaccessible information on molecular structure and intramolecular dynamics. Structural information is particularly interesting for weakly bound molecular clusters. In this way a microscopic understanding of solvation and complexation processes is expected in the near future.

References

15.1   P.P. Sorokin, J.R. Lankard: IBM J. Res. Develop. **10**, 162 (1966)
15.2   F.P. Schäfer, W. Schmidt, J. Volze: Appl. Phys. Lett. **9**, 306 (1966)
15.3   V.S. Letokhov, V.P. Chebotayev: *Nonlinear Laser Spectroscopy* (Springer, Berlin, Heidelberg 1977)
15.4   E. Riedle, H.J. Neusser, E.W. Schlag: J. Chem. Phys. **75**, 4231 (1981)
15.5   F. Biraben, M. Bassini, B. Cagnac: J. de Phys. **40**, 445 (1979)
15.6   E. Riedle, R. Moder, H.J. Neusser: Opt. Commun. **43**, 388 (1982)
15.7   L.S. Vasilenko, V.P. Chebotayev, A.V. Shishaev: JETP Lett. **12**, 113 (1970)
15.8   E. Riedle, H.J. Neusser: J. Chem. Phys. **80**, 4686 (1984)
15.9   T.W. Hänsch, B. Couillaud: Opt. Commun. **35**, 441 (1980)
15.10   J.H. Callomin, T.M. Dunn, I.M. Mills: Phil. Trans. Roy. Soc. London A **259**, 499 (1966)
15.11   H. Sieber, E. Riedle, H.J. Neusser: J. Chem. Phys. **89**, 4620 (1988)
15.12   G. Herzberg: *Molecular Spectra and Molecular Structure, III. Electronic Spectra and Electronic Structure of Polyatomic Molecules* (van Nostrand Reinhold, New York 1966)
15.13   D.H. Levy: Ann. Rev. Phys. Chem. **31**, 197 (1980)
15.14   Th. Weber, A. von Bargen, E. Riedle, H.J. Neusser: J. Chem. Phys. **92**, 90 (1990)
15.15   Th. Weber, E. Riedle, H.J. Neusser, E.W. Schlag: Chem. Phys. Lett. (1991) in press
15.16   Th. Weber, H.J. Neusser: J. Chem. Phys. **94**, 7689 (1991)

Edward W. Schlag   Hans Jürgen Neusser

16. Electro-optic and Photoconductive Sampling of Ultrafast Photodiodes with Femtosecond Laser Pulses

J. Kuhl, M. Klingenstein, M. Lambsdorff, J. Rosenzweig, C. Moglestue and A. Axmann

With 8 Figures

Photoconductive and electro-optic sampling of the photocurrent of GaAs metal –semiconductor–metal photodiodes reveal the influence of field screening and optical phonon scattering on the frequency bandwidth of these detectors. The experimental results agree with theoretical predictions of a self-consistent two-dimensional Monte Carlo calculation.

16.1 Introduction

Contemporary electronic and optoelectronic devices can generate and process electrical signals with rise and decay times in the 0.3–10-ps regime – well beyond the time resolution of conventional electronic test equipment which is still limited to > 10 ps. Examples of such high speed devices are photoconductive switches, metal–semiconductor–metal (MSM) photodiodes or GaAs FETs with submicrometer gates. During the last decade photoconductive switching [16.1–3] and electro-optic sampling [16.4–6] utilizing picosecond and sub-picosecond optical pulses have been developed as powerful tools to investigate the frequency response of such ultrafast devices directly in the time domain.

It has been possible to extend the electronic measurement capabilities by 1 to 2 orders of magnitude compared to conventional electronics through the application of these new optical techniques. Thus, the rapid advances in the generation of ultrashort optical pulses achieved in the 25 years since the invention of the dye laser have had a steadily increasing impact on the progress of semiconductor electronics technology, rendering possible the development of novel and faster electronic devices.

In the first part of this chapter the present status of these contactless optical test techniques will briefly be reviewed. The subsequent experimental and theoretical analysis of the carrier transport in MSM GaAs diodes strikingly demonstrates the considerable potential of these methods for the character-ization of ultrafast optoelectronic devices with picosecond time resolution.

16.2 Photoconductive Switches

The development of ultrafast photoconductive switches has been pioneered by *Auston* [16.1–3, 7]. The switch is fabricated on a semiconducting film with high

Topics in Applied Physics, Vol. 70
Dye Lasers: 25 Years Ed.: Dr. Michael Stuke
© Springer-Verlag Berlin Heidelberg 1992

defect density. Suitable materials are crystalline or polycrystalline Si films damaged by heavy bombardment with Si or O ions. Free electrons and holes photoexcited in the conduction and valence band by absorption of an ultrashort optical pulse generate photoconductivity only for a very short time interval after excitation since the carriers are rapidly trapped and localized at deep defect levels. The free-carrier lifetime in the extended band states of damaged Si material can be reduced for instance to about 500 fs by means of ion irradiation at a dose of 10^{14}–10^{15} ions/cm^2. If 100-fs optical pulses, which can be routinely generated by a colliding pulse mode-locked (CPM) ringlaser [16.8] or a hybridly mode-locked dye laser at a repetition rate of 100 MHz, are used for excitation, the rise time of the photoconductivity follows the integrated pulse shape and the decay is limited by the free-carrier trapping time. The shape of the current pulse measured at the switch contacts is obtained by convoluting the photoconductivity response of the semiconducting film with the RC-limited response of the device. Thus, the switch has to be mounted in a broadband microstrip line or a coplanar transmission line in order to avoid excessive broadening of the electrical pulse.

Whereas the time resolution attainable with microstrip lines is limited to several picoseconds, properly designed coplanar transmission lines permit signal rise and decay times of a few hundred fs corresponding to a bandwidth of almost 1 THz. For our studies we used the so-called sliding contact configuration (Fig. 16.1), which was originally designed by *Ketchen* et al. [16.9].

The coplanar transmission line structure consists of two parallel 5-μm-wide and 0.5-μm-thick Al lines separated by 10 μm and fabricated by optical lithography and lift-off processing on thin films of silicon-on-sapphire (SOS, thickness 0.6 μm). The electrical pulses are generated by switching the conductivity of the charged coplanar line which is connected to a dc bias of 25 V via local excitation with 100-fs laser pulses of a colliding-pulse mode-locked dye laser (photon energy 2 eV) focused between the lines. Photoexcitation of electrons and holes in the semiconducting substrate varies the conductivity of the gap by

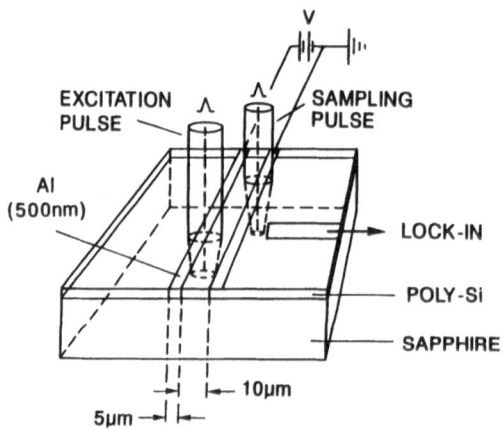

Fig. 16.1. Photoconductivity cross-correlation measurement on coplanar transmission lines

orders of magnitude, thus leading to the excitation of a current pulse propagating along the line. The duration of this current pulse is adjustable within wide ranges by proper choice of the ion irradiation dose. The shape of the generated electrical pulse can be analyzed by cross-correlating the photoconductive response of two switches.

In the structure depicted in Fig. 16.1 this is accomplished by illuminating the gap that connects the transmission line to an electrical probe (lock-in amplifier with a current sensitive preamplifier) with a synchronized 100-fs optical pulse derived from the same laser. The pulse shape is sampled by measuring the photocurrent with the lock-in amplifier as a function of the delay between the optical excitation and the sampling pulse. Figure 16.2 (left-hand side) shows the cross-correlation of the shortest electrical pulse, which we obtained from a switch with a Si^+ irradiation dose of $3 \times 10^{14}\,cm^{-2}$. Deconvolution of this cross-correlation trace leads to an electrical pulse duration as short as 0.5 ps. This pulse duration increases almost linearly to about 10 ps if the ion implantation dose is reduced to $3 \times 10^{12}\,cm^{-2}$ (Fig. 16.3). Increase of the ion dose beyond 3 $\times 10^{14}\,cm^{-2}$, however, does not lead to further pulse shortening.

For many practical applications of photoconductive switches to investigations of high-speed electronic and optoelectronic devices, it will be indispensable to integrate the switch and the electronic circuit on the same substrate. The fact that most high-speed discrete devices and integrated circuits are based on GaAs has stimulated intense research during recent years to develop high-speed photoconductive materials which can be easily integrated with GaAs devices. Very recently we have succeeded in generating 0.5 ps pulses from GaAs after

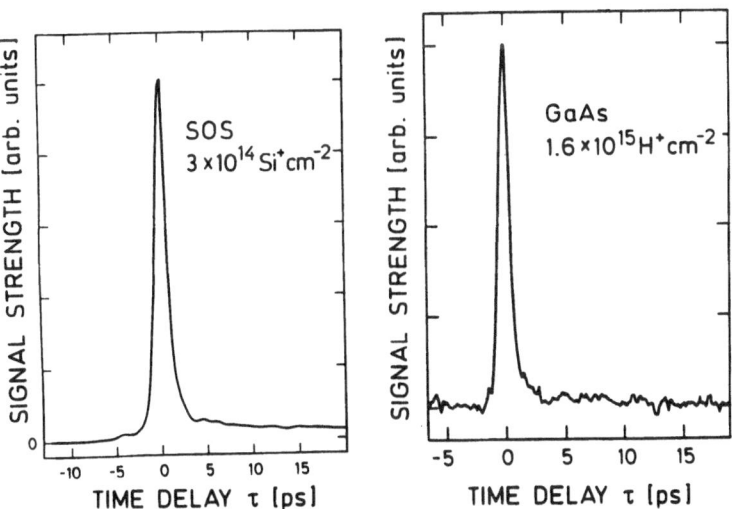

Fig. 16.2. Photocurrent cross-correlation curves for silicon-on-sapphire with total irradiation dose of $3 \times 10^{14}\,Si^+\,cm^{-2}$, FWHM: 1 ps (left) and GaAs with total proton irradiation dose of 1.6 $\times 10^{15}\,cm^{-2}$, FWHM: 1 ps (right)

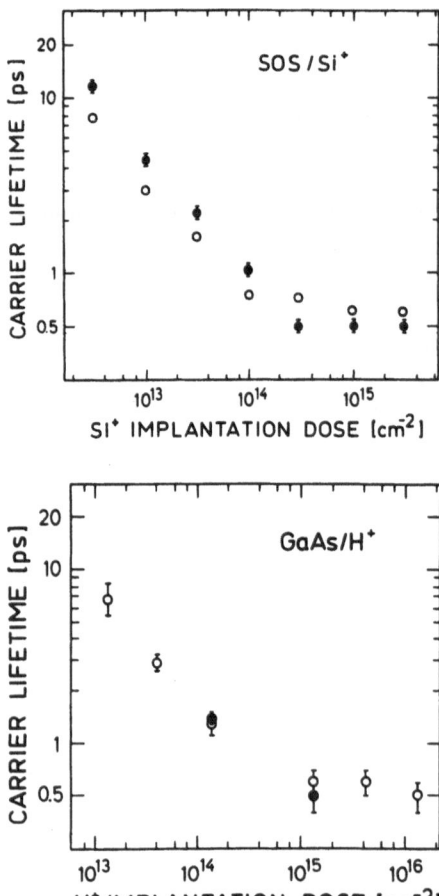

Fig. 16.3. Carrier lifetime vs. ion implantation dose for SOS (upper part) and GaAs (lower part) photoconductive switches derived from time-resolved reflectivity ($\bigcirc$) and photoconductivity experiments ($\bullet$)

bombardment with protons at a dose of 1.6×10^{15} cm^2 [16.10] (Fig. 16.2). The dependence of carrier lifetime on proton implantation dose for GaAs is shown in Fig. 16.3 (lower part). At low implantation levels the carrier lifetime depends significantly on the irradiation dose reflecting the increasing number of structural defects. Above an implantation dose of 1.37×10^{15} H$^+$ cm^{-2}, however, the lifetime remains constant at a value of ~ 0.5 ps. This lifetime saturation is most probably caused by amorphization of GaAs. Comparable speed has been reported for thin crystalline GaAs films epitaxially grown onto a GaAs substrate at relatively low temperatures ($\sim 200^\circ$C). In addition, this novel material should have a 10–100 times higher response than conventional (radiation damaged) photoconductive materials [16.11]. 480-fs pulses and a high sensitivity have been also observed with CdTe grown by metal-organic chemical vapor deposition on sapphire and quartz substrates as the photoconductive material [16.12]. The authors claim however, that this material, can also be easily grown on GaAs substrates.

16.3 Electro-optical Sampling

Electro-optical sampling exploits the Pockels effect, i.e. the variation of birefringence in electro-optically active crystals by an electric field, for the detection and measurement of electrical transients. The change of the polarization state of linearly polarized light passed through the crystals provides a quantitative measure of the electric field strength. For easy analysis of the polarization the electro-optic crystal is mounted between crossed polarizers. Thus, the variation of the polarization is transformed into a varying intensity of the transmitted beam which can be easily monitored by a photodiode and a lock-in amplifier which is tuned to the frequency of a chopper inserted into the optical beam generating the electrical signal via a photoconductive switch.

Figure 16.4 depicts the scheme for an electro-optic sampling set-up. The crucial component is a small LiTaO$_3$ crystal polished to the shape of a small pyramid cut at the top to form a surface area as small as $50 \times 50\ \mu m^2$. This plane is pressed against the coplanar transmission line so that the crystal penetrates into the fringing electric field between the two lines. This noninvasive electro-optic wafer probe permitting testing of high-speed integrated circuits was originally designed by *Valdmanis* [16.13]. 100-fs output pulses from a CPM laser are used as the optical probe for the electric field.

The plane of polarization of each optical pulse passing through the LiTaO$_3$, is rotated and the angle of rotation is proportional to the field strength, i.e. the corresponding current amplitude of the electrical pulse on the line. The respective intensity variation is monitored with a slow photodiode which does not need to resolve the individual optical pulses. Picosecond and subpicosecond time resolution is accomplished by varying the temporal delay of the optical probe pulse with respect to the electrical pulse on the line. Electro-optic sampling with 100-fs pulses offers a time resolution of 300 fs which exceeds the present limit attainable with photoconductive switches by almost a factor of 2.

Besides having higher time resolution, electro-optic sampling provides the further advantage that the electrical pulse shape can be easily sampled at any

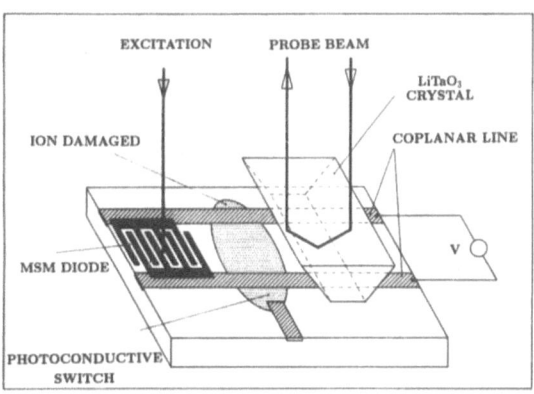

Fig. 16.4. Experimental setup for characterization of MSM photodiodes by photoconductive or electro-optic sampling with ultrashort laser pulses

position within an extended circuit by simply shifting the probe tip to the corresponding place. In contrast, photoconductive switching allows signal analysis only as discrete positions where a photoconductive switch has been integrated in the circuit design. It is interesting to note that GaAs, which is an important material for the fabrication of most high frequency electronic components is also electro-optically active. This property opens the possibility to utilize the electro-optic effect in the substrate material for sampling of ultrashort voltage and current pulses [16.14].

16.4 Characterization of MSM Photodiodes

16.4.1 MSM Schottky Photodetectors

Metal–semiconductor–metal (MSM) GaAs Schottky diodes consisting of a planar multi-finger electrode pattern on a high purity semiconductor substrate are sensitive, high speed photodetectors for multigigabit optical data processing [16.15, 16]. Narrow Schottky contact spacing which results in rapid carrier sweep-out after photoexcitation and a small active area in order to minimize the capacitance of the device are the key design features.

The diodes which have been fabricated by means of electron beam lithography on bulk LEC-GaAs using Ti/Pt/Au as the Schottky contacts had an active area of $10 \times 10 \, \mu m^2$, and typically 0.5–1-μm-wide fingers with 0.5–1.0 μm separation between them. The sensitivity is 0.2 A/W at 3 V and 825 nm.

For sampling of the diode response, the photodiode was integrated with a photoconductive switch on the same GaAs chip which was locally ion damaged with a total dose of $4.11 \times 10^{14} \, H^+ \, cm^{-2}$ (Fig. 16.4). The GaAs carrier lifetime was thus reduced to 0.8 ps [16.10]. The transmission line consists of two 5-μm-wide stripes with 10 μm spacing. The 5-μm-wide sampling strip is located at 2.5 mm distance from the diode. Besides photoconductive sampling, we also applied electro-optic sampling with the small LiTaO$_3$ tip at a distance of 100 μm from the diode. Excitation of the diode and photocurrent sampling were performed using the 70-fs optical pulses derived from a dispersion-compensated colliding-pulse mode-locked dye laser (wavelength 620 nm, repetition rate 116 MHz). The time resolution was determined to be 0.8 ps and 0.3 ps for photoconductive and electro-optic sampling, respectively.

Here we present a direct comparison of experimental and theoretical results of the electron and hole transport properties in these diodes as a function of the geometrical structure and bias of the device. Our analysis is focused, in particular, on the influence of space-charge effects originating from photo-generated electrons and holes at higher laser pulse intensities on the velocity of the carrier extraction. Finally, we try to answer the question whether the response speed of these detectors is limited by the intrinsic physical carrier transport times or determined by the RC time constant associated with the capacitance of the diode structure.

16.4.2 Monte Carlo Simulation of the Transport

The response of the MSM diode to a 70-fs pulse is calculated by two-dimensional Monte Carlo (MC) simulation of the carrier transport [16.17–19]. This method represents a self-consistent solution of Boltzmann's transport and Poisson's field equations, in both space and time. The latter is iteratively solved every 40 fs over a rectangular uniform mesh of 10 nm. An unintentional background doping density of 5×10^{13} (n-type) is assumed.The scattering mechanisms taken into account to solve Boltzmann's equations are optical and acoustic phonon, intervalley and impurity scattering. The model includes the Γ, X, and L, minima of the conduction band and the heavy-hole, light-hole and split-off valence band. Carrier–carrier scattering and external parasitic effects (capacitance) have not been included in this simulation. During illumination lasting 70 fs and terminating at $t = 0$ fs, electron–hole pairs were generated with a depth distribution

$$n(y) = N_0 \exp(-ay) , \tag{16.1}$$

where N_0 represents the density at the surface and $a = 4.279 \times 10^4 \, \mathrm{cm}^{-1}$ the absorption coefficient. The subsequent motion of these photogenerated electrons and holes to the anode and cathode, respectively, under the influence of the electric field and the stochastic scattering events is then calculated for successive 40-fs time intervals. The transport is simulated in a self-consistent way since the repeated solution of Poisson's equation after every 40 fs takes into account screening of the electric field due to the space-charge fields originating from the spatial separation of electron and hole distributions. The output current is determined by counting the electrons and holes which are absorbed during each 40-fs interval by the anode and cathode, respectively.

The transport simulation has to be restricted to 10^4 artifical "superparticles" because of the limitations on computer capacity and time. When solving Boltzmann's equation every superparticle is regarded as a single charged carrier. For the recalculation of the electric field by means of Poisson's equation and for estimation of the current, one superparticle is considered to consist of many real charged particles and the number depends on the light pulse intensity.

The upper part of Fig. 16.5 depicts the temporal evolution of the electron ($\bigcirc$) current, the hole ($\times$) current and the total ($\bullet$) current predicted by this model for an MSM diode with a finger separation of 1.5 μm and a bias voltage of 2 V at a relatively modest excitation level of $5.1 \times 10^{15} \, \mathrm{cm}^{-3}$. The time dependence of the total current ($\bullet$) exhibits a first decay (1/e-time 8 ps) which can be attributed to the electron sweep-out and a second slow decay (1/e-time 30 ps) which results from the hole sweep-out. The signal rise time and the full width at half maximum are 4 ps and 11 ps, respectively.

Figure 16.6 (left) shows the influence of the excitation density on the photocurrent pulse shape. The MC calculation predicts an initially faster decay and a significantly longer tail of the photocurrent if the carrier density is increased to $5.1 \times 10^{16} \, \mathrm{cm}^{-3}$. This behavior is due to effective screening of the

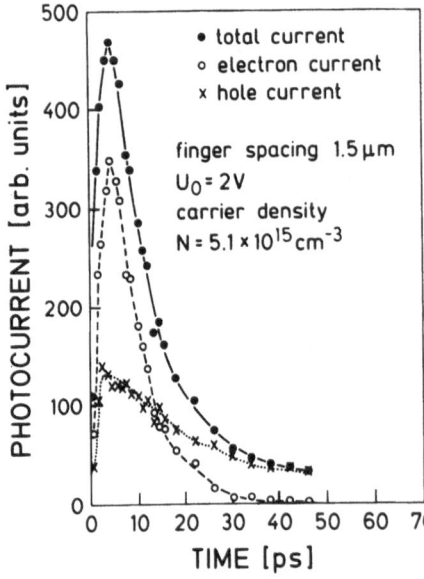

Fig. 16.5. Monte Carlo simulation of the electron ($\bigcirc$), hole ($\times$) and total current ($\bullet$) in the MSM photodiode (finger separation 1.5 μm, bias 2 V) for a modest excitation density of 5.1 $\times 10^{15}$ e–h pairs per cm^3

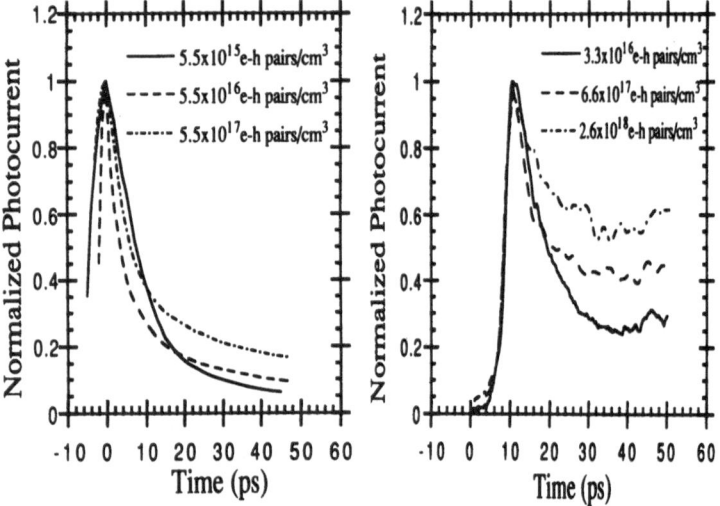

Fig. 16.6. Theoretical variation of the photocurrent output pulse shape with excitation density (left) and response of a 0.75-μm MSM photodiode to excitation with varying e–h pair density (right)

external field by the space-charge fields created by the separation of the electron and hole populations inside the device. If the excitation intensity is increased by another factor of 10, the initial peak is broadened again and the long tail becomes much stronger. Under this condition, the transport properties are totally governed by screening effects which lead to an almost complete compensation of the external field across the entire space between anode and cathode after 2 ps. In this situation the collection rates of electrons and holes are almost the same. The output current decays to 30% of its peak value after 12 ps and then forms a long tail which extends over a few 100 ps.

It should be noted that the initial decay of the signal at higher intensities reflects the collapse of electric field between anode and cathode rather than depletion of carriers. These disturbing screening effects become important at even lower densities if the electric field is smaller. A diode with 0.5–1-μm-finger distance should thus be operated at a bias of a few volts.

16.4.3 Experimental Results

Figure 16.6 (right) shows the response of a 0.75-μm diode (bias 3 V) for several pulse intensities corresponding to electron–hole pair densities of 3.3 $\times 10^{16}$ cm^{-3}, 6.6×10^{17} cm^{-3}, and 2.6×10^{18} cm^3. The signal sampled by the photoconductor has a rise time of less than 4 ps independent of the excitation level. The rapid decay of the signal peak becomes initially faster with increasing excitation intensity as predicted by the MC simulation as a consequence of field screening and is significantly slowed down at the highest intensity. In addition, the signal reveals the formation of a continuously growing long tail with increasing intensity. By and large, the experimental curves qualitatively confirm the expected strong influence of space-charge effects on the carrier transport velocity.

In Fig. 16.7 we present a comparison of a measured current pulse shape with theoretical predictions. The time dependence of the calculated photocurrent (solid line) is shown together with the experimental photocurrent curve measured by electro-optic sampling (diode finger separation 0.5 μm, bias 2 V).

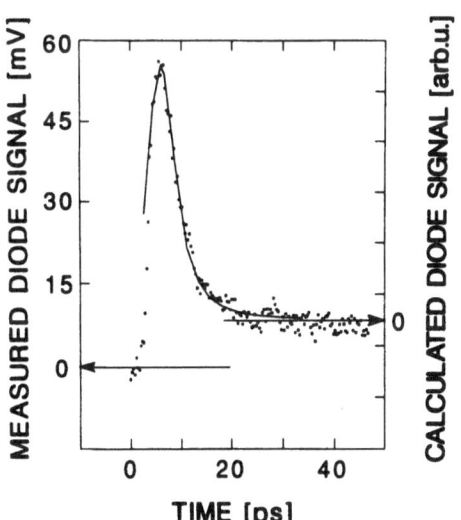

Fig. 16.7. Comparison of the experimental photocurrent curve (●) measured by electro-optic sampling and the Monte Carlo calculation of the total current (solid line) in an MSM photodetector (finger separation 0.5 μm, bias 2 V, excitation: 5.1×10^{15} e–h pairs per cm^3)

222   *J. Kuhl* et al.

Excellent agreement between experiment and theory can be observed except for a long tail in the measured photocurrent which is not predicted by the theory and whose origin is not yet understood. The measured rise time of the photocurrent amounts to 1.7 ps (10–90%), the first rapid decay to 4.5 ps and the second slow decay to 10 ps.

In order to identify the response-limiting processes we investigated the temperature dependence of the current output pulses [16.20]. For these experiments the diode (0.75 µm finger spacing, 4 V bias, $N_0 = 3.5 \times 10^{16}$ cm^3) was mounted in a He cryostat and the photocurrent pulse shape was analyzed by photoconductive sampling. Figure 16.8 depicts the measured FWHM of the current pulses (triangles) vs. temperature. The pulse duration decreases from 10.8 ps at 300 K to 6.5 ps at 100 K, remains almost constant between 100 K and 50 K and increases again to 11.1 ps if the temperature is decreased to 10 K. The ratio of the finger distance to the FWHM of the current pulse duration (shown as black dots) represents an approximate measure for the saturation value of the carrier drift velocity.

The variation of the time resolution of the photoconductive sampling with temperature due to multiple trapping of carriers in shallow traps can be excluded and the shortening of the signal between 300–100 K can definitely be attributed to an accelerated depletion of the diode after photoexcitation. This experimental finding is explained by the increase of the electron drift velocity at lower temperatures because of a significant reduction of intra- and inter-valley phonon scattering. The phonon scattering rates drop in accordance with the decreasing population of LO, LA, and TA phonon modes with temperature. The corresponding temperature dependence of the electron drift velocity can be described by a simple model which takes into account phonon absorption and emission rates depending on the Bose distribution of phonons. In our analysis we assumed a weighted average phonon energy of 30 meV. The solid line in Fig. 16.8 is obtained by a fit of this model to the experimental data where the

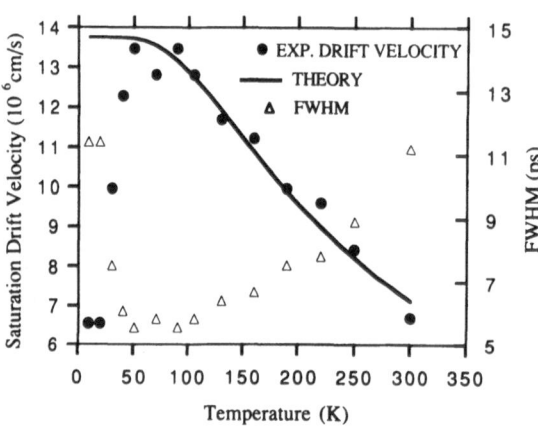

Fig. 16.8. FWHM of the photocurrent pulses (triangles, right scale), measured (black dots) and calculated (solid line) saturation drift velocity of electrons in the MSM diode (left scale) vs. temperature

saturation drift velocity v_d at $T = 0\,\mathrm{K}$ is the only fit parameter. The value of v_d $= 1.4 \times 10^7\,\mathrm{cm/s}$ agrees fairly well with the published data for GaAs.

The excellent description of the experimental data by the phonon scattering model in the temperature range 50–300 K identifies carrier–phonon interaction as the limiting mechanism for the transport velocity. The model of intra- and inter-valley phonon scattering predicts, however, an almost constant electron drift velocity for $T < 70\,\mathrm{K}$. The rapid increase of the response time for $T < 50\,\mathrm{K}$ can be attributed to ionized impurity scattering at low temperatures. The low temperature experiments demonstrate that the time resolution is not influenced by the diode's intrinsic capacitance, C, which was estimated to be 6 fF [16.21] corresponding to an RC time constant of 600 fs (transmission line impedance $R \sim 100\,\Omega$) about a factor of 3 below the shortest rise times. Consequently, the photocurrent rise and fall times are determined by carrier transit time effects.

Fourier transformation of the experimental photocurrent curve in Fig. 16.7 yields a frequency bandwidth of 50 GHz (3 dB point), if the long tail is neglected. This bandwidth is significantly lower than the value of 105 GHz reported by *van Zeghbroeck* et al. [16.22]. The origin of the discrepancy is not clear. The Monte Carlo calculation predicts, however, a nonnegligible influence of field screening on the response current for the values of the intensity and bias voltage given in [16.22].

16.5 Conclusion

We have demonstrated that the electron and hole contribution of the photocurrent in GaAs MSM photodiodes after excitation with an ultrashort light pulse are nearly separated in time. The response speed at low excitation densities is limited by optical phonon scattering. Field screening effects become important at higher intensities or low bias voltages. In order to prevent space-charge induced deterioration of the response the detector should be operated at high bias levels ($> 2\,\mathrm{V}$) and the optical pulse energy has to be limited to 100 fJ. These requirements, which have been ignored in previous theoretical and experimental work, will be surely achieved in future high-bit-rate optical communication systems. Ensemble Monte Carlo calculations support our experimental data.

References

16.1   D.H. Auston: Picosecond Photoconductors, in: *Picosecond Optoelectronic Devices*, ed. by C. H. Lee (Academic, New York 1984)
16.2   D.H. Auston: Ultrafast Optoelectronics, in *Ultrashort Laser Pulses and Applications*, ed. by W. Kaiser, Topics Appl. Phys; Vol. 60 (Springer, Berlin, Heidelberg 1988)
16.3   D.H. Auston: Picosecond Photoconductivity: High Speed Measurements of Devices and Materials, in *Semiconductors and Semimetals*, Vol. 28, ed. by R. B. Marcus (Academic, New York 1990)

16.4   J.A. Valdmanis, G. Mourou: Subpicosecond Electrical Sampling and Applications, in *Picosecond Optoelectronic Devices*, ed. by C. H. Lee (Academic, New York 1984)

16.5   J.A. Valdmanis, G. Mourou: Electro-Optic Sampling: Testing Picosecond Electronics, Laser Focus/Electro-Optics, Feb. 1986, p. 84 and March 1986, p. 96

16.6   J.A. Valdmanis: Electro-Optic Measurement Techniques for Picosecond Materials, Devices, and Integrated Circuits, in *Semiconductors and Semimetals*, Vol. 28, ed. by R. B. Marcus (Academic, New York 1990)

16.7   D.H. Auston: IEEE J. Quantum Electron. **QE-19**, 639 (1983)

16.8   R.L. Fork, B.I. Greene, C.V. Shank: Appl. Phys. Lett. **38**, 671 (1981); J. A. Valdmanis, R. L. Fork, J. P. Gordon: Opt. Lett. **10**, 131 (1985); J. A. Valdmanis, R. L. Fork: IEEE J. Quant. Electron. **QE-22**, 112 (1986); M. Nakazawa, T. Nakashima, H. Kubota, S. Seikai: Opt. Lett. **12**, 681 (1987); H. Kubota, K. Kurokawa, M. Nakazawa: Opt. Lett. **13**, 749 (1988)

16.9   M.B. Ketchen, D. Grischkowsky, T.C. Chen, C.-C. Chi, I.N. Duling, III, N.J. Halas, J.-M. Halbout, J.A. Kash, G.P. Li: Appl. Phys. Lett. **48**, 751 (1986)

16.10   M. Lambsdorff, J. Kuhl, J. Rosenzweig, A. Axmann, Jo. Schneider: Appl. Phys. Lett. **58**, 1881 (1991)

16.11   F.W. Smith, H.Q. Le, V. Diadiuk, M.A. Hollis, A.R. Calawa, S. Gupta, M. Frankel, D.R. Dykaar, G.A. Mourou, T.Y. Hasiang: Appl. Phys. Lett. **54**, 890 (1989)

16.12   M.C. Nuss, D.W. Kisker, P.R. Smith, T.E. Harvey: Appl. Phys. Lett. **54**, 57 (1989)

16.13   J.A. Valdmanis: Electron. Lett. **23**, 1308 (1987)

16.14   B.H. Kolner, D.M. Bloom: IEEE J. Quantum Electron. **QE-22**, 69 (1986).

16.15   W. Roth, H. Schumacher, J. Kluge, H.J. Geelen, and H. Beneking: IEEE Trans. Electronic Devices **32**, 1034 (1985)

16.16   M. Ito and O. Wada: IEEE J. Quantum Electron. **QE-22**, 1073 (1986)

16.17   C. Moglestue: IEEE **CAD-5**, 326 (1986)

16.18   C. Moglestue, J. Rosenzweig, J. Kuhl, M. Klingenstein, M. Lambsdorff, A. Axmann, Jo. Schneider, J. Hülsmann: J. Appl. Phys. **70**, 2435 (1991)

16.19   M. Lambsdorff, M. Klingenstein, J. Kuhl, C. Moglestue, J. Rosenzweig, A. Axmann, Jo. Schneider, J. Hülsmann, H. Leier, A. Forchel: Appl. Phys. Lett. **58**, 1410 (1991)

16.20   M. Klingenstein, J. Kuhl, J. Rosenzweig, C. Moglestue, A. Axmann: Appl. Phys. Lett. **58**, 2503 (1991)

16.21   G.D. Alley: IEEE Trans. Microwave Theory and Techniques **MTT-18**, 1028 (1970)

16.22   B.J. van Zeghbroeck, W. Patrick, J.-M. Halbout, P. Vettiger: IEEE Electron Device Lett. **9**, 527 (1988)

Jürgen Kuhl

17. Ultrafast Spectroscopy of Plasmas Generated by Very Intense Femtosecond Dye Laser Pulses

D. von der Linde

With 6 Figures

During the last two decades dye lasers have played an outstanding role in the generation of ultrashort optical pulses. Owing to the large frequency width of the optical transitions in organic dye molecules, dye lasers can produce pulses with a duration well below 10^{-13} s. The output energy of a dye laser oscillator can be increased by using dye laser amplifiers. The characteristic energy fluence for gain saturation in dye molecules is typically a few 10^{-3} J/cm^2. Thus a relatively small-scale dye laser amplifier with an output beam cross section of ≈ 1 cm^2 is capable of providing laser pulses with a few millijoules of optical energy. Properly designed dye laser amplifiers produce high-quality, diffraction-limited output beams which can be readily focused to a spot diameter of just a few microns. The peak intensity of optical radiation that can be attained in this way is of the order of 10^{17} W/cm^2.

The electric field strength corresponding to such an extremely high intensity is close to 10^{10} V/cm, which is comparable to the strength of the Coulomb field in atoms. Any material exposed to such high light intensity will be ionized during the interaction. For example, solid material will be turned into a plasma in a time given by the duration of the laser pulse, say, a hundred femtoseconds or less. This time is too short for significant transport of mass and energy to occur. Therefore the density of the generated plasma is approximately equal to the density of solid material; the average energy of the electrons is expected to be a few hundred electron volts.

During the last few years there has been rapidly growing interest in plasmas generated by intense femtosecond pulses. One of the exciting prospects is that these plasmas emit short bursts of X-rays [17.1, 2] and that it may be possible to develop a point source of subpicosecond X-ray pulses [17.3] by tightly focusing femtosecond optical pulses on a suitable target. Encouraged by the rapid progress in X-ray optics [17.4] one may envision that ultrashort X-ray emission from a microplasma is collected and focused on a sample for new types of applications or experiments. Still more challenging would be the realization of proposed new X-ray laser schemes [17.5, 6] which make use of laser plasmas produced by high-energy femtosecond laser pulses.

A great deal of research remains to be done before these challenging applications become a reality. The detailed nature of the interaction of extremely intense, ultrashort laser pulses with matter and the properties of the generated plasmas are not yet well understood. A typical new feature of femtosecond laser plasmas is that the thickness of the plasma generated on the

Topics in Applied Physics, Vol. 70
Dye Lasers: 25 Years Ed.: Dr. Michael Stuke
© Springer-Verlag Berlin Heidelberg 1992

surface of a solid target is only a small fraction of the wavelength. By contrast, the scale length of plasmas generated by picosecond or nanosecond laser pulses is comparable with or larger than the wavelength.

Much of the previous work on plasmas produced by femtosecond laser pulses was concerned with the fundamental optical properties [17.7–13] and the characterization of the X-ray emission [17.14–20]. This chapter deals with a number of new observations on femtosecond laser plasmas. Section 17.2 describes the spectra of the optical second harmonic of light reflected from the plasma, which show a wealth of interesting structures. Time-resolved pump–probe experiments with relatively intense collinear probe pulses reveal striking correlations between the time dependence of the optical reflectivity and the second harmonic (Sect. 17.3), which may be indicative of ponderomotive interaction of the probe pulse with the plasma.

Section 17.4, finally, describes noncollinear pump–probe experiments with very weak probe pulses which clearly resolve the transformation of the solid into a plasma by the ultrashort excitation pulse, and the subsequent expansion dynamics of the highly excited material.

17.1 Experimental

Intense femtosecond laser pulses were generated by amplifying the output pulses of a standard colliding pulse, passively mode-locked (CPM) dye laser operating near 620 nm [17.21]. An energy gain of more than 10^6 was obtained by passing the femtosecond dye laser pulses through a 6-pass, bow-tie-type preamplifier pumped at 10 Hz repetition rate by 4 mJ, 8 ns pulses from a frequency-doubled Nd:YAG laser. The final amplifier stage was an axicon amplifier pumped with the remaining 250 mJ of second harmonic from the Nd:YAG laser. The output energy of the axicon is approximately 0.5 mJ. Due to dispersive pulse broadening during amplification the pulse width is ≈ 350 fs, and the amplified pulses possess considerable frequency chirp. The pulse width can be reduced to ≈ 60 fs by passing the pulses through a dispersive delay line consisting of optical prisms. However, to avoid spatial and spectral distortions of the pulses caused by the nonlinearity of the refractive index of the prism material the intensity must be kept below a certain limit. Therefore the output beam of the axicon amplifier must be expanded to a rather large beam diameter (≈ 1.5 cm) before being passed through the prism sequence. Some of the experiments discussed below were done with the uncompressed, chirped pulses of 350 fs.

The maximum pulse energy on target was $E_p = 0.5$ mJ for the uncompressed pulses, and $E_p = 0.3$ mJ for the compressed pulses. For an achromatic focusing lens with $f = 2.5$ cm a focal spot diameter $D = 3$ μm was measured (full width at half maximum, FWHM). Assuming a Gaussian spatial profile, the peak intensity of the femtosecond dye laser pulses is calculated using the expression $I_p = 0.88 \, (E_p/D^2 t_p)$. One obtains $I_p = 1.4 \times 10^{16}$ W/cm^2 for the uncompressed pulses ($t_p = 350\ \text{fs}$), and $I_p = 4.9 \times 10^{16}\ \text{W/cm}^2$ for the compressed pulses ($t_p = 60\ \text{fs}$). The actually measured spatial profile has approximately 20% more energy in the wings than a Gaussian. Thus there is a corresponding overestimate of the peak intensity when a formula based on a Gaussian distribution is used.

Amplified spontaneous emission (ASE) was controlled using spatial filters and saturable absorbers. The ASE energy fluence is typically 0.2% of the total fluence and corresponds to an intensity of less than $10^9\ \text{W/cm}^2$.

A schematic diagram of the experimental arrangement is shown in Fig. 17.1. The incident femtosecond laser pulses were focused onto the surface of the target using a suitable objective lens. The targets were placed in a vaccum chamber with a pressure of 10^{-3} mbar. The reflected light was recollimated and passed through a beam splitter to separate the fundamental (ω) and the second harmonic (2ω). For the measurement of the spectra near ω and 2ω an optical multichannel analyzer coupled to a 0.5-m grating spectrometer was used. In the time-resolved pump–probe experiments only the *integrated* reflected fundamental and second harmonic was measured using a silicon photodiode (ω) or a photomultiplier (2ω).

Three different types of experiments were performed:

1) Single-beam experiments with p-polarized excitation pulses. Here the general characteristics of the reflected light and the spectra were measured (ω and 2ω).
2) Collinear time-resolved pump–probe experiments with s-polarized pump and p-polarized probe pulses. To overcome background signals due to the reflected pump pulse, the probe pulses had to be relatively strong in the collinear geometry.
3) Noncollinear pump–probe experiments in which the pump pulse could be blocked by a beam stop. In this case very weak probe pulses could be used.

Simple glass substrates served as targets. The samples were rapidly raster scanned to provide a fresh sample surface for each pulse. The noncollinear pump–probe experiments required particular care to maintain spatial overlap of the micron-size pump and probe focal spots and to keep the target surface in the focal plane during scanning. To meet the stringent requirements of spatial overlap and focusing, a common objective lense with $f = 6$ cm was to focus both the pump beam and the probe beam (focal diameter $\approx 7.2\ \mu\text{m}$).

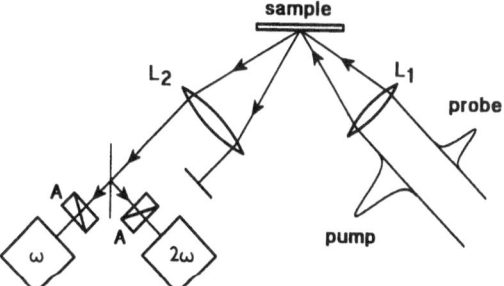

Fig. 17.1. Schematic diagram of experimental arrangement: L_1–focusing lens. L_2–recollimation lens. A: polarization analyzers. ω and 2ω: detectors or spectrometers for the measurement of the reflected fundamental and second harmonic

17.2   Characteristics of the Reflected Light

The properties of the reflected pulses changed dramatically when the incident intensity was increased. At peak intensities around 10^{16} W/cm^2 the observed reflectivity was much greater than that of the unperturbed glass sample. The characteristics of the reflected light are consistent with the formation of a thin plasma layer with an extremely steep gradient of the electron density. For example, with a p-polarized light pulse of 350 fs duration the reflectivity as a function of angle of incidence had a minimum near 50°. Model calculations suggest that the characteristic scale length L describing the electron density gradient is approximately 0.05 λ under these conditions. This number can be used for a crude estimate of the expansion velocity of the plasma, $v_{\text{exp}} \approx L/t_{\text{p}}$ = 0.05$\lambda/t_{\text{p}} \approx 10^5$ m/s, where $t_{\text{p}} = 350$ fs is the pulse duration.

A typical frequency spectrum of the reflected pulse is shown by the solid curve in Fig. 17.2 (p-polarization, 40° angle of incidence). The dashed line is the spectrum of the incident pulse. It can be seen that the reflected spectrum is essentially blueshifted by $\Delta\lambda = 0.5$ nm. If interpreted as a Doppler shift from an expanding plasma, the Doppler velocity is calculated to be 1.5×10^5 m/s. This value is in reasonable agreement with the expansion velocity estimated from the angular dependence of the optical reflectivity. It is clear, however, that these estimates are rather crude and involve temporal and spatial averages over the intensity distribution of the laser pulse.

A striking observation in these experiments was that a relatively large amount of reflected second harmonic was generated with p-polarized excitation pulses. Energy conversion of up to 10^{-4} was observed. Second harmonic generation with s-polarization was far less effective. The reflected second harmonic was predominantly p-polarized and increased strongly with angle of incidence. More importantly, the structure of the second harmonic spectra changed significantly when the angle of incidence was changed. Examples of second harmonic spectra are given in Fig. 17.3a–c. The dashed lines represent

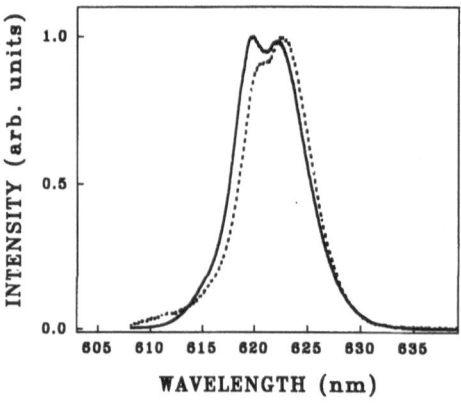

Fig. 17.2. Example of a spectrum of the reflected light (solid curve). Dashed curve: spectrum of the incident light. Angle of incidence: 40°; excitation intensity: $\approx 10^{16}$ W/cm^2; pulse duration: 350 fs

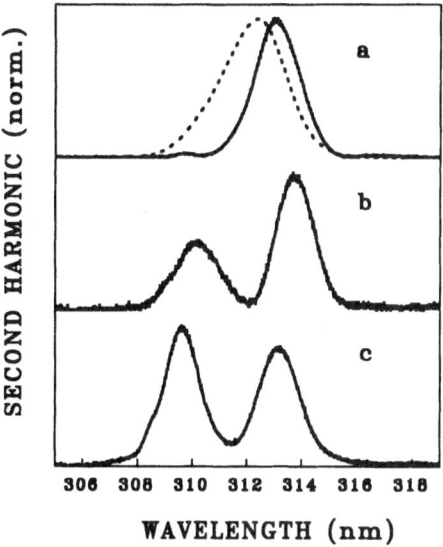

SECOND HARMONIC (norm.)

306   308   310   312   314   316   318

WAVELENGTH (nm)

Fig. 17.3. Spectra of the reflected second harmonic (solid curve). Dashed curve: reference second harmonic spectrum from a KDP crystal. Excitation intensity: $\approx 10^{16}$ W/cm²; pulse duration 350 fs; angle of incidence: (a) 63°; (b) 42°; (c) 35°

second harmonic reference spectra measured with a thin KDP crystal (phase-matching bandwidth $\Delta\lambda \approx 50$ nm). For large angles of incidence the second harmonic was essentially shifted to the red, e.g., $\Delta\lambda = +0.9$ nm, but there was also a very weak blueshifted component (Fig. 17.3a, $\varphi = 63°$). The relative strength of the blue component increased with decreasing angle of incidence (Fig. 17.3b, $\varphi = 42°$). In Fig. 17.3c the blueshifted second harmonic exceeds the red component. Note that the maxima of the spectra in Fig. 17.3 are normalized to unity. When the actual signals are compared it turns out that the redshifted component decreases strongly with decreasing angle of incidence, whereas the blue component is approximately constant.

The observed structure of the second harmonic spectra, in particular the pronounced redshifted component, cannot be explained by the Doppler effect. Expansion of the plasma away from the surface is expected to lead to an overall blue shift. Moreover, the values of the observed shifts of the second harmonic are too large and inconsistent with the observed Doppler shift of the reflected fundamental.

One of the possible mechanisms of second harmonic generation in an inhomogeneous plasma is scattering from longitudinal electron plasma oscillations [17.22]. It has been suggested that the observed second harmonic shifts could be due to coupling of electron–plasma oscillations with ion-acoustic modes [17.23]. This coupling would give rise to second harmonic radiation with frequency shifts determined by ion-acoustic frequencies. The mechanism could account for the magnitude of the observed frequency shifts of the second harmonic, but a model explaining the structure of the second harmonic spectra is still lacking.

17.3   Collinear Pump–Probe Experiments

To study the formation of the plasma and its subsequent time evolution, experiments with two collinear, orthogonally polarized laser pulses with a variable time delay were performed. The s-polarized leading pulse served as excitation, and the delayed p-polarized pulse was used as a probe. The reflected pump pulse was suppressed with a polarizer. Generally the second harmonic background due to the pump was weak, but considerations of the signal-to-background ratio required the use of probe energies at least 1/8 of the pump energy. On the other hand, the collinear geometry relieved to some extent the difficulties of controlling the spatial overlap of pump and probe.

In the experiments the reflected fundamental and the second harmonic were measured simultaneously. Striking correlations in the variation of the reflectivity and the second harmonic with delay time were observed. An example is shown in Fig. 17.4 for a pump–probe energy ratio of 8:1. The probe pulse intensity corresponds to 4.4×10^{14} W/cm². For such high intensity a plasma is generated by the probe pulse alone. The reflectivity for negative delay time,

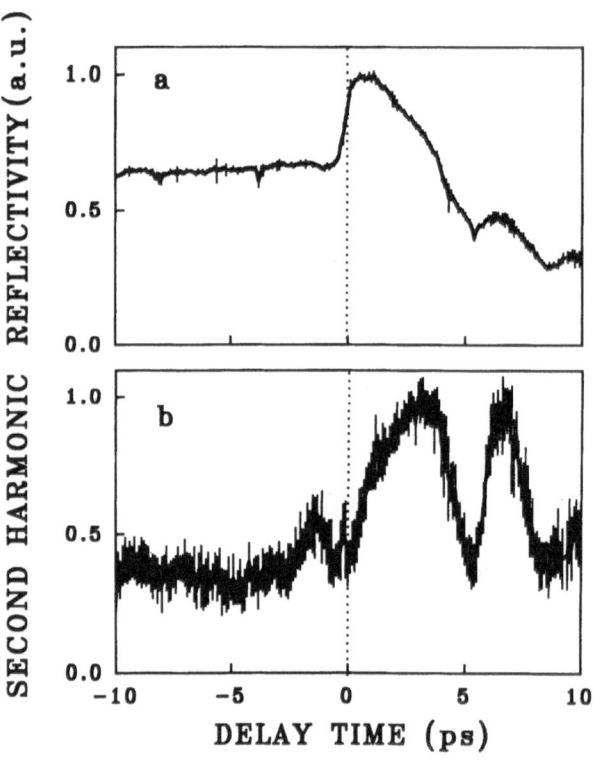

Fig. 17.4. Collinear pump–probe experiments: (**a**) reflectivity (arbitrary units) and (**b**) second harmonic of the probe pulse as a function of delay time. Excitation intensity: 3.5×10^{15} W/cm²; probe intensity: 4.4×10^{14} W/cm²; angle of incidence: 48°

therefore, is the self-reflectivity from the plasma generated by the probe pulse (Fig. 17.4a). The sharp increase of the reflectivity near zero delay time is due to plasma formation by the pump pulse proper. The subsequent decay over several picoseconds signifies the expansion of the plasma, being caused by optical absorption in the growing plasma layer. Note the distinct reflectivity minima near 5 ps and 8.5 ps.

The corresponding second harmonic is shown in Fig. 17.4b. It can be seen that the second harmonic rises from the background to a first maximum at 3 ps. The point to be noticed here is that there are several maxima and mimima, which are clearly correlated with the observed temporal structure of the reflectivity. The background for large negative delay is due to the sum of the second harmonic generated by the pump pulse and the probe pulse independently. The small maximum near -2 ps corresponds to the situation in which pump and probe have interchanged their roles.

It will be shown in the next section that the periodic modulation of the reflectivity and the second harmonic disappear when the intensity of the probe pulse is reduced. This suggests that the effect may be caused by some nonlinear interaction of the probe pulse with the expanding plasma, when high probe intensity is used.

The following tentative explanation is offered for the periodic modulations. Interference between the incident and the reflected probe light gives rise to a standing wave pattern in the expanding plasma layer. For probe pulse intensities corresponding to several times 10^{14} W/cm^2 substantial ponderomotive forces act on the plasma, which can be comparable with the pressure forces. As a result the plasma density profile is modified during the interaction with the probe pulse and assumes a spatial modulation which tracks the light intensity of the standing wave pattern [17.24]. This mechanism can lead to a self-consistent redistribution of the plasma density and of the electric fields. It is possible that during the expansion the system passes repeatedly through spatially resonant configurations in which the absorption of electromagnetic energy is enhanced. This would lead to corresponding periodic changes of the reflected light.

This model could also explain the observed pronounced maxima of the second harmonic. Formation of resonant structures in the plasma profile would be accompanied by periodic enhancements of the electric fields in the plasma which in turn would drive the source current responsible for second harmonic generation to a maximum in a resonant configuration.

17.4   Non-collinear Pump–Probe Experiments

The principal advantage of a noncollinear pump–probe beam geometry is that it is possible to suppress the reflected pump beam completely by means of a beam stop. Very much weaker probe pulses can be used in this case because background signals from the strong pump pulse, which place a lower limit on the probe pulse energy, are absent. It is readily possible to detect the reflection and the second harmonic even if the probe pulse energy is more than two orders of magnitude less than the pump pulse energy.

In the noncollinear experiments the probe pulse intensity was 7.1×10^{13} W/cm^2. Figure 17.5 shows the result of a time-resolved reflectivity measurement of a plasma generated by a 60 fs, s-polarized pump pulse $(8.5 \times 10^{15}$ W/cm$^2)$. The vertical scale in Fig. 17.5 is the measured absolute reflectivity. For negative delay times the optical reflectivity is measured to be $R_p < 1\%$, which is the reflectivity of the unperturbed glass substrate for p polarization for an angle of incidence of $\approx 50°$. In this experiment a sharp, clean rise of the reflectivity to a maximum of $R_p \approx 20\%$ is observed. The measured rise time is 100 fs. A reflectivity minimum occurred around 250 fs. The inset in Fig. 17.6 shows the decay of the reflectivity on a time scale of several picoseconds.

The reflectivity data demonstrate three important points:

1) Modulations of the decay curve as observed in the experiments with more intense probe pulses (Fig. 17.4) are absent.
2) The unperturbed reflectivity observed for negative delay (probe pulses lead) indicate that the probe pulses alone do not generate a plasma.
3) The sharp rise of the reflectivity within 100 fs and the absence of any reflectivity pedestal clearly indicate that the pump pulse interacted with the unperturbed solid and not with a plasma. A "preplasma" [17.25] could be generated if the pump pulse had a poor peak-to-background intensity ratio.

The second harmonic of the probe pulse measured under the same conditions is shown in Fig. 17.6. A relatively slow rise of the second harmonic to a

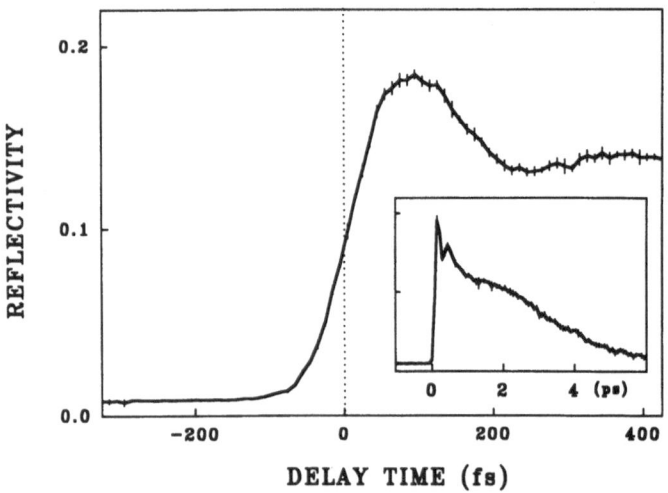

Fig. 17.5. Noncollinear pump–probe experiments: reflectivity (absolute units) of the probe pulse vs. delay time. Inset: reflectivity data on an extended time scale. Angle of incidence: 50°; excitation intensity: 8.5×10^{15} W/cm^2; probe intensity: 7.1×10^{13} W/cm^2; pulse duration: 65 fs

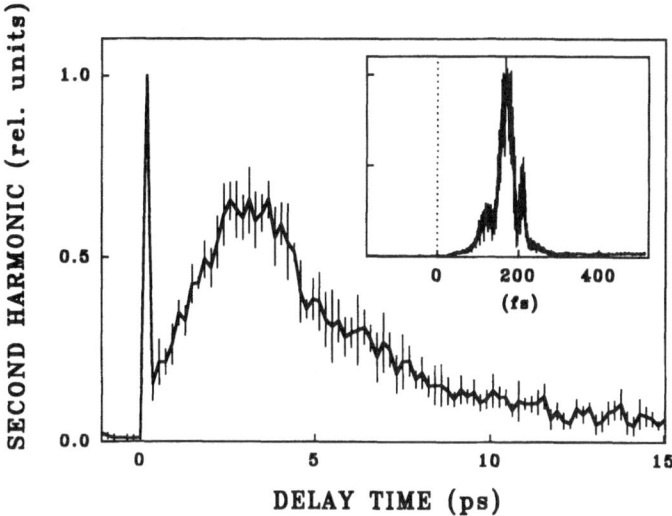

Fig. 17.6. Noncollinear pump–probe experiments: Second harmonic of the probe pulse vs. delay time. Inset: Structure of the sharp maximum near 200 fs on an expanded time scale. In the main part of the figure the second harmonic corresponding to the maximum at 200 fs is not fully resolved. It is four times stronger than the second maximum at 3 ps. The experimental parameters are the same as in Fig. 17.5

maximum near 3 ps, followed by a smooth decay over several picoseconds was observed. The strong modulations which were observed with the more intense probe pulses (Fig. 17.4) have disappeared. Note, however, the extremely sharp spike of second harmonic observed at a delay time of 200 fs. This structure is shown on an expanded time scale in the inset of Fig. 17.6. The maximum at 200 fs exceeds the maximum at 3 ps by a factor of 4. The measurement also reveals an interesting temporal fine structure corresponding to variations of the second harmonic on a time scale of about 50 fs.

Second harmonic generation in a rapidly expanding, strongly inhomogeneous plasma is a rather complicated problem. There are a number of interacting processes which could be responsible for the observed changes of the second harmonic., e.g., the changes of the plasma density profile, changes of the electron temperature, occurrence of resonant absorption, the phase-matching conditions, etc. Because a thorough theoretical analysis is still lacking, only tentative interpretations can be given.

The decay of the second harmonic at long times may be due to an increase of optical absorption with increasing geometrical thickness of the plasma, which has also been invoked as an explanation of the decay of the optical reflectivity. Alternatively, the rise and decay of the second harmonic with increasing plasma thickness (scale length) could also be caused by changes of the phase-matching conditions associated with the expansion of the plasma.

The most conspicous feature of the second harmonic is the sharp spike at 200 fs. This structure is possibly associated with the onset of resonant absorption [17.26, 27] when the plasma has reached a suitable thickness. Preliminary model calculations suggest that this condition may occur for scale length corresponding to several percent of the wavelength. With an expansion velocity of about 10^7 cm/s this would correspond to a few hundred femtoseconds, in agreement with the observed onset of the second harmonic. However, the short duration and the temporal fine structure are not explained as yet.

17.5   Conclusions

Focused intensities corresponding to electrical field strengths comparable with the fields in atoms can be achieved with femtosecond pulses from small-scale dye laser systems. Solid material exposed to such extremely intense ultrashort pulses is immediately ionized and transformed into a plasma.

In this chapter a series of femtosecond optical measurements on plasmas generated by ultrashort laser pulses have been described. In particular, it was shown that efficient optical second harmonic generation takes place and that the spectra of the second harmonic exhibit interesting structure.

The evolution of the plasma during the first few picoseconds could be measured with the help of time-resolved pump–probe techniques. With relatively intense probe pulses distinct temporal correlations in the evolution of the optical reflectivity and the second harmonic were observed. It was suggested that these effects can be attributed to perturbations of the plasma density profile caused by the probe pulse through ponderomotive forces.

An important result of the time-resolved measurements was that there is clear evidence that the laser excitation pulses actually interact with the intact solid and not with a preplasma. The solid material is transformed into a plasma in approximately 100 fs, the time being determined by the laser pulse duration.

Femtosecond time-resolved spectroscopy of laser-produced plasmas is still in an early stage. The experiments described here have revealed many intriguing phenomena. Time-resolved spectroscopy combined with theoretical efforts to explain the observed effects are likely to contribute to a better understanding and eventually realization of the exciting application of microplasmas produced by femtosecond laser pulses.

Acknowledgements. The author is indebted to H. Schüler, T. Engers and H. Schulz for their excellent experimental work. Special credit is due to H. Schulz for pointing out the relevance of ponderomotive effects. Support of this work by the Deutsche Forschungsgemeinschaft is gratefully acknowledged.

References

17.1   D. Kühlke, U. Herpers, D. von der Linde: Appl. Phys. Lett. **50**, 1785 (1987)
17.2   O. R. Wood, W. T. Silvast, H. W. K. Tom, W. H. Knox, R. L. Fork, C. H. Brito-Cruz, P. I. Maloney, C. V. Shank: J. Opt. Soc. Am. **B 13**, 24 (1987)

17.3   M. M. Murnane, H. C. Kapteyn, R. W. Falcone: IEEE J. Quantum Electron. **QE 25**, 2417 (1989)
17.4   N. M. Ceglio: J. X-Ray Sc. Techn. **1**, 7 (1989)
17.5   F. P. Schäfer: J. Phys., Coll. C6, Suppl. 10, **47**, 149 (1986)
17.6   M. Steyer, F. P. Schäfer, S. Szatmari, G. Kühnle: Appl. Phys. **B 50**, 265 (1990)
17.7   H. M. Milchberg, R. R. Freeman, S. C. Davey, R. M. More: Phys. Rev. Lett. **61**, 2364 (1988)
17.8   J. C. Kieffer, P. Audebert, M. Chaker, J. P. Matte, H. Pepin, T. W. Johnston, P. Maine, D. Meyerhofer, J. Delettrez, D. Strickland, P. Bado, G. Mourou: Phys. Rev. Lett. **62**, 760 (1989)
17.9   O. L. Landen, D. G. Stearns, E. M. Campbell: Phys. Rev. Lett. **63**, 1475 (1989)
17.10 P. Mulser, S. Pfalzner, F. Cornolti: In *Laser Interaction with Matter*, ed. by G. Verlade, E. Minguez, J. M. Perlado (World Scientific, Singapore 1989) p. 142
17.11 H. M. Milchberg, R. R. Freeman: J. Opt. Soc. Am. **B 6**, 1351 (1989)
17.12 R. Fedosejevs, R. Ottmann, R. Sigel, G. Kühnle, S. Szatmari, F. P. Schäfer: Phys. Rev. Lett. **64**, 1250 (1990)
17.13 R. Fedosejevs, R. Ottmann, R. Sigel, G. Kühnle, S. Szatmari, F. P. Schäfer: Appl. Phys. **B 50**, 79 (1990)
17.14 D. G. Stearns, O. L. Landen, E. M. Campbell, J. H. Scofield: Phys. Rev. **A 37**, 1684 (1988)
17.15 S. E. Harris, J. D. Kmetec: Phys. Rev. Lett. **61**, 62 (1988)
17.16 G. Kühnle, F. P. Schäfer, S. Szatmari, G. D. Tsakiris: Appl. Phys. **B 47**, 361 (1988)
17.17 J. A. Cobble, G. A. Kyrala, A. A. Hauer, A. J. Taylor, C. C. Gomez, N. D. Delamater, G. T. Schappert: Phys. Rev. **A 39**, 454 (1989)
17.18 H. W. K. Tom, O. R. Wood II: Appl. Phys. Lett. **54**, 517 (1989)
17.19 M. M. Murnane, H. C. Kapteyn, R. W. Falcone: Phys. Rev. Lett. **62**, 155 (1989)
17.20 M. M. Murnane, H. C. Kapteyn, M. D. Rosen, R. W. Falcone: Science **251**, 531 (1991)
17.21 H. Schulz, D. von der Linde: SPIE Proceedings 1268 (1990) p. 30
17.22 N. G. Basov, Yu. A. Zhararenkov, N. N. Zorev, G. V. Sklizkov, A. A. Rupasov, A. S. Shikhanov: *Heating and Compression of Thermonuclear Targets by Laser Beam* (Cambridge University Press, Cambridge 1986)
17.23 T. Engers, W. Fendel, H. Schüler, H. Schulz, D. von der Linde: Phys. Rev. **A 43**, 4564 (1991)
17.24 K. Baumgärtel, K. Sauer: *Topics on Nonlinear Wave–Plasma Interaction* (Birkhäuser, Basel 1987)
17.25 C. H. Nam, W. Tighe, E. Valeo, S. Suckewer: Appl. Phys. **B 50**, 275 (1990)
17.26 N. G. Denisov: Sov. Phys. JETP **4**, 544 (1957)
17.27 V. L. Ginzburg: *The Properties of Electromagnetic Waves in Plasmas* (Pergamon, New York 1964)

Dietrich von der Linde

18. Dye Laser Spectroscopy of Isolated Atoms and Ions in Liquid Helium

G. zu Putlitz and M. R. Beau

With 5 Figures

Alkaline earth atoms and ions in superfluid helium have been investigated by means of dye laser spectroscopy. The observed line shifts and broadenings provide microscopic insight into the structure of such point defects. The results are compared with the data of low pressure shifts in gaseous helium extrapolated to liquid helium density, which reveals new effects due to the liquid environment. Moreover these impurities can be used as new tools for the investigation of the superfluid itself, for instance coloring of vortices or trapping in electric and magnetic traps.

These studies were only made possible by the unique properties of the dye laser, such as high spectral power density, wide range tunability and unsurpassed beam quality.

18.1 Effects of Pressure on Spectral Lines

The study of pressure shifts of atomic and ionic transitions due to collisions with buffer gas atoms have a long tradition as a tool for the examination of interatomic potentials [18.1]. As a result of the interaction of buffer gas or matrix atoms (e.g. rare gas atoms) and the probe atoms (e.g. alkali or alkaline earth atoms), the energy levels evolve as a function of the internuclear distance r (Fig. 18.1). From the theoretical point of view calculations of this kind are very sophisticated, since in principle they require the solution of the Hamiltonian of some tens of electrons in the Coulomb field of two nuclei. However, in a reasonable approximation one can extract the main contributions of the resulting adiabatic potentials by means of simplifying assumptions. The attractive long range interaction can be approximated by the van der Waals force ($\propto r^{-6}$), which depends on the polarizability of the atoms. The repulsive part – the so-called Pauli force – has its roots in the nonorthogonality of the atomic wavefunctions. In the case of probe ions rather than atoms, the attractive part is dominated by the monopole–interaction with an r^{-4} radial dependence.

Of course, a more refined analysis of the effects described above can be made bearing in mind that the interaction between the two colliding partners is electric in nature, resulting in a mixing of wavefunctions. Calculations of this type have been carried out for simple systems, for example Na-He.

Topics in Applied Physics, Vol. 70
Dye Lasers: 25 Years Ed.: Dr. Michael Stuke
© Springer-Verlag Berlin Heidelberg 1992

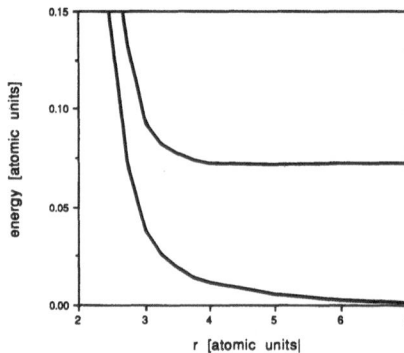

Fig. 18.1. Theoretical interaction energies for ground and excited states of Na-He [18.2]

From the experimentalist's point of view the valence electron explores the configuration coordinate during an optical transition, resulting in a frequency shift and broadening. Consequently, pressure effects provide a sensitive test for the electronic environment that the probe atom is experiencing.

For a given interatomic separation r the excitation frequency is altered by $\Delta v(r) = \Delta v_e(r)$-$\Delta v_g(r)$, where $\Delta v_e(r)$ and $\Delta v_g(r)$ denote the shifts for the excited and ground states, respectively. The net effect can be obtained by an average of $\Delta v(r)$ weighted by the distances r realized under experimental conditions. This average is of course different for different collision energies, in other words, it makes the effect of shifts and widths temperature dependent. The resulting line shape is obtained by the appropriate calculation of the frequency distribution of $\Delta v(r)$.

For the low density regime (a few millibars) the impact damping theory has been developed [18.3]. The scattering process is regarded under the aspect of different internuclear separations for each impact parameter. Shifts and widths exhibit a linear dependence as a function of pressure. Experimental results for sodium in different rare gases are shown in Fig. 18.2. The general behavior of the shift can be qualitatively understood in terms of the expressions mentioned above. For helium, which has a small electric polarizability, the repulsive Pauli force dominates and the slope for the pressure shift is positive. In contrast to this behavior the heavy noble gases bear a large polarizability and hence the attractive contributions lead to a frequency decrease of the spectral lines.

For higher densities collisions of three and more particles become important and the statistical theory applies [18.5]. In this model an equilibrium distribution of the buffer gas atoms around the alkali is assumed:

$$P(r) = c \cdot \exp\left(-\frac{V(r)}{kT} \right) \cdot r^2$$

where $V(r) = h \cdot \Delta v_g(r)$ is the potential for the rare gas atoms in the presence of the alkali and h, k and T denote the Planck constant, Boltzmann constant and temperature, respectively. The evaluation of the line shape is then straightfor-

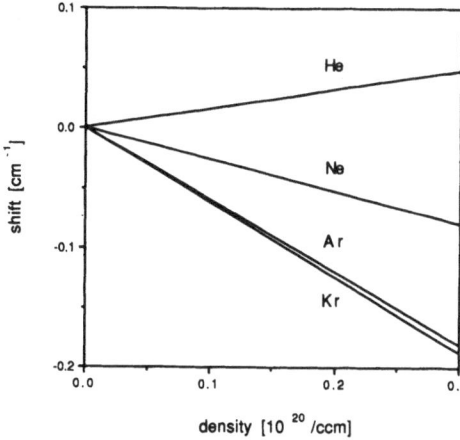

Fig. 18.2. Pressure shifts of the sodium D_1 line in different buffer gases. (Data from [18.4])

ward. In general, one observes a linear regime with a steeper slope compared to the low-pressure data.

Because of the simultaneous line broadening at higher pressures, precision is limited to small nonlinear effects. It should be noted that the hyperfine pressure shifts (much more accurately measurable by means of optical pumping and radio frequency methods) have revealed a quadratic term at higher buffer gas densities [18.6, 7].

With this introduction to the interaction of single ions or atoms (preferentially alkali or alkaline earth atoms) with buffer gas particles one may wonder what happens if the density is increased to a pressure equivalent to a liquid (~ 800 bar). In addition, the interaction between the constituent atoms of the matrix plays an important role.

18.2 Impurities in Liquid Helium

Liquid helium is a very interesting fluid as such, particularly because of its phase transition from the fluid to the superfluid state and its unique phase diagram. As a consequence, point defects in liquid helium have been of considerable interest in the past. However, until now no universal technique had been available to implant probe atoms and ions of any kind into the liquid. For this reason, mostly electrons, helium ions and metastable helium atoms have been studied, which can be created by radioactive sources or electron bombardment within the matrix itself.

The model developed for electrons and metastable helium atoms inside the liquid treats the ground (g) and excited (e) states separately [18.8, 9]. Starting from $h \cdot \Delta v_{e,g}(r)$, the pair potential of the impurity and a helium atom, the total energy shift due to the surrounding liquid is obtained by weighting with a density function ϱ

$$E_{e,g}(R, \alpha) = \int h\Delta v_{e,g}(r) \cdot \varrho(r, R,\alpha) \, d^3r \ .$$

The density function implicitly assumes a bubble structure, with R denoting the radius of the void and a profile parameter α describing the smoothness of the density transition:

$$\varrho(r, R, \alpha) = \begin{cases} 0, & \text{for } r < R \\ \varrho_0 \cdot \{1 - [1 + \alpha \cdot (r - R)] \cdot \exp[-\alpha \cdot (r - R)]\} & \text{otherwise.} \end{cases}$$

To obtain the total energy of the point defect, the energy of liquid distortion E_b has to be considered. Here

$$E_b = E_s + E_{pv} + E_{vk}$$

with the surface energy $E_s = \sigma \cdot 4\pi \cdot R^2$ the pressure volume work $E_{pv} = p \cdot \frac{4}{3}\pi \cdot R^3$ and the volume kinetic energy $E_{vk} = \hbar^2/8 \, M \int d^3r (\nabla \varrho)^2/\varrho$. The first two expressions are obvious with R and σ denoting the surface tension and the external pressure, respectively. The third term (M is the mass of a helium atom) results from the increase of the liquid's kinetic energy due to the density gradient at the bubble margin [18.8, 9]. To find the absolute minimum as a function of the liquid distribution, the total energy shift $E_t(R, \alpha) = E_{e,g} + E_b$ has to be minimized with respect to R and α. Results for the barium atom are illustrated in Fig. 18.3.

This derivation so far is general and can be applied partially to the classical probes in superfluid helium: the positive ion (He$^+$) and the negative ion (e$^-$). In the case of the electron bubble, however, the situation is somewhat different. Because there is no initial (atomic) wavefunction which is perturbed, it is necessary to parametrize it as a Gaussian or a hydrogenic wavefunction. For

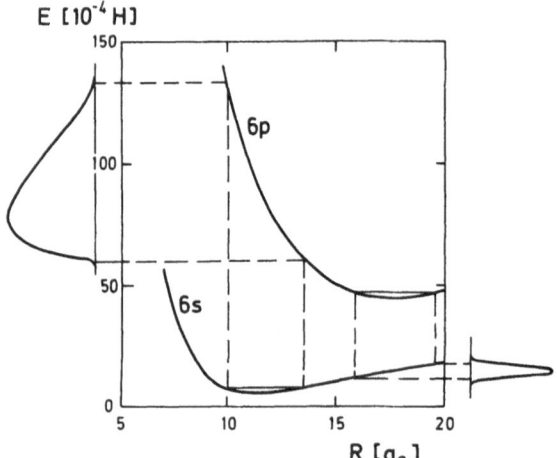

Fig. 18.3. Configuration co-ordinate diagram of the barium ground and excited states in liquid helium. The ordinate is the energy shift expressed in 10^{-4} atomic units; the abscissa depicts the bubble radius. For clarity the excited state is shifted by 30×10^{-4} a.u.

this excess electron the energy E_g is composed of the kinetic energy and the Pauli force. The latter can be approximated by the optical pseudo-potential [18.10]. The total energy of the defect has to be minimized with respect to R and α, as well as with respect to the parameters describing the electron wavefunction. It turns out that the void exhibits macroscopic dimensions with a diameter of about 32 Å and is indeed energetically favored compared to a delocalized (Bloch wave) electron state.

The He^+ ion on the other hand experiences a short range Pauli force as a result of its strongly localized electron. So the monopole–dipole interaction of the charge carrier and the surrounding liquid becomes dominant. If one converts the interaction energy into pressure, the solidification pressure is easily exceeded and a solid crust is formed around the ion core. Therefore the name "snowball" has become established in the literature.

From this concept it is clear that the valence electron(s) play(s) a key role for the nature of the defect structure. The theoretical concept of the optical pseudopotential for alkaline and alkaline earth ions [18.11] has supported this point of view. For the alkaline ions with a closed shell configuration a snowball structure has been deduced, similar to that of the helium ion. In the case of the alkaline earth ions however, the Pauli force (due to the extended valence electron) pushes the liquid apart and a bubble-like structure was derived. As a result, new interesting phenomena are anticipated, which go beyond the scope of pressure effects in the gas phase.

From the expected large matrix shifts and broadenings it is obvious that the use of hollow cathode lamps doped with the corresponding element is very inconvenient due to the small spectral overlap of their emission characteristics. Hence white light sources narrowed by appropriate filters were used in the past. The advent of the dye laser [18.12, 13] added a completely new dimension to such experiments. With its high spectral power density and its tunability to the desired wavelength the dye laser has proven to be the ideal light source for such investigations.

Understanding the interaction of atoms or ions with liquid helium offers the possibility of using these atoms and ions as probes for the properties of the superfluid.

The superfluid, discovered 53 years ago by Kapitza and Allen *et al.*, still draws the attention of many scientists [18.14]. Important problems remain unresolved, e.g. the quantitative explanation of various critical velocities, a microscopic theory of rotons and the dynamics of quantum circulations called vortices, to mention but a few [18.15, 16]. This can be illustrated by an example: The observation of the mass flow between two reservoirs connected by a small channel yields a surprising result. The flow velocity is nearly independent of the applied pressure and increases with decreasing diameter of the capillary. Early theories predicted well the dependence on the diameter, but the values for the critical velocities were one magnitude too large [18.16].

As has been shown in previous studies, the motion of electrons and helium ions can reveal interesting properties of both the defects and the surrounding liquid. For example, proof of the existence of the roton resulted from the investigation of the electron mobility as a function of temperature [18.17]. It was found that the mobility μ varies as the inverse roton density $\mu(T) \propto n_r^{-1}$ and could therefore be used to determine the roton energy gap.

In the superfluid helium the circulation $\oint v \cdot ds$ is quantized and takes the values $\kappa = n \cdot h/M$ with an integer n, the Planck constant h and the mass of a helium atom M. As a consequence the velocity of the superfluid varies inversely with the distance from the vortex core. A vortex ring is established when a vortex line is bent into a ring. Mobility studies in the high field regime demonstrate a critical velocity at which an ion nucleates a vortex ring and is subsequently captured by the ring. This shows the suitability of ions for the investigation of the superfluid.

For a long time it has been a great challenge to prove the existence of Landau's critical velocity, at which rotons should be created. In 1976 it was shown that rotons are created only in pairs [18.18]. This selection rule is still unexplained and could lead to a new understanding of these elementary excitations.

Despite all these successes it is not possible to deduce the intrinsic structure of the impurities only from mobility measurements. There are many open questions, for example, the interpretation of the roton-limited mobility [18.19, 20]. Therefore it is clear that for a test of microscopic theories optical spectroscopy of impurities in superfluid helium is a very promising technique, as has been proven in solid state physics. To make a long story short: Ions and atoms in superfluid helium accessible to optical spectroscopy can contribute to both an understanding of the defects and the investigation of the superfluid itself.

18.3 Experiments

In order to study single atoms and ions immersed in superfluid helium, several severe problems have to be solved. First of all the ions and atoms in question have to be created in the gas phase close to the surface of the liquid. The experimental chamber holding the sample is immersed in helium in a cryostat at a temperature of 1.4 K. The methods employed so far are evaporation from a specially insulated oven with subsequent ionization by a discharge and sputtering in the focus of a strong laser beam. The example chosen can be seen in Fig. 18.4. The light of a 1 MW pulsed nitrogen laser has been focused on a metal specimen above the liquid level, resulting in a dense plasma in front of the metal surface. With an efficiency of about 10^{-3} ions are sputtered and subsequently drawn into the liquid by means of an electric field. Guided by this field the ion cloud moves downwards into the optical volume and is excited by the dye laser (perpendicular to the plane of the figure). The spectral distribution of the fluorescence light is analyzed by a 1/2 m monochromator and detected by a red-sensitive photomultiplier (S20 photocathode) covering a spectral range from 200

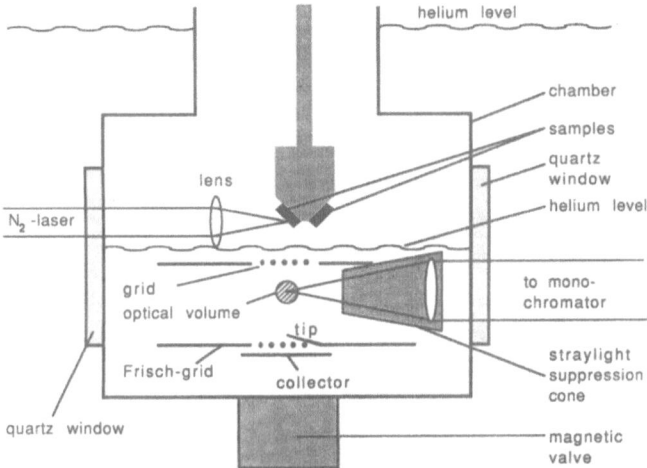

Fig. 18.4. Schematic view of the experimental cell immersed in liquid helium

to 800 nm. Special attention must be devoted to the liquid level below the sample because it is crucial for the implantation of the ions. Therefore it is measured by a superconducting wire and regulated by a magnetic valve to a precision of 1 mm. The ion density achievable in the superfluid is rather low $(10^8\,\mathrm{cm}^{-3})$ and special care must be taken to avoid stray light from the excitation source. The unsurpassed beam quality of a dye laser is extremely helpful in this regard.

For the study of atoms, the ions have to be neutralized inside the liquid. For this purpose a field emission tip is installed at the lower end of the helium bath. A thin tungsten wire is etched electrochemically to a very sharp tip with a bending radius of about $0.2\mu m$. If a potential of several kilovolts is applied to the wire, electrons are emitted into the liquid [18.21], where they can recombine with the injected ions. During this process many lines of the atomic spectrum can be observed, from the ultraviolet to the infrared. These spectra yield valuable information about the recombination process itself, the initial population of the excited states and the relaxation of the helium environment due to the neutralization of the ions [18. 22, 23]. A striking feature of the recombination spectra is the absence of transitions emerging from states situated less than 1.8 eV below the ionization limit. However, it should be pointed out that these missing lines as well as the observation of all other transitions can be well explained by the tunnel model developed in our group. During the recombination process an electron (forming its own bubble) tunnels into the ionic Coulomb potential and occupies an atomic-like state in the liquid. Together with the subsequent relaxation of the fluid into the new equilibrium configuration, this "1.8 eV limit" is predicted well [18.24, 25].

Ions neutralized by electrons to form atoms are not affected by the electric field anymore and exhibit dwelling times of some hundred milliseconds in the

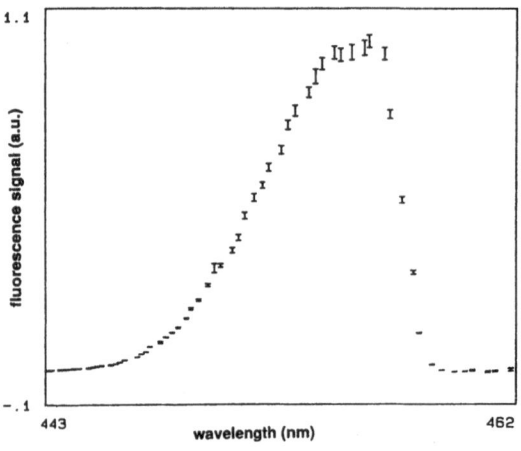

Fig. 18.5. Excitation spectrum of strontium in liquid helium at 1.5 K. One sees clearly the asymmetry predicted by theory (Fig. 18.3)

optical volume. They can be studied conveniently by laser induced fluorescence and laser excitation spectroscopy. An example of such a spectrum is given in Fig. 18.5.

18.4 Spectroscopic Results

Some observations of the excitation bands for alkaline earth ions and atoms are summarized in Table 18.1. The shifts and widths of the transition wavelengths are on the order of a few nanometers, which shows the small perturbation of the atomic states in liquid helium, similar to the case of pressure shifts and in contrast to the much larger effects in other noble gases. In agreement with theory the excitation spectra are significantly asymmetric, which will be discussed later. The experimental data are first compared with the theory outlined above. In our group, configuration coordinates for the pair potentials are computed with the pseudopotential formalism, which represents an ab initio calculation without adjustable parameters [18.26]. The spectral line shape is obtained by evaluating Franck-Condon factors, as illustrated in Fig. 18.3. In the case of excitation the line shift is determined by the distance of the configuration coordinates at the minimum of the ground state potential. The line width and asymmetry are derived from the slope and the curvature of the upper potential curve. Both the theoretical and experimental transitions are much more altered in width and frequency than the emission lines (compare Tables 18.1 and 18.2). This qualitative agreement supports the bubble model for the impurities investigated so far (see also Fig. 18.5). Despite the fact that the numerical agreement is not yet convincing, it should be pointed out that the theoretical shifts possess the correct sign and order of magnitude. Moreover, the increase in shifts and widths as a function of atomic number scales as predicted. Further

Table 18.1. Excitation bands for alkaline earth atoms and ions in liquid helium at a temperature of 1.5 K. The wavelength shifts of the atomic lines correspond to 50% rise to maximum on the red side

| Free atom | | Atom in liquid helium | | | |
| | | Experimental | | Theoretical | |
Element	λ [nm]	Shift [nm]	FWHM [nm]	Shift [nm]	FWHM [nm]
Ca	422.67	-2.65 ± 0.5	4.75 ± 0.5	-21.5	5.6
Sr	460.73	-3.3 ± 0.5	5.95 ± 0.5	-25.0	6.3
Ba	553.55	-4.3 ± 0.5	8.5 ± 0.5	-33.5	7.6
Ba^+	493.41	-11.4 ± 0.5	4.4 ± 0.5	-6.8	2.7

Table 18.2 Comparison of the fluorescence of various elements in liquid helium to the gas phase shifts and broadenings (from 18.28) extrapolated to liquid helium density

| Free atom | | Atom in liquid helium | | Pressure effects scaled | |
Element	λ [nm]	Shift [nm]	FWHM [nm]	Shift [nm]	FWHM [nm]
Mg	285.17	3.6(2)	1.5(2)	—	3.2(2)
Ca	422.67	0.63(7)	0.81(4)	0.32(3)	9.2(2)
Sr	460.73	0.60(16)	0.9(2)	0.0(1)	12.0(3)

improvements are expected by replacing the pseudopotential method used here by pair potentials derived from molecular calculations [18.27].

An interesting approach is to compare the data with the low pressure measurements in the gaseous phase scaled to the density of superfluid helium. Those data are available for the fluorescence of alkaline earth atoms [18.28]. If the low pressure shifts and broadenings are extrapolated to the density of liquid helium (2.2×10^{22} cm^{-3}) one observes a much smaller broadening but a larger shift for the liquid compared to the gas phase prediction. Qualitatively these effects can be explained as follows: During a scattering process the gaseous, "hot" helium atoms explore the whole potential curve corresponding to their kinetic energy, leading to a comparably broad frequency distribution in absorption and emission. In contrast, liquid helium atoms stay much closer to the potential minimum and probe only that part of the interaction curve that is given by zero point motion (Fig. 18.3). In general, pair potentials for the ground and excited states evolve in different ways, and as a result the energy difference as a function of the internuclear distance is composed of positive and negative contributions. As discussed above, the gas phase atoms probe the whole configuration coordinate, in contrast to the liquid atoms, which examine only the neighborhood of one point of the interaction curve. So the resulting shift is greatly reduced in the case of high temperature pressure effects. In addition to this, in the case of the liquid, the host–host atom interaction becomes important. The bubble energy E_b (mainly the surface tension) results in a decrease of the bubble radius, which in turn tends to increase the difference.

18.5 Conclusions and Outlook

Laser spectroscopy of isolated atoms and ions immersed in superfluid helium has proven to be a successful tool for the investigation of point defects in the liquid environment. Comparison of line shifts in superfluid helium with low pressure gas phase shifts extrapolated to the liquid density show the similarity of the two effects. However, the liquid exhibits new phenomena due to the low temperature and to the helium–helium interaction. Optical spectroscopy of such impurities can also be used as a new tool for the investigation of the superfluid itself. For example, electrons captured in a vortex can be ejected by means of an optical transition [18.29]. Similar behavior should be observable with foreign ions or atoms and offers the possibility of investigating this effect in more detail. This "photoejection" is also sensitive to nonradiative transitions and should permit the discovery of the excitation bands of calcium and strontium ions, which are unobserved so far.

A completely new regime of experiments is envisaged if it becomes possible to store ions and atoms in the center of a vortex line (which has been demonstrated in the case of electrons [18.30]) or in suitable ion traps immersed in the liquid. Laser excitation followed by spatially resolved detection of the fluorescence light could be used to track the vortices and observe their dynamics. Critical velocities in microscopic channels (mentioned in Sect. 18.2) have been treated recently [18.31]. It was found that at least three vortices entering the channel can sustain a vortex tangle leading to dissipation. It would be a great challenge to observe these vortices by the the proposed coloring technique.

Storage of ions and atoms within liquid helium using electric and magnetic traps is also of great interest; it was demonstrated recently by observing the trapping of antiprotons in liquid helium [18.32]. The motion of such ions and atoms can be studied by optical detection. Such measurements could reveal critical velocities and ionic effective masses in a well-defined environment in order to cross-check the data obtained from plasma frequency excitation [18.33].

Clearly there is a wealth of future experiments to be carried out for a large variety of ions and atoms which can now be injected into superfluid helium by the universal technique described here. However, it should be stressed on this occasion that none of our experiments could have been promoted to real beauty without the invention of the tunable dye laser by F. P. Schäfer, P. P. Sorokin and their colleagues 25 years ago.

Acknowledgement. This work is supported by the Deutsche Forschungsgemeinschaft, Grant No. PU 22/26-1.

References

18.1   See for example: B. Wende (ed.): Proc. 5th Int. Conf. Spectral Line Shapes, Berlin 1980 (de Gruyter, Berlin 1981); K. Burnett (ed.): Proc 6th Int. Conf. Spectral Line Shapes, Boulder 1982 (de Gruyter Berlin 1983)

18.2   M. Krauss, P. Maldonado, A. Wahl: J. Chem. Phys. **54**, 4944 (1971)
18.3   H. M. Foley: Phys. Rev. **69**, 616 (1946)
18.4   J. F. Kielkopf: J. Phys. B. **13**, 3813 (1980)
18.5   L. B. Robinson: Phys. Rev. **117**, 1275 (1960)
18.6   E. S. Ensberg, G. zu Putlitz: Phys. Rev. Lett. **22**, 1349 (1969)
18.7   T. Crane, D. Casperson, P. Crane, P. Egan, V. W. Hughes, R. Stambaugh, P. A. Thomson, G. zu Putlitz: Phys. Rev. Lett. **27**, 474 (1971)
18.8   K. Hiroike, N. R. Kestner, S. A. Rice, J. Jortner: J. Chem. Phys. **43**, 2625 (1965)
18.9   A. P. Hickman, W. Steets, N. F. Lane: Phys. Rev. B **12**, 3705 (1975)
18.10   Y. M. Shih, C. Woo: Phys. Rev. A **8**, 1437 (1973)
18.11   M. W. Cole, R. A. Bachmann: Phys. Rev. B **15**, 1388 (1977)
18.12   F. P. Schäfer, W. Schmidt, J. Volze: Appl. Phys. Lett. **9**, 306 (1966)
18.13   P. P. Sorokin, J. R. Lankard: IBM J. Res. Devel. **10**, 162 (1966)
18.14   R. J. Donelly: Sci. Am. 66 (Nov 1988) 66; Phys. Today (Feb. 1987), Special issue: ^{3}He and ^{4}He
18.15   A. L. Fetter: In *The Physics of Liquid and Solid Helium*, ed. by K. H. Bennemann, J. B. Ketterson (Wiley, New York 1976)
18.16   D. R. Tilley, J. Tilley: *Superfluidity and Superconductivity* (Adam Hilger, New York 1990)
18.17   L. Meyer, F. Reif: Phys. Rev. **110**, 279 (1958)
18.18   D. R. Allum, R. M. Bowley, P. V. E. McClintock: Phys. Rev. Lett. **36**, 1313 (1976)
18.19   K. W. Schwarz: Adv. Chem. Phys. **33**, 3 (1975)
18.20   W. I. Glaberson, W. W. Johnson: J. Low Temp. Phys. **20**, 313 (1975)
18.21   A. Phillips, P. E. V. McClintock: Philos. Trans. R. Soc. London A **278**, 271 (1975)
18.22   H. Bauer, M. Beau, J. Fischer, H. J. Reyher, J. Rosenkranz, K. Venter: Physica B **165**, 137 (1980)
18.23   M. Beau, J. Fischer, H. J. Reyher, H. Schreck: To be published
18.24   M. Beau: Ph.D. Thesis (University of Heidelberg, Heidelberg 1989)
18.25   M. Beau, G. zu Putlitz: To be published
18.26   H. Bauer, M. Beau, B. Friedl, C. Marchand, K. Miltner, H. J. Reyher: Phys. Lett. A **146**, 134 (1990)
18.27   J. Pascale: Phys. Rev. A **28**, 632 (1983)
18.28   N. J. Bowman, E. L. Lewis: J. Phys. B **11**, 1703 (1978); G. U. Zhuvikin, N. P. Penkin, L. N. Shabanova: Opt. Spectrosc. **42**, 134 (1977); J. M. Farr, W. R. Hindmarsh: J. Phys. B **4**, 568 (1971)
18.29   C. C. Grimes, G. Adams: Phys. Rev. **B41**, 6366 (1990); J. A. Northby, T. M. Sanders: Phys. Rev. Lett. **18**, 1184 (1967)
18.30   E. J. Yarmchuk, R. E. Packard: J. Low Temp. Phys. **46**, 479 (1982)
18.31   K. W. Schwarz: Phys. Rev. Lett. **64**, 1130 (1990)
18.32   M. Iwasaki et al.: Phys. Rev. Lett. **67**, 1246 (1991)
18.33   C. J. Mellor, C. M. Muirhead, J. Traverse, W. F. Vinen: Proc. LT 18. Jpn. J. Appl. Phys. **26**, 381 (1987) and references therein

Gisbert zu Putlitz   Michael Beau

Topics in Applied Physics Founded by Helmut K. V. Lotsch